THE MAN
IN THE ICE 3

Veröffentlichungen des Forschungsinstituts für Alpine Vorzeit
der Universität Innsbruck 3

Herausgegeben von H. Moser, W. Platzer, H. Seidler, K. Spindler

The Man in the Ice
Volume 3

K. Spindler, H. Wilfing, E. Rastbichler-Zissernig,
D. zur Nedden, H. Nothdurfter (eds.)

Human Mummies
A Global Survey of their Status and the Techniques of Conservation

SpringerWienNewYork

Univ.-Prof. Dr. Hans Moser
Rector of the University of Innsbruck, Innsbruck, Austria

Univ.-Prof. Dr. Werner Platzer
Institute of Anatomy, University of Innsbruck, Innsbruck, Austria

Univ.-Prof. Dr. Horst Seidler
Institute of Human Biology, University of Vienna, Vienna, Austria

Univ.-Prof. Dr. Konrad Spindler
Institute of Pre- and Protohistory, University of Innsbruck, Innsbruck, Austria

Dr. Harald Wilfing
Institute of Human Biology, University of Vienna, Vienna, Austria

Mag. Elisabeth Rastbichler-Zissernig
Institute of Prehistoric Alpine Research, University of Innsbruck, Innsbruck, Austria

Univ.-Prof. Dr. Dieter zur Nedden
Radiological Department II of the University Hospital Innsbruck, Innsbruck, Austria

Dr. Hans Nothdurfter
Regional Authorities for Bozen, Bozen, Italy

This work is sponsored by
Bundesministerium für Wissenschaft, Forschung und Kunst in Wien
Kulturabteilung der Stadt Wien
Südtiroler Landesregierung

This work is subject to copyright.
All rights are reserved, whether the whole or part of the material is concerned, specifically those of translation, reprinting, re-use of illustrations, broadcasting, reproduction by photocopying machines or similar means, and storage in data banks.

© 1996 Springer-Verlag/Wien
Printed in Austria

Cover: Artificial mummy of Francesco Ferdinando d'Avalos (1489–1525), marquis of Pescara and captain general of the troops of Charles V, who won the famous battle of Pavia against the French king François I in 1525: detail of the skull with horizontal craniotomy filled with wool (Basilica of San Domenico Maggiore, Naples; photo: Istituto di Anatomia Patologica, Università di Pisa)

Data conversion and printing: Druckhaus Grasl, A-2540 Bad Vöslau
Binding: Fa. Papyrus, A-1100 Wien

Printed on acid-free and chlorine-free bleached paper

With 226 partly coloured Figures

Libary of Congress Cataloging-in-Publication Data

Human mummies : a global survey of their status and the techniques of
    conservation / K. Spindler ... [et al.] eds.
        p.   cm. – (The man in the ice, ISSN 0947-3483 ; v. 3)
    Includes bibliographical references.
    ISBN 3-211-82659-9 (alk. paper)
    1. Mummies–Collection and preservation. 2. Mummies–Conservation
and restoration. 3. Human remains (Archaeology) I. Spindler,
Konrad, 1939–   . II. Series
GN293.H85   1996
393'.3–dc20   96–7335
                                                            CIP

ISSN 0947-3483

ISBN 3-211-82659-9 Springer-Verlag Wien New York

Preface

On 15–17 September, 1993, Innsbruck, Austria, hosted the International Mummy Symposium. This does not mean that beautiful North Tyrol was the setting for a gathering of the world's most prominent mummies themselves, but rather the exciting discovery of a Late Neolithic glacial mummy released from the ice of the Ötztal Alps provided the focus of attention for numerous scholars from many different parts of the world to come together to address various questions relating to mummified human remains.

Normally researchers studying the remains of historical or prehistoric human bodies will at best have bony substance to work on. It is rarely the case that soft parts and internal organs are preserved in various states of conservation. Where that is the case we speak of human mummies, and they can be the product of either natural or artificial mummification. Natural mummification may occur when a corpse is placed in ice, salt or a bog, while artificial mummification can take the form of embalming and/or evisceration, for example. It is also possible for both types of processes to be involved in the preservation of human remains. As a rule the mummies that have been discovered in the past had been deliberately deposited, in the most frequent case in graves but also at sacrificial sites and shrines, or at disposal sites following judicial killing.

Only under extremely rare conditions does mummification of a human body occur following an accident or natural disaster, all the more so as the victim's relatives would normally launch a search to find the body so as to provide a proper burial in keeping with the funeral customs of the society involved.

The discovery of such a mummy is of outstanding importance as this is the only type that offers real insights into the cultural background, whereas the mummies discovered in graves or tombs are merely a source of information on the burial customs of that society. In this context the bodies of the Hallstatt miners discovered in the salt mines of Hallstatt and Hallein in Upper Austria come to mind, although those finds – made as they were several centuries ago – have since been lost to modern research. Another remarkable case, that of the pre-Columbian miner from Restauradora Mine near Chuqincamata in northern Chile, is the result of impregnation with copper salts, and the mummy became an attraction as "Copper Man" at various fairs around the country. As he was found with a complete set of miner's tools, the mummy offers a unique insight into the life and working conditions of an Indio miner of the first millennium AD. Even so, the mummified remains comprise only the skeleton with a completely rigid covering of skin, whereas the other soft parts have not survived.

In contrast, mummification in ice, and especially in the permafrost, can produce much better results. The three English sailors lost on the Franklin expedition who were exhumed on Beechey Island in the Canadian Arctic a few years ago, for example, seemed almost to be sleeping in their graves.

Under such circumstances it is little short of a miracle that, more than 5000 years after his death, the mummy of a man was found on 19 September 1991 in an excellent state of preservation (cf. "The Man in the Ice", vol. 1, 1992, and vol. 2, 1995).

It is a great honour for the University of Innsbruck that the Regional Government of the Autonomous Province of South Tyrol as the owner of the find has entrusted it with the task of conserving and studying this unique find, although it was clear from the start that the complexity of the work involved would necessitate the establishment of an international and interdisciplinary project on a hitherto unheard-of scale (cf. K. Spindler et al., Preface. In: "The Man in the Ice", vol. 2, 1995, Vff). It was also apparent that the University alone would be neither willing nor able to accept sole responsibility for the success of such a demanding assignment.

For that reason it was decided at a relatively early date to invite internationally prominent experts on mummification to attend a conference in Innsbruck with the primary aim of addressing aspects of conservation and defining the state of the art in the study of human mummies.

An attempt was accordingly made to contact those research workers throughout the world who have been involved in work on major mummy finds. The reaction was overwhelming, and the invitation to come to Innsbruck was accepted almost without exception, so that a truly high-powered programme could be drawn up for the symposium. The majority of the papers presented, plus a number of additional important reports, form the subject of this, the third volume in the "Iceman" series.

The official programme of the symposium was followed by a closed session attended by the various speakers plus a number of other experts, who addressed specific questions relating to the mummy from the Hauslabjoch.

The Innsbruck symposium was organised by the students and co-workers of the Institute of Prehistory and Early History, and the staff of the Institute of Prehistoric Alpine Research at the University of Innsbruck. To them we owe a sincere vote of thanks for an exciting conference and for the harmonious spirit of the days we spent together in the early autumn of 1993.

We are also grateful to all those institutions and individuals who – with help and advice, editorial skills, funds and moral support – have supported the Innsbruck symposium and publication of the proceedings. We hope we will be forgiven for not listing the many names involved; the addressees of our deep gratitude will know who they are. A name omitted from a list would be the greater offence.

Springer-Verlag Wien New York has taken great pains to give this, the third volume in the "Iceman" series the fine format it deserves. Thank you.

Innsbruck, January 1996 *Konrad Spindler*

Contents

General aspects

Seipel, W.: Mummies and ethics in the museum ... 3
Lewin, P. K.: Current technology in the examination of acient man ... 9
Hunt, D. R., Hopper, L. M.: Non-invasive investigations of human mummified remains by radiographic techniques ... 15
Rae, A.: Dry human and animal remains – their treatment at the British Museum ... 33

Egyptian mummies

Seipel, W.: Research on mummies in Egyptology. An overview ... 41
Maekawa, S., Valentin, N.: Development of a prototype storage and display case for the Royal Mummies of the Egyptian Museum in Cairo ... 47

Asiatic mummies

Wang, B.-H.: Excavation and preliminary studies of the ancient mummies of Xinjiang in China ... 59
Yamada, T. K., Kudou, T., Takahashi-Iwanaga, H., Ozawa, T., Uchihi, R., Katsumata, Y.: Collagen in 300 year-old tissue and a short introduction to the mummies in Japan ... 71

Arctic mummies

Zimmerman, M. R.: Mummies of the Arctic regions ... 83
Notman, D., Beattie, O.: The palaeoimaging and forensic anthropology of frozen sailors from the Franklin Arctic expedition mass disaster (1845–1848): a detailed presentation of two radiological surveys ... 93
Hart Hansen, J. P., Nordqvist, J.: The mummy find from Qilakitsoq in northwest Greenland ... 107

South American mummies

Allison, M. J.: Early mummies from coastal Peru and Chile ... 125
Arriaza, B.: Preparation of the dead in coastal Andean pre-ceramic populations ... 131
Aufderheide, A. C.: Secondary applications of bioanthropological studies on South American Andean mummies ... 141
Horne, P. D.: The Prince of El Plomo: a frozen treasure ... 153

European mummies

Brothwell, D.: European bog bodies: current state of research and preservation 161

Daniels, V.: Selection of a conservation process for Lindow Man 173

Rodríguez-Martín, C.: Guanche mummies of Tenerife (Canary Islands): conservation and scientific studies in the CRONOS Project. 183

Fornaciari, G., Capasso, L.: Natural and artificial 13th–19th century mummies in Italy 195

Ascenzi, A., Bianco, P., Nicoletti, R., Ceccarini, G., Fornaseri, M., Graziani, G., Giuliani, M. R., Rosicarello, R., Ciuffarella, L., Granger-Taylor, H.: The roman mummy of Grottarossa 205

Fulcheri, E.: Mummies of Saints: a particular category of Italian mummies 219

Kaufmann, B.: Mummification in the Middle Ages 231

Kaufmann, B.: The corpse from the Porchabella-glacier in the Grisons, Switzerland (community of Bergün) . . 239

Iceman research: current events

Spindler, K.: Iceman's last weeks 249

Bereuter, T. L., Lorbeer, E., Reiter, C., Seidler, H., Unterdorfer, H.: Post-mortem alterations of human lipids – part I: evaluation of adipocere formation and mummification by desiccation 265

Bereuter, T. L., Reiter, C., Seidler, H., Platzer, W.: Post-mortem alterations of human lipids – part II: lipid composition of a skin sample from the Iceman 275

Makristathis, A., Mader, R., Varmuza, K., Simonitsch, I., Schwarzmeier, J., Seidler, H., Platzer, W., Unterndorfer, H., Scheithauer, R.: Comparison of the lipid profile of the Tyrolean Iceman with bodies recovered from glaciers 279

Kralik, C., Kiesl, W., Seidler, H., Platzer, W., Rabl, W.: Trace element contents of the Iceman's bones. Preliminary results 283

Weisgram, J., Splechtna, H., Hilgers, H., Walzl, M., Leitner, W., Seidler, H.: Remarks on the anatomy of a mummified cat regarding the extent of preservation 289

General aspects

Mummies and ethics in the museum

W. Seipel

Kunsthistorisches Museum, Vienna, Austria

On 31. 10. 1848 the troops of Prince Alfred Windischgrätz, lying outside the city walls of Vienna, trained their artillery on the Inner City. In their attempt to restore law and order to the capital they also damaged large parts of the Imperial Castle. The most disastrous hits of all – as it turned out – were the ones that damaged the rooms in the attic of the Imperial Library on the Josefsplatz, where the zoological collection of the Imperial Family was stored. The artillery bombardment and the resulting fire destroyed many crates and boxes filled with ethnological exhibits. They also destroyed – and this is of particular interest in our context – the "Four representatives of humanity down to the last speck of dust" as a contemporary description put it.

To explain what the "Four representatives of humanity" are all about, let us not anticipate events but go back to the year 1721. This is the year Angelus Soliman, as he would later be known in Vienna, was born somewhere in the region of Eritrea as the son of an African prince. Aged seven, he was abducted in the course of a tribal feud and sold on the slave-market. After some time spent in the service of an African master he was eventually brought by boat to Messina in Sicily, where he was given an excellent education in the house of a wealthy and apparently also kindhearted marquise. On being baptised he took the name of Angelo Soliman, by which he was known from then on. Soon his refined conduct and demeanour became well known and he came to the notice of the Imperial general Prince Lobkowitz, who resided in Sicily and was a frequent guest at the house of the marquise in Messina. Prince Lobkowitz engaged a teacher and Angelo learnt German in seventeen days. He was made the prince's permanent companion, accompanying him on his travels as well as into battle. Legend has it that Prince Lobkowitz bequeathed Angelo in his will to Prince Wenzel Liechtenstein, whose service he entered in 1755. He also accompanied Wenzel Liechtenstein on various travels, for example to Parma or to Frankfurt for the Imperial Coronation of Josef II in 1764. In 1768 Soliman, described as the "moor in the service of Prince Wenzel zu Liechtenstein, born in Africa of non-catholic parents", was married in church. This seems to have led to a break with Liechtenstein. In 1773, the nephew and heir of Prince Wenzel Liechtenstein, Prince Franz Liechtenstein, took Soliman into his service and made provisions for the yearly payment of the sum of 600 Gulden to Soliman until his death.

In 1783 Angelo became a freemason. He was received into the recently founded masonic lodge "True Concord". This is, of course, the lodge to which Mozart and Haydn also belonged. He went into retirement in 1783 and lived without any major illnesses until 1796, when he died in his seventieth year after suffering a stroke while out in the street. According to the death register of the Schottenstift (in Vienna) he was buried "on the 23rd of November in the cemetery of Währing" (then a little village outside the city walls of Vienna).

It is at this point, of course, that his "biography" begins to concern us. After all, a later report on Angelo Soliman set down the following facts:

1. "that in 1796 on the order of Emperor Franz II his body was flayed

2. that his skin was then mounted onto a wooden model in such a way as to imitate the appearance of Angelo Soliman in a life-like manner, and then publicly displayed for ten years

3. that this skin mounted onto the wooden model, in other words the tangible form of our brother Angelo Soliman, was consumed by fire amid deafening noise after 52 years".

What had happened after the death of Angelo Soliman? This is recorded in the minutes of the Austrian Academy of Sciences from the year 1856. They show us that the careful preparation of the corpse of Angelo Soliman was, in fact, undertaken by the director of the Physikalische Kabinett on the express wish of Emperor Franz II himself. The Physikalische Kabinett was at that time united with the Zoological Museum and here the exhibit was displayed. The artist or rather conservator who undertook the work was the sculptor Franz Thaller, the actual preparation tock place in a coach house in the courtyard of the k.k. (Imperial) Library. A description tells us that "Angelo Soliman was shown in a standing position, the right foot moved back and the left arm moved forward; he wore a belt around the loins and a crown on his head, both made from alternating red, blue and white ostrich feathers. Arms and legs were each decorated with a string of white glass-pearls and he wore a large long pleated necklace". He was displayed in the

"fourth room of the left wing; this is a room which is decorated in the style of a tropical forest with shrubs, pools of water and reeds. It contains a waterpig, a tapir, some musk-rats and several singing birds. In the same room just left of the exit is a glass-case painted in green colour, the door of which is covered with a curtain made from green material; the inside of this glass-case is painted red. It was in this cupboard that Angelo Soliman was kept. He was shown to the visiting public by the warden just before the visitors left the room".

Even after the rearrangement of the collection in 1802 the cupboard containing the moor's remains was not moved. In fact, he was joined by a present from the King of Naples, a preparation of a negro girl, and later by the figures of an African zoo-warden from Schönbrunn and of an African hospital porter. Even without a detailed analysis of the preparation technique used, it should be noted that contemporary descriptions emphasize how life-like Soliman appeared. Although several representations of Angelo Soliman while still alive are known, no pictoral records of his exhibition in the Zoological Museum exist.

Before we continue the story of this "spectacle", allow me a jump in time: to the year 1992. During the preparations for the Summer Olympic Games in Barcelona a strong protest by participants from African and Asian countries took place. This was caused by the exhibition of a stuffed and prepared figure of an African negro in a little communal museum not far from Barcelona – the name of which had better not be mentioned. This was deeply wounding to the African athletes' sense of honour. A boycott of the Olympic Games was threatened by the African participants unless the offending exhibit was removed before the opening of the Games. After the museum's initial and somewhat incomprehensible refusal to comply a compromise was reached; the exhibit was removed from public display for the duration to the Games.

At this point one should mention those human exhibits which have always aroused the particular interest of many visitors to museum, and which are often used in the titles of exhibitions, on posters or in the plots of so-called horror films to capture the public's imagination and curiosity: Egyptian mummies. Let me start with a few historical remarks. The appreciation of Egyptian mummies did not initially derive from interest in their appearance, their history or the historical, let alone human messages they contained. Instead it was the incorrect identification of certain substances used in the embalming-process with pitch or bitumen, considered a panacea since the 11th century. Hence the derivation of the word "mummy" from the Arab word "mumiya", which means asphalt, bitumen, pitch. It was used in medicine, especially to purify and to stanch bleeding, but also in the treatment of many other ailments. As natural forms of bitumen were rare and therefore very expensive, it was already suggested by Arab scholars and doctors of the 11th century to use the resin in Egyptian mummies instead. We know today that this substance (bitumen) is not found in Egypt, and that it was used only during the Graeco-Roman period – and very rarely at that – in the process of mummification. Nonetheless, hundreds if not thousands of mummies were cut up in the search for the precious bitumen, ground to powder, used in trinctures and sold all over the civilised world. Already in 1203 the Arab writer Abd-el-Latif wrote: "For half a Dirham I acquired three heads filled with this bitumen-substance." The demand for mummy-powder or suitable mummy parts remained unabated from the Middle Ages to the Renaissance, even though one of the most famous doctors of antiquity, Abisen, had warned about using these substances. To satisfy this almost insatiable demand for mummies in the West, the production of hundreds of fake mummies became commonplace. One did not even shy away from mummifying the bodies of newly deceased persons and turning them into mummy-medicine. It was still possible to obtain mummy-powder at quite a number of chemists in Europe right up to the end of the 19th century. It was thought to be particularly useful in cases of stomach disorder.

The public in the West only became seriously interested in mummies and in the bodies they contained after Napoleon's expeditions to Egypt. Particularly the litarary descriptions by the French writers Théophil Gautier, Flaubert or Balzac fired the public's imagination with many of those romantic notions about Egyptian mummies that still surround them today.

Napoleon's expedition to Egypt had brought back mummies which were then stored in the Louvre. Due to their subsequent slow disintegration and the unpleasant smell emanating from them, the French king Charles X ordered them to be buried a second time, this time in the gardens of the Louvre near the Collonade by Perrault. It is one of the ironies of history that in 1830 exactly the same spot was chosen for the burial of revolutionaries who had fallen on the barricades. In 1840, they were exhumed and reburied with great pomp on the Place de la Bastille. One can therefore assume that quite a few of the Egyptian mummies were transferred from the site of their second burial together with the heroes of the Revolution, to find their final resting-place on the Place de la Bastille!

From the 1830's onwards it became popular in England to undertake the unwrapping and dissection of Egyptian mummies publicly in front of paying crowds. Doctor T. J. Pettigrew was particularly famous. He used the large auditorium of Charing Cross Hospital for his "operations". This soon became too small and he had to use an even larger exhibition hall. His rival was a man called Athanasi who advertised his exhibitions in the following manner: "Giovanni Athanasi has the pleasure to inform the public that the most interesting mummy ever found in Egypt will be unwrapped in the large assembly room of Exeter Hall on the evening of Monday the 10th of April at 5 p.m. Tickets including a description of the mummy can be purchased from Giovanni d'Athanasi at number 3 Wellington Street, Strand. A limited number of seats have been reserved immediately next to the table on which the mummy will lie, for the price of 6 Shillings each". It reflects the spirit of the age that Alexander 10th Duke of Hamilton had his own magnificent mausoleum built in his garden, and after his death on 18. 8. 1852 was mummified by Pettigrew and buried in an Egyptian sarcophagus. Many more such bizarre stories with examples from different countries of the "enlightened Occident" could easily be recounted.

But let us return to the land of the mummies, to Egypt. From 1871 onwards, certain circumstances arose which indicated that tomb-robbers had found a large number of mummies, parts of which were turning up in antique shops. Eventually it was possible to discover their hiding place, a tomb dating from the 21st Dynasty, and all the royal mummies it contained were brought to the Egyptian Museum in Cairo. In 1891 and 1898 more mummies were discovered, and in 1902 they were also brought to the Egyptian Museum. Here they were exposed, together with the mute witnesses to the history of Egypt, to the prying eyes of the visitors until 1929, when it was decided (out of a feeling of reverence) to remove them from the public showrooms to an Arab mausoleum.

Let us stop here for a moment and return to Angelo Soliman. At much the same time as trade in mummies thrived in Egypt, he became a focal point of interest in Vienna. This, however, did not happen with universal consent. The surviving records and minutes of the ministry concerned show that the only daughter of Soliman, Josephine, repeatedly and insistently asked "for the skeleton and the skin of her father to be handed over to her for burial". In spite of several petitions in 1796 her request remained unanswered. Eventually, she turned to the archbishop of Vienna to ask for his intervention on her behalf. As this document still survives in the Wiener Fürsterzbischöfliche Archiv we can quote a few telling lines from it:

"Concerning this particular case we permit ourselves a few remarks. It is the custom of civilised peoples, in fact it is demanded by affluence, decorum and modesty, not to expose the human body to the gazing eye. Instead it is covered in life by clothes and in death by the earth. Only the great good for humanity derived from the dissection of the and the anatomical experiments with the bodies by doctors may be seen as giving sufficient grounds for an exemption from this rule. The exhibition of a moor as a handsome curiosity merely for the gratification of the coveting eye and curiosity is not excusable."

Little needs to be added to this deeply humane and scientifically sound opinion. Except of course the fact – and here we reach the very heart of our problem – that it seems almost impossible to strike the right balance between public curiosity and scientific interests – particularly in spectacular cases, as, for example, in the case of the Iceman ("Ötzi").

Already in 1959 it was thought desirable to exhibit the royal mummies in the Egyptian Museum in Cairo once more. But the development of mass-tourism, the use of flash-lights and cameras caused the truly disgraceful spectacle which everyone who visited the Egyptian Museum in the early 1960's, and in particular room 52 containing the royal mummies, will recall with sadness and disgust. One could not help observing the sad spectacle of the base, voyeuristic and sensationalist treatment of the tangible representatives of three thousand years of Egyptian history. It took President Anwar Sadat to order again in the 1970's the closing of the rooms containing the mummies. A decision for which everyone advocating a humane and dignified treatment of these physical remains of what was once a human individual must be grateful.

Let us now cast a swift glance at the "Code of Ethics" issued by ICOM. Let me quote two passages from this to try to define our problem:

"By definition a museum is an institution in the service of society and of its development, and is generally open to the public (even though this may be a restricted public in the case of certain very specialised museums, such as certain academic or medical museums, for example):" (2.6)

"Subject to the primary duty of the museum to preserve unimpaired for the future the significant material that comprises the museum collections, it is the responsibility of the museum to use the collections for the creation and dissemination of new knowledge, through research, educational work, permanent displays, temporary exhibitions and other special activities. These should be in accordance with the stated policy and educational purpose of the museum, and should not compromise either the quality or the proper care of the collections. The museum should seek to ensure that information in displays and exhibitions is honest and objective and does not perpetuate myths and stereotypes." (2.8)

"Where a museum maintains and/or is developing collections of human remains and sacred objects these should be securely housed and carefully maintained as archival collections in scholarly institutions, and should always be available to qualified researchers and educators, but not the morbidly curious. Research on such objects and their housing and care must be accomplished in a manner acceptable not only to fellow professionals but to those of various beliefs, including in particular members of the community, ethnic or religious groups concerned. Although it is occasionally necessary to use human remains and other sensitive material in interpretative exhibits, this must be done with tact and with respect for the feeling for human dignity held by all peoples." (6.7)

The last paragraph in particular undoubtedly reflects the sentiments already expressed in the archbishop's intervention on behalf of Josefine Soliman. We can only agree with this attitude wholeheartedly. In my opinion the public display of mummies or parts of mummies or of human preparations is intrinsically inhuman and quite inappropriate to our cultural conciousness – regardless of their historical and scientific importance, their state of conservation or public interest. The recently opened Österreichisches Kriminalmuseum in Vienna displays the head of a female delinquent from the 19th century. Even taking the particular character of the museum into account, this seems incomprehensibly tastless and only helps to increase the public's primitive urges for gruesome thrills. There can be no doubt that the clearly stated recommendations of the "Code of Ethics" issued by ICOM must be applied if the "Man from the Hauslabjoch" is ever exhibited. The press coverage so far has already shown how difficult it is for scientists to secure some form of dispassionate reporting concerned more with the interests of science than with morbid curiosity. The many pictures of the disgraceful treatment accorded to the "Man from the Hauslabjoch", reprinted in dozens of articles in the press, have created a very bad impression. This should not be increased by the further profanation a public display would entail. Whether it is Angelo Soliman, an Egyptian pharaoh or the "Man from the Hauslabjoch", these bodies deserve some remnant of human dignity which we must try to retain for them, irrespective of their academic and historic testimonies. After all, the humane treatment and respect for the human remains of our past also honours the sciences concerned. Whether it is archaeology, the history of medicine, prehistory or anthropology – in the end they are all united in their search for the identity of man, found not only in all the achievements of art and culture but also in the physical remains of human existance. It is one of the duties of museums to demonstrate this. While it is part of their responsibility to make the results of academic research accessible to a wider public, they must also exercise restraint in order to protect the timeless dignity of man in life and in death.

Summary

Interest in Egyptian mummies dates back well into the past. In the 12th and 13th century, Arabs believed Ancient Egyptian mummies to have healing powers, and this lead not only to the destruction of thousands of mummies, it was also the reason why the belief in the medical importance of Egyptian mummies remained unquestioned in Europe until the end of the 19th century. In addition, numerous mummies were brought to Europe in the wake of Napoleon's expedition to Egypt, which gave a new dimension to the public's. Since 1830, the public unwrapping of mummies from Egypt became a popular entertainment in London. Proceeding from the fate of the African servant of count Lobkowitz and count Liechtenstein, who was mummified after his death and publicly exhibited in the show-cabinets of the Natural History Rooms, and the criticism already voiced then, this paper believes that, taking the existing guidelines on ethics issued by ICOM into account, mummies should not be exhibited publicly in exhibitions or museums, except for purely scholarly reasons. This, of course, applies not only to Ancient Egyptian mummies, but also to the "Man from the Hauslabjoch".

Zusammenfassung

Das Interesse an ägyptischen Mumien reicht weit in die Vergangenheit zurück. Die Heilkräfte, die die Araber im 12. und 13. Jhdt. den altägyptischen Mumien zuschrieben, führten zur Zerstörung von Tausenden Mumien, und bis zum Ende des 19. Jhdt.s hielt man in Europa am Glauben an ihre medizinische Bedeutung fest. Im Gefolge der Ägyptenexpedition Napoleons wurden zahlreiche Mumien nach Europa gebracht und weckten enormes Interesse bei der Bevölkerung. Ab 1830 war das öffentliche Auswickeln von ägyptischen Mumien in London geradezu eine Volksbelustigung. Ausgehend von dem Schicksal des schwarzen Dieners der Grafen Lobkowitz und Liechtenstein, der nach seinem Tod in Wien 1796 mumifiziert und in den Schauvitrinen der naturgeschichtlichen Sammlung ausgestellt wurde – was schon damals auf Kritik stieß – kommt der Autor zum Schluß, daß in Übereinstimmung mit den Ethikrichtlinien der ICOM Mumien nicht in Ausstellungen oder Museen öffentlich zur Schau gestellt werden sollten, es sei denn für rein wissenschaftliche Zwecke. Das gilt natürlich nicht nur für die altägyptischen Mumien, sondern auch für den „Mann vom Hauslabjoch".

Résumé

L'intérêt pour les momies d'Égypte a des origines de longue date. En effet déjà au cours du XIe et XIIe siècle les arabes étaient convaincus des pouvoirs thérapeutiques des momies d'Égypte ce qui provoqua non seulement la destruction de milliers de momies mais fût aussi la raison pourquoi elles gardèrent leur importance pour la médecine européene jusqu'à la fin du XIXe siècle. Par la suite une grande quantité de momies fût transférée en Europe suite aux expéditions de Napoléon en Égypte ce qui enrichit l'intérêt du public d'une nouvelle dimension: en 1830 naquit à Londres le deux de société de défaire les bandeaux des momies. Le sort du serviteur africain des comtes Lobkowitz et Liechtenstein, décédé à Vienne en 1796, momifié et exposé dans le cabinet du musée d'histoire naturelle fût déjà à l'epoque l'objet de critiques. Aujourd'hui les normes étiques du conseil international des musées icom prevoyent que les momies ne soyent pas exposées au public dans des musées sauf pour des raisons purement scientifiques. Cela ne s'applique pas seulement aux momies d'égypte mais aussi à la «momie du glacier».

Riassunto

L'interesse per le mummie dell'antico Egitto risale a tempi assai lontani. Le virtù terapeutiche attribuite dagli arabi alle mummie egizie soprattutto nell' 11. e nel 12. secolo furono la causa non solo della distruzione di migliaia di mummie, ma fecero sì che l'importanza della mummia nel campo della medicina fosse valida anche in Europa fino alla fine del 19. secolo. Oltre a ciò, in seguito alla spedizione di Napoleone, un gran numero di mummie furono trasferite in Europa, per cui l'interesse del pubblico fu accresciuto di una nuova dimensione. Togliere le bende alle mummie che provenivano dall'Egitto era divenuto a Londra nel 1830 si può dire un gioco di società. Il destino di un servente negro del conte Lobkowitz e Lichtenstein, originario dell'Africa e morto a Vienna nel 1796, che dopo la sua morte fu mummificato ed esposto al pubblico nelle sale del Museo di storia naturale, fu già allora oggetto di critica: in base alle direttive etiche del consiglio internazionale dei musei ICOM attualmente in vigore, si sostiene l'opinione che le mummie non debbano venire esposte al pubblico nell'ambito di mostre o in musei, se non per motivi di carattere esclusivamente scientifico. Ovviamente questa conclusione è valida non solo per le mummie egizie, ma anche per «l'uomo di Hauslabjoch».

Correspondence: Dr. Wilfried Seipel, Kunsthistorisches Museum, Burgring 5, 1010 Vienna, Austria.

Current technology in the examination of ancient man

P. K. Lewin

Hospital for Sick Children, Toronto, Ontario, Canada

In recent years there has been an explosive outburst of research in the quest for information about the life, disease and death of Ancient man.

This research has been focussed on the numerous artificially and naturally preserved bodies found not only in cemeteries, but in sites as varied as the bogs of Denmark and England, and the permafrost of Northern Canada and Siberia, and of course the "Iceman" found recently in a glacier in the Ötztal Alps, on the border between Austria and Italy.

Unfortunately many of the investigations have been undertaken in a circus-like atmosphere, with autopsies of "mummies" being conducted under the glare of media lights, resulting in the destruction of invaluable human remains, with often very little scientific information, if any, being obtained.

However, the ethical examination of ancient human remains by a recognized scientific team is important, because a multidisciplinary approach can often determine the cause of the demise of those examined in the context of their geographic and cultural environment. This type of information is needed to compile data on the status of health and disease of ancient man. These investigations can often determine the causes of death of preserved individuals, whether by trauma or diseases such as infection or cancer. The dietary status of the individuals can also be evaluated. Such information can then be compared with that of present day populations, both locally and world wide, and this knowledge will be most useful in determining the history of diseases that afflict mankind today. With newer scientific techniques, particularly in molecular biology and genetics, data may be found which may prove to be important in predicting deleterious genetic drifts in certain populations, the adaptation to various infectious diseases and the increasing virulence of others. In addition, the examination of ancient man may give us an indication of the effect of environmental factors, particularly that of pollution, on the health of present mankind.

With this in mind, I was amazed when I first examined, in 1966 (1), skin from an ancient Egyptian mummified hand. The tissues were well preserved at the ultrastructural level (Figure 1), confirming similar observations on Egyptian mummified tissues by light microscopy performed by Armand Ruffer in the late 1880's (2). Electron microscopy demonstrated intact cell membranes, nuclei, cellular organelles including mitochondria, intact fibres like collagen and fibrin. By inference, the demonstration of intact organelles and chemical structures would indicate that the underlying genetic materials in the cells are at least in part viable and intact.

Although I tried in 1984 (3) to extract DNA (deoxyribonucleic acid) from mummified material for cloning, I was unsuccessful. However, viable genetic DNA fragments from an Egyptian mummy was extracted and then cloned by Paabo in 1985 (4). This important new technique can be further enhanced by the possibility of extracting minute quantities of DNA from human tissues including hair, and amplifying these minute quantities by the use of the polymerization chain reaction (PCR). Once amplified into respectable quantities, complementary genetic markers such as the HLA tissue type, can be extracted and compared to those of present day individuals. Once these techniques are perfected, the investigative routes are limitless and can determine the exposure of ancient man to various infectious agents. The connectedness of people from wide-spread regions can also be determined through genetic analysis.

With the permission of President Sadat of Egypt in 1979, and the assistance of Professor Mourad A. Sherif of Cairo, Dr. Donald Hopkins and I were allowed to examine minute scrapings from the skin of the mummy of Ramses the Fifth in the royal mummy collection in the Cairo museum. Ramses the Fifth is covered over by a generalised pimple-like rash which is also noted on his face and body (Figs. 2A & 2B), which has always been suspected to be due to smallpox. We were only allowed to collect minute scrapings which had had pealed off his neck, and the specimens were sent on to the United States Center for Disease Control in Atlanta, Georgia, and examined there by Drs. Erskine Palmer and James Nakano. The specimens were found to be non-infective, but on electron microscopy, smallpox-like particles were noted (Figure 3). The immunoprecipitation test for

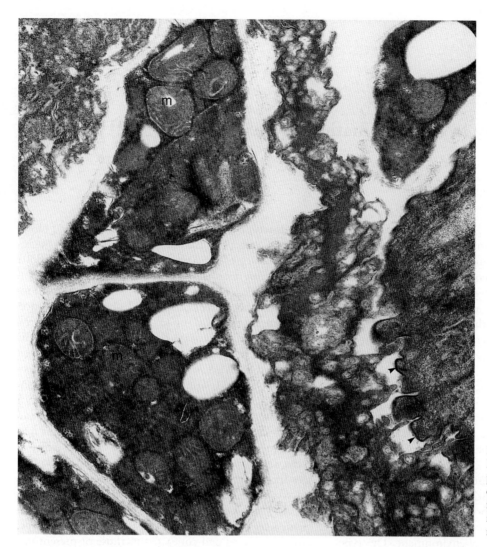

Fig. 1. Electron microscopy of Egyptian mummified skin (approx. 3000 years old). Epidermal cells (arrowheads) are separated from the dermal cells containing mitochondria (*m*) by a basement membrane. ×35,000

smallpox was also positive. This indicates that smallpox did occur in ancient Egypt, and was probably one of the ancient plagues quoted in the Bible (5, 6).

Although the smallpox specimen from ancient Egypt was not viable, it is probable that the smallpox virus and possibly other infectious agents, like that of the influenza pandemic of 1918, are still viable and present in some bodies preserved in the permafrost here in the Canadian arctic and in Siberia. It is therefore imperative that all archeological, anthropological or paleopathological expeditions working in areas of permafrost, including those in glacial areas, should always wear gloves and handle biological and human remains with especial care, using aseptic techniques. This will not only prevent contamination of specimens, but also be useful in preventing any infectious particles infecting the hand by way of cuts, etc. (7).

The newer radiological imaging methods have revolutionised the examination of ancient man and animals, and I was involved in 1977 with Dr. Harwood Nash in pioneering the use of Computerized X-Ray Tomography (CT Scan) for application in the field of archeology, anthropology and medical archeology. We first examined the naturally mummified brain of "Nakht", a 14 year old weaver who died about 3200 years ago in ancient Egypt. The brain, although shrunk, was well preserved, and CT Scan demonstrated intact ventricles and some differentiation between grey and white areas (8). This was followed by a whole body CT scan of the priestess Djemetesankh, who died about 2500 years ago. The mummy is still preserved intact in her coffin (9).

In the technique used for CT scan, the object to be examined is exposed to a series of sequential x-rays at different angles, the x-ray source rotating around the object in a clockwise direction. The varying relationships of the x-ray images on the internal structures are then resolved in a computer, to yield a digital representation which can be translated into different densities of colour to produce detailed two-dimensional images in various planes of objects examined.

A

B

Fig. 2. Ramses V (collection of Royal Mummies, Cairo Museum). (A, top) Note blisters on right cheek near the jaw. (B, right) Note blisters on left shoulder. Courtesy of Dr. Don Hopkins, Chicago

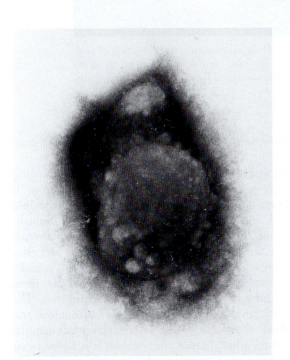

Fig. 3. Electron microscopic examination of skin from Ramses V demonstrating a smallpox like particle (×100,000)

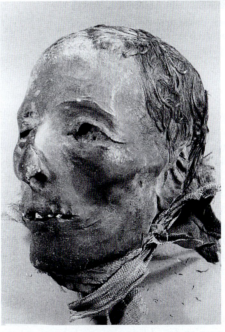

Fig. 4. Ancient Egyptian mummified head (Greco-Roman period)

In 1987, John Stevens, Judy Trogadis and I (10) extended these procedures to produce the first three-dimensional images from the sequential data obtained from CT scans. Thus, with these methods, we can visualize the mummy in an intact coffin, electronically peel off the skin, and examine the underlying bones and body contents, such as the brain (Figures 4–6). Some of the newer imaging techniques, such as magnetic resonance imaging (MRI), which measures the magnetic resonance of the hydrogen proton molecule of water in a powerful magnetic field, is not useful in archeology because of the dryness of most mummified material. Newer MRI systems, using phosphorus or other elements, may produce very detailed tomographic images in the future. With the increasing resolution of fine sections of the body and organs, imaging

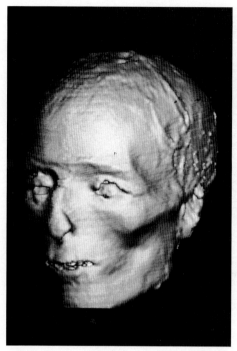

Fig. 5. Computer image reconstruction of the Egyptian mummified head seen in Fig. 4 (Lewin, Stevens and Trogadis)

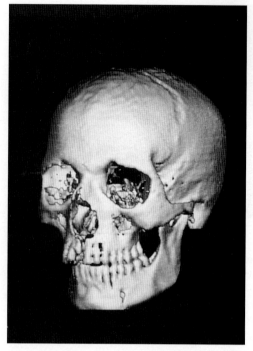

Fig. 6. 3D digital computer reconstruction of skull from an Egyptian mummified head (Fig. 4) of the Greco-Roman period (Lewin, Stevens and Trogadis)

histopathology can be undertaken to diagnose pathology in tissues.

Thus images produced by these new techniques provide morphological and pathological data previously available only through destructive measures. These techniques therefore provide valuable methods for allowing ancient specimens to be kept intact for posterity (11).

The exhumation and handling of any ancient or mummified individual always has to proceed with humility and due deference, particularly in regard to the local community whose ancestors are being investigated. Permission, and even participation of the local residents or villagers, is to be encouraged. Such actions facilitate the communication between the archeological team and the local community. In addition to careful archeological planning, it is important that such teams have an anthropologically trained advisor to supervise and participate in the examination of human remains in graves or other sites. The team members should be prepared to document either photographically or by video, the in situ presentation of such remains and also document the method used to remove the human or biological remains. Those handling biological material should always practice aseptic techniques by wearing gloves, to prevent contamination of dried up bodies by soiled hands. Even more precautions should be taken in relation to frozen remains, because of the remote concern that possible viable infectious material such as smallpox may be present.

If portable imaging units are available, all biological material should be x-rayed prior to examination. If specimens are to be taken, sterile instruments should be used, and the sterile specimens should be immediately sent for microbiological and virological investigations. Samples should also be sent for light and electron microscopy, and lastly if pieces of skin, muscle, hair, or any other organs or tissues are available, these should be collected for possible DNA extraction. The scientific advisor should be available to monitor and distribute these specimens. If the preserved material came from a hot, humid climate then it is advised that these specimens be kept in a dry air-tight container filled with nitrogen or other inert gas. Frozen material should be preserved in the frozen state in a container with high humidity.

The only method that I totally disagree with is the use of ionizing agents, such as gamma or X-rays to preserve biological materials. This method was applied to Ramses II to prevent fungus infection. Such irradiation has in my view destroyed the mummy of Ramses II (12), particularly as the doses were so large that it destroyed the molecular configuration of proteins and DNA, thus preventing future chemical analysis or DNA extraction for genetic investigation or cloning.

In summary the ethical investigations of preserved human remains offers us an insight into the kinds of disease, the causes of death and the type of nutrition of ancient man. Such information will increase our understanding of the diseases that affect mankind today.

Acknowledgement

This paper is dedicated to all my colleagues and assistants who, over the years, have tolerated and assisted me in my exotic extracurricular activities.

Summary

The ethical investigation of preserved human remains offers us an insight into the kinds of disease, the causes of death and the type of nutrition of ancient man.

The author describes his pioneering efforts in the field of electron microscopy of Egyptian mummified tissues and the use of the latest imaging techniques in medical archeology. These CT scans and 3-dimensional images, which in many cases demonstrate very detailed structure, will obviate the necessity of autopsy, and so preserve these specimens for posterity.

With the advent of genetic (DNA) extraction from the tissues of ancient man and the possibility of generating large amounts of copies of the genetic material, we are now in a position to compare the genetic structure of ancient and modern man.

Lastly, a warning note is issued about the dangers of possible viable agents, such as smallpox, being preserved on bodies in the permafrost, and the precautions that should be taken to deal with this.

Zusammenfassung

Die anthropologische Untersuchung erhaltener menschlicher Reste gibt uns einen Einblick in Krankheiten, Todesursachen und Ernährung des Urmenschen.

Der Autor beschreibt seine ersten Versuche auf dem Gebiet der Elektronenmikroskopie von mumifizierten Geweben aus Ägypten und die Anwendung der neuesten Bildtechniken in der medizinischen Archäologie. Diese Computertomographien und dreidimensionalen Bilder, die in vielen Fällen sehr detaillierte Strukturen zeigen, ersetzen eine Autopsie und erhalten somit diese Funde für die Nachwelt.

Durch die Anwendung der Methode der genetischen Extraktion (DNA) aus erhaltenen Geweberesten und die Möglichkeit, daraus eine große Anzahl von Kopien des genetischen Materials zu erzeugen, sind wir nun in der Lage, die genetischen Strukturen mit jenen des modernen Menschen zu vergleichen.

Schließlich wird noch vor möglichen Gefahren bezüglich keimbarer Agens gewarnt, wie z. B. Pocken, die an Körpern aus dem Permafrost noch erhalten sind, und es werden Vorsichtsmaßnahmen vorgeschlagen, die gegebenenfalls angewandt werden sollten.

Résumé

La recherche éthique des restes humains nous offre la possibilité de faire des études sur des différentes maladies, des raisons pour la mort et le type de nutrition des hommes anciens.

L'auteur décrit ses premiers efforts dans le domaine d'examen par un microscope électronique d'un tissu momifié égyptien et ainsi l'utilisation des dernières nouveautés de la technique vidéo dans le domaine de l'archéologie médicale. Ces résultats de la tomodensitométrie et les images à trois dimensions, qui montrent assez souvent des structures très détaillées, remplacent l'autopsie et présèrvent ainsi ces spécimens pour la postérité.

Avec l'invention de l'extraction génétique des tissus (l'A.D.N.) d'un homme ancien et la possibilité d'en faire un grand nombre de reproduction du matériel génétique nous sommes en mésure de comparer la structure génétique de l'homme ancien avec celle de l'homme moderne. A la fin on signale le danger possible des agents viables, comme la variole, qui ont été préservés sur les corps dans le permafrost ainsi que les précautions qu'il faudrait prendre.

Riassunto

La ricerca etica su resti umani ci offre la possibilitá di effettuare studi sulle varie malattie, le ragioni che hanno portato alla morte ed il tipo di alimentazione dell'uomo primitivo. L'autore descrive i suoi primi sforzi nell'ambito dell'analisi con il microscopio elettronico di un tessuto mummificato egiziano, videotecnologie piú avanzate tridimensionali, che evidenziano spesso strutture molto dettagliate in grado di sostituire l'autopsia, conservando cosí i ritrovamenti per le generazioni future. Con l'intervento dell'estrazione genetica di tessuto, il DNA, da un uomo primitivo e la possibilitá di riprodurlo in gran quantiá, siamo in grado di paragonare le strutture genetiche dell'uomo primitivo con quelle dell'uomo moderno. Infine si ricorda il pericolo di possibili agenti variabili come la varicella, che si sono conservati sui corpi nel permafrost, come anche le precauzioni che si dovrebbero prendere.

References

1. Lewin P. K. Paleo-electron microscopy of mummified tissue. Nature, 1967: 213, 416–417.
2. Ruffer M. A. Studies in the Paleopathology of Egypt (Ed. Moodie R. L.) 1921 University of Chicago Press.
3. Lewin P. K. DNA extraction for cloning from ancient Egyptian mummified tissue (Preliminary report). Paleopathology Newsletter Sept. 1984: 47,5.
4. Paabo S. Molecular cloning of ancient Egyptian mummy DNA. Nature 1985: 314, 644–645.
5. Lewin P. K., "Mummy" riddles unravelled. Bulletin, Microscopical Soc. Can. 1984: 12, 4–8.
6. Lewin P. K. Technological innovation and discoveries in the investigation of ancient preserved man. Human Paleopathology, current synthesis and future options. (Eds. Ortner D. J. and Aufderheide A. C.) 1991, 90–91. Smithsonian Institution Press, Washington.
7. Lewin P. K. Mummified frozen smallpox: Is it a threat? J. A. M. A. 1985: 253, 3095.
8. Lewin P. K. and Harwood-Nash D. C., X-Ray Computed axial tomography of an ancient Egyptian brain. IRCS (Inter-

national research communication system, Med.Sci.) 1977: 5.78.
9. Lewin P. K. Whole body CT scan of an Egyptian mummy. Paleopathology Newsletter (Abstract) April 1978, T7–T8.
10. Lewin P. K., Trogadis J. E. and Stevens J. K. Three dimensional reconstruction from serial X-ray tomography of an ancient Egyptian mummified head. Clinical Anatomy 1990: 3, 215–218.
11. Burns G. and Lewin P. K. Eco-archeometry, the development of environmental sciences in archeology and paleopathology. Journal of the Irish Colleges of Physicians and Surgeons 1992: 21(2) 98–100.
12. Lewin P. K. The use of modern technology in medical archeology. Bull. et Mém. de la Soc. D'Anthrop. de Paris 1981, t. 8 série XIII, 339–341.

Correspondence: Dr. Peter K. Lewin, The Hospital for Sick Children, 555 University Avenue, Toronto, Ontario M5G 1X8, Canada.

Non-invasive investigations of human mummified remains by radiographic techniques

D. R. Hunt and **L. M. Hopper**
Department of Anthropology, National Museum of Natural History, Washington, D.C., U.S.A.

Introduction

Radiography has been used as a tool for mummy research practically since its discovery by Roentgen. Within the same year as Roentgen's discovery, Konig (1896) reported on an x-ray image taken of a mummy in Germany. The next year, Sir Flinders Petrie used x-rays to study a mummy at the British Museum but was constrained by the size and weight of the equipment and only imaged the feet. In 1897 in Vienna, a doctor named Block x-rayed a whole mummy for a medical study. Eminent Egyptologist, Georg Ebers studied these findings (El Mahdy, 1989: 75). Tuthmoses IV was x-rayed by Khayat in 1903 and the films were studied by G. Eliot Smith. Smith determined the remains were much younger than the age estimated from translated writings (Smith, 1914). The results of this report began a long debate concerning historical estimates of Pharaonic age versus the biological age of Thutmoses. Were the historical estimates wrong? Was Smith wrong in his assessment? Or could the mummy have been mis-identified?

As radiographic technology advanced and its value became more widely recognized, in depth radiologic studies of mummies became more common. R. Moodie radiographically recorded and reported on the Egyptian and Peruvian mummified remains at the Field Museum, Chicago in 1931. Over the course of several years Frans Jonckheere and P. H. K. Gray studied and reported on several hundred mummies using radiographs as one of their main means of investigation (Jonckheere and Gray, 1951). All the mummies at the City Museum in Liverpool, England were x-rayed in 1966 by Gray and Slow (1968). A landmark study using radiography was done in 1967 at the Cairo Museum in a joint venture between the University of Michigan School of Dentistry, University of Alexandria and the Cairo Museum. The first publication from this expedition, *X-Raying the Pharaohs* (Harris and Weeks, 1973) became a very popular book for both the student and amateur Egyptologist. A more technical publication, *X-Ray Atlas of the Royal Mummies* (Harris and Wente, 1980) is an excellent report of this professional research.

The next generation of radiological technology, Computerized Tomography (CT or CAT scan) was developed rather recently in the late 1960's. Since its inception, CT has also been used as a research tool on Egyptian mummies. CT images prove to be invaluable in ascertaining more exact information about mummies which plain film images cannot provide (Pahl, 1986). Studies have been performed in Philadelphia in 1973, as part of the Manchester Project in 1975 (David, 1979; 1988), and in Toronto in 1977, Boston in 1984 (D'Auria, et al., 1988), Minneapolis in 1983 (Notman, 1986), Lausanne in 1984 and Lyons in 1985 (Goyon and Josset, 1988). The Lyons mummy was believed to be a sailor (evidenced by the rigging still attached to the sailcloth used as a outer wrapping) and mummified in the mid or lower class style (Herodotus II,85 in Budge, 1925: 205) so most of the internal organs were still present.

Several mummies from museums in Minnesota were CT'ed in 1983 (Notman, 1986). One particularly interesting case was Lady Tashat who had fractures and strangely positioned skeletal elements. Plain film x-ray indicated broken bones and the presence of a adult second skull between the legs of the mummy, but the superimposing of many of the skeletal elements made diagnosis difficult. CT scans produced first in Minneapolis and then a second scanning at the Mayo Clinic, Rochester clarified the presence of damaged ribs and vertebrae scattered in the chest cavity and the left upper arm and clavicle broken downward. It is hypothesized that the second skull may have been placed between the legs in a later rewrapping of the mummy. Egyptian priests commonly repaired mummies that had been vandalized by grave-robbers. In this instance, they may have inadvertently (or advertently) put the skull within the mummy wrappings for its 'reburial'.

Because of technological advances in computation and data storage abilities of CT machines, Three-dimensional imaging is now possible and is becoming more widely available. The mummy Tabes from the Museum of Fine Art, Boston may have been the first reported 3-D head reconstruction (D'Auria et al., 1988). Since that time, this technique has been used for various 3-dimensional reconstructions of mummified remains, Egyptian and otherwise. Indiana University School of Medicine in 1990 studied the female mummy Wenuhotep using 3-dimensional CT with normal findings for a 550–400 B.C. mummy (Braunstein and Contes, 1992). Using more sophisticated computer software, such as presented by zur Nedden and Wicke (1992) on the Hauslabjoch "Iceman", permits the use of data gathered from the CT scan to reproduce the skull or other CT window features in wax or polymer using laser image cutters.

Nuclear Magnetic Resonance Imaging (NMR or MRI) has become another new technique for diagnostic radiology in the 1980's, however, it is inapplicable in mummy investigations due to the lack of active hydrogen (the necessary proton source) in mummified remains. This inability to use MRI was demonstrated when attempts were made to use it in research on Lady Tashat at the Mayo Clinic in 1983 (Notman, 1986). Using latest generation MRI and new enhancing techniques, an attempt to generate images of the NMNH child mummy was done at GWMC during the present study. The efforts were again fruitless. Until other proton ions are available as resonators, dried tissue remains will not successfully produce images. However, the capabilities of CT imaging are now sophisticated enough to image structures (especially in static 'patients') in vertical, horizontal and in 3-dimensional formats as mentioned above.

The study of these four mummies was instigated by renovation of the Egyptian Culture section of the Western Civilization Hall exhibit at the National Museum of Natural History, Smithsonian Institution. It was decided to change one of the existing displays on burial practices into a tomb scene with a mummy surrounded with associated funerary items and artifacts of the same time period. This change is part of overall updating of the information presented in the Egyptian displays, but another impetus for the renovation came from public demand for mummy exhibits after all NMNH mummies were removed in 1991 with the closing of the Biological Anthropology-Human Variation/Evolution Hall.

To dispel the Natural History Education Office's criticism that Physical Anthropology exhibits are freak shows, coordination with the Ethnology and Archeology Division curatorial staff is ongoing to produce an exhibit which will portray the mummy in the most realistic of archeological settings, associated with period artifacts. The display will also integrate the religious and cultural significance of the mummification practice in Egypt. And present an in depth scientific investigation of the mummy to educate the public on the methods and technology used in physical anthropological research concerning human mummified remains and the significance of this research while preserving the integrity of the specimen.

Fig. 1 (top). General Electric SPX Fluroscopic and Radiographic Unit with Spot Film Camera at the George Washington University Medical Center Department of Radiology. Angie Dopkowski aligning NMNH-385664 for cross-table lateral projection. Fig. 2 (right). General Electric GE9800 Quick HiLight with Scintillation Ceramic Detectors at the George Washington University Medical Center Department of Radiology. NMNH-126790 heading into scanner

Materials and methods

Four intact Egyptian mummies from the Department of Anthropology, National Museum of Natural History, were chosen for study in a joint venture with the Washington University Medical Center Radiological Department using latest generation radiological equipment. The anthropology department's objective in this study was to increase the knowledge of and the general and specific information about each of these mummies without damaging their integrity. The provenance for these specimens is minimal and therefore this type of research is especially helpful to acquire lacking information, such as age, sex, cultural chronology and possible individual characteristics. Results from the investigation will be added to the museum catalog information to aid in future use of these specimens for research and exhibition purposes.

Head Diagnostic and Forensic Radiological Technologist, Lisa M. Hopper and staff focused on manipulations and modifications of radiologic methodology to produce clear, accurate and minimally distorted images. Their research results demonstrated the capabilities and limitations of their present equipment for forensic as well as diagnostic radiological analysis and the abilities of the equipment to be manipulated in problematic imaging, particularly with radio-opacities, superimpositioning and contrast.

Fluoroscopy and spot film imaging using a General Electric SPX Fluoroscopic and Radiographic Unit with Spot Film Camera (Fig. 1) was performed on Mummy 381234 and 126790 as an initial investigation into the type of problems which might be encountered inside the mummies. Complete body plain film radiography was performed on all four mummies using the General Electric SPX F & R Unit with Spot Film Camera. Standard anterior/posterior and crosstable radiographic procedures were employed. Conventional protocol was followed to avoid distortion due to tube to film or object to film alignment. Modifications for dry tissue radiography were made by varying milliamperage seconds (mAs), kilovoltage (kVp), screen speed choice and a variety of film cassettes (Cahoon, 1974). These variables were relatively consistent for the four specimens, suggesting some uniformity within technique deviations, just as there are relative standards for living diagnostic radiography. Adjustments through density variations were made by using filters. Technique variation specific to each mummy are discussed in the case studies below.

Computer Tomography (CT) was accomplished using a General Electric GE 9800 Quick HiLight with scintillation ceramic de-

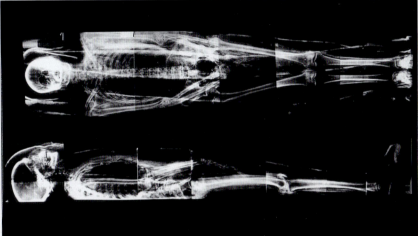

Fig. 3 (left). Mummy NMNH-126790. Cartonnage is typical of Ptolemaic period. Fig. 4 (top). Composite anterior-posterior and lateral plain film radiographs of NMNH-126790. Note cartonnage imaging and modern opaque pin and nail artifacts

Fig. 5. Mummy NMNH-126790 situated in its sarcophagus base

Fig. 6. External sarcophagus case of NMNH-126790 with painted face, false beard and blue painted hair

tectors (Fig. 2). In all CT scans, body boundaries were marked and reconstruction diameter set at 20 cm (Bushong, 1993: 408–427). The linear attenuation coefficient was set for dense bone. All images were reconstructed with standard algorithm set for bone windows.

Automated film processing by a Picker MB6 was used to maintain emulsion and processing uniformity.

Particular difficulties were experienced when radiographing each specimen. The fragile nature of the subjects prohibited rotation or movement of extremities, thus producing super-impositions with the spine, humeri, hands and femora. Problems in image interpretation in the plain film x-rays was usually clarified by CT images. Additionally, Mummy 126790 was permanently affixed to its sarcophagus by resins requiring an additional photo-radiographic restriction in the lateral projections. Special filters were used to penetrate the sarcophagus to produce quality images without compromising clarity by over-penetrating the less dense structures above the sarcophagus. This procedure was accomplished using a Clear-Pb Lateral Decubitus Filter and was done for this case and also subsequently Mummy 381234, because regular contrast variations were suboptimal. This compensation filter corrected spine and humerus imposition exposure while attenuating radiation to the skull and ribs, thus minimizing "image burn out". Without the use of this filter, 50 % of all x-rays taken would have been suboptimal.

Contrast was a continual challenge. Most contrast problems were rectified by the above described filters or adjustments of kVp. Some resin areas were so dense that there was inadequate penetration with conventional x-ray even with the use of filters or bucky devices and changes in mAs and kVp. Diagnostic imaging through these resins was generally produced in CAT scans.

Another area of challenge was object film distance (OFD). Besides the normal amount of distortion of object to film distance, the sarcophagus in Mummy 126790 required a greater OFD, thus additionally distorting some images. This anode heel effect (Bushong 1993) was a limiting factor in the short-focus radiography on large film size used in this study. To correct the elongation and distortion of some radiographs, we shortened the exposure field to 14 inches as opposed to 17 inches simply by turning the cassette. To improve image sharpness and reduce irradiation fog, detail screens were used.

Case study #1 – Mummy NMNH-126790 (Figs. 3 & 4)

LOCATION: LUXOR, EGYPT
COLLECTOR: HONORABLE S.S. COX (Constantinople, Turkey)
ACCESSION NO: 17401
ACCESSION DATE: 05/14/1886
RELATIVE DATE: Ptolemaic Period (305–52 B.C.)
RADIOCARBON DATE: not performed

Catalog comments

National Museum of Natural History Report for 1886, Pt II., p. 50. states; "An Egyptian mummy in excellent state of preservation and obtained at Luxor, in Upper Egypt, by Honorable, S. S. Cox, U.S. Minister of Turkey, was presented by him to the Museum. This Mummy measuring 5′ 6″ in height is delicately proportioned, and is altogether a very good specimen."

Specimen information

AGE: 40–50 years (from skeletal aging)
LENGTH: 170.5 cm
SEX: Male (from skeletal features)
CONDITION: Good–fully wrapped, in Sarcophagus, with cartonnages

Some conservation of the mummy was done by the Smithsonian's Anthropology Laboratory in 1932. No meaningful hieroglyphics are present on the sarcophagus, breastplates, or wrappings to identify the individual or dynastic period. The method of wrapping, sarcophagus manufacture, position of the hand over the groin area and design of the cartonnages are indicative of the Ptolemaic period (304–30 B.C.). Associating artifacts in the archeology collections are also indicative of the Ptolemaic period.

Radiologic studies completed: Fluoroscopy, Plain film radiographs, CAT scans (Tables 1 a & 1 b).

Table 1 a. Exposure guide for plain film radiographs of NMNH-126790

Region	Proj.	Decub. Filter	mAs	kVp	Screen	FFP	Grid
Skull	AP	NO	20	70	Detail	54″	Bucky
Skull	LAT	YES	10	70	Detail	44″	8:1
Chest	AP	NO	20	70	Detail	54″	Bucky
Chest	LAT	YES	16	80	Detail	44″	8:1
Abdomen	AP	NO	20	76	Detail	54″	Bucky
Abdomen	LAT	YES	16	80	Detail	44″	8:1
Pelvis	AP	NO	32	74	Detail	54″	Bucky
Pelvis	LAT	YES	25	80	Detail	44″	8:1
Femur	AP	NO	50	62	Detail	54″	Bucky
Femur/Pen. Resin	AP	YES	50	70	Detail	54″	Bucky
Femur	LAT	YES	20	70	Detail	44″	NO
Knee/Bilat Knee	AP	NO	40	62	Detail	54″	Bucky
Superimposed Tibia/Fibula	LAT	NO	20	70	Detail	44″	8:1
(Bilat) Tibia/Fibula	AP	NO	40	62	Detail	54″	Bucky
Superimposed Feet/Bilat Feet	LAT	NO	20	70	Detail	44″	8:1
	AP	NO	40	62	Detail	54″	Bucky
Superimposed	LAT	YES	20	70	Detail	44″	8:1

Table 1 b. CT scan transaxial image measurements for NMNH-126790

Region	Slice (mm)
Head	10 mm
Chest	20 mm
Abdomen	10 mm
Lower extremity	10 mm

Physical assessment

This mummy is completely wrapped with original cartonnages still present on the front of the mummy and it is fixed into its associating sarcophagus (Fig. 5). The sarcophagus is made of wood in the shape of a wrapped mummy. The areas where the wood pieces have been fitted together are filled with a granular paste-like substance and covered with pitch. The exterior is painted with a black color wash. A depiction of face and hair are fashioned on the lid with a stucco substance being the final cover over the wood carving. The face is brown with open eyes and false beard. The beard and the hair are colored with a faded blue paint (Fig. 6). The lid is secured onto the base by six tongue and groove alignment wood plates, the plates dowelled into the base. There are stylized border paintings in a gold color on the feet and lower leg region of the lid. A rectangular border contains hieroglyphic writing, but the translation is not meaningful or identifiable.

The cartonnages are still in good shape except for some wear damage to the feet covering suffered whenever the sarcophagus lid was removed. The quality of the cartonnages is moderate, the gold leaf covering the deities faces is not well crafted and rather haphazardly applied. Representations on the plates are standard for Ptolemaic Period, depicting mummification and burial ceremonies but in a stylized manner, not specific as would have been portrayed in earlier Dynastic artwork.

Since the mummy is completely wrapped, all physical anthropological observations are made using radiographs. One of the more striking features of the lateral head radiograph is the outline presence of the cartonnage mask covering the face (Fig. 7). The supraorbital eminence is distinctive on the skull and the mental eminence is also defined as is the mastoid process and nuchal line, which is actually a torus. All these traits are male features Bass 1987: 81–82; Krogman and İşcan, 1986: 191–193). The sella is not clear on the lateral radiograph and examination of the CT slice at the sella/cribriform level, reveal these bony features are in fact missing, indicating damage to these structures, most probably from brain removal through the nose, a practice not as commonly done in later periods. The two radio-opaque artifacts in the lateral projection are modern finishing nails in the sarcophagus installed at some earlier time. Other artifacts are straight pins used during the 1932 conservation of the mummy to hold the cartonnages onto the front of the mummy.

The dentition is worn, the occlusal surfaces flat and crowns low. There is anti-mortem tooth loss of the maxillary 1st molars with alveolar resorption in process and the mandibular 1st and 2nd molars with alveolar resorption essentially completed. Alveolar resorption around other teeth suggest gum recession and possible periodontal disease. There do not appear to be any abscesses present, however. There is no indication of cranial sutures in either the lateral or A-P head radiographs other than a remnant opacity along the sagittal line suggesting an age greater than 40–45 yrs (Meindel and Lovejoy, 1985). The mid-thoracic and lumbar regions are showing minor osteophytic activity and the acetabulum exhibits slight lipping consistent with an individual in between 30–40 yrs depending on physical activity level. The thyroid is opaque on the lateral x-ray identifying ossification of the cartilage. The extent of the ossification is far enough along to be visible on x-ray and would be indicative of an individual over 40 yrs of age (Černy, 1984). All of these attributes and the level of dental wear and loss would put this individual most likely between 40–45 yrs at death.

The arms of the mummy are along the side of the body with the hands outstretched, side by side and laying over the groin. The pelvis is male-like with a narrow subpubic angle, triangular and narrow pelvic inlet, narrow sciatic notch and short pubic bone length (Bass, 1987: 201–206, Stewart 1979: 104–110). The overall bone size and morphology is also consistent with a male individual. No observable pathologies, traumas or remarkable features are present in the legs or feet of this mummy.

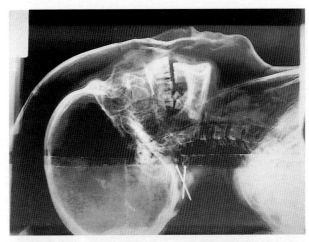

Fig. 7. Lateral cranial radiograph of NMNH-126790 with radio-opaque ghost of cartonnage mask over the face. Projection shows anti-mortem loss of molar dentition and noticeable tooth wear. Note also ossified hyoid and thyroid cartilage

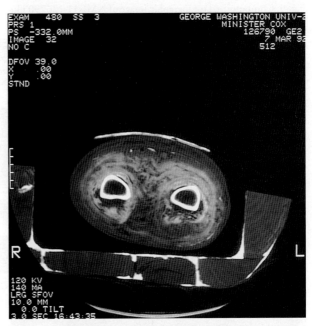

Fig. 8. Computerized tomographic image of NMNH-126790 sectioned through the mid-femur region showing dehydrated soft tissues and layers of wrappings

Tomographic images illustrate mummification procedures with linen packing in the abdominal region, layering of wrapping on the body and the extremities and solidified pools of resins used in preservation. Cross-section views indicate dehydrated muscle groups still present in the abdomen, arms and particularly legs (Fig. 8). One of the most interesting aspects of the CT images is the clear reproduction of the woody growth rings in the boards of the sarcophagus. These features are probably not useable for dendrochronology in their present imaging. However, this may be an indication that computerized tomography could be used as a means to study wood structure and chronology alleviating the need to make cross-sections of valuable archeological artifacts.

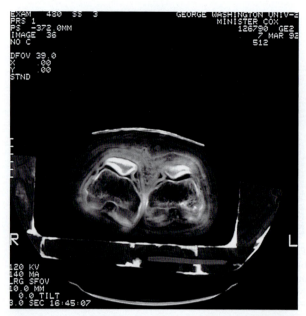

Fig. 9. Computerized tomographic image of NMNH-126790 illustrating the imaging of the woody tree rings in the sarcophagus

The construction of the sarcophagus can also be visualized in the CT scans (Fig. 9). There are three major wood pieces that were used to construct the sarcophagus, all tongue and grooved together with the cracks filled in with material (granular material and pitch) to finish out the exterior. The quality of the workmanship is moderate, since the mummy does not fit well into the encasement, and there is evidence of carving along the sides to make the individual fit. This modification is consistent with the Ptolemaic Period since in the later periods, standard cases were made and then the mummy made to fit the sarcophagus.

Physicians' report

Plain film study

The bony skull is unremarkable with no evidence of fracture or bony abnormality. The sella is normal. The frontal sinuses are well aerated. Left orbit is difficult to see, right orbit appears normal. The cervical vertebral bodies and intervertebral spaces are well maintained. There is no subluxation. The facet joints are unremarkable. The thoracic vertebral bodies and intervertebral bodies are well maintained and well aligned. The ribs are without evidence of fractures. The clavicles scapulae and humeral heads show no evidence of previous trauma, bony lesions or degener-

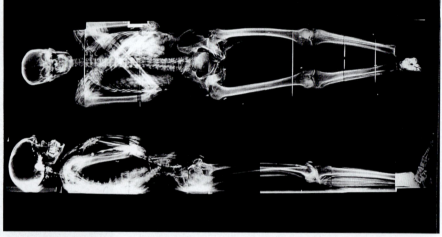

Fig. 10 (left). Mummy NMNH-381234. Fig. 11 (top). Composite anterior-posterior and lateral plain film radiographs of NMNH-381234

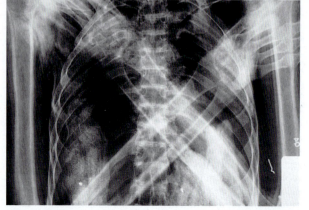

Fig. 12. Anterior-posterior chest radiograph of NMNH-381234 showing rolls of linen packing in the pleural cavity

Fig. 13. Parineal area of NMNH-381234 where tissue was removed to expose pubic symphyses. Packing is also observable in the anus region

ative change. The AC joints are mildly narrowed. The humeri, radii, and ulnae are normal with no evidence of fracture. The wrists and hands are hard to visualize because of overlying artifact. However, there is widening of the right radial scaphoid joint. The carpals, metacarpals, and phalanges are unremarkable and without evidence of fractures. The sternum is unremarkable. No intrathoracic organs present. Multiple overlying artifacts including pins are visual. The lumbar and sacral spine show no vertebral body abnormalities and joint spaces are well maintained. There is no evidence of subluxation. The S1 joint is not well visualized. The symphysis pubis is slightly narrowed. The left and right hip show minimal narrowing and no degenerative change. The femoral necks are intact. The intra-abdominal and pelvis organs have been removed. The femoral heads appear normal and there is no evidence of fracture or bony lesions within the femurs. Knee anatomy is normal except for some mild narrowing of the right joint. The tibiae and fibulae show no abnormalities or fractures. The gross morphology of the ankle and foot appear normal and without trauma except for some slight narrowing of the right ankle joint, but evaluation of these features are difficult because of overlying artifact.

CT examination

This mummy is wrapped with an extensive amount of cloth and still resting in a wooden coffin. The cervical and thoracic spine appear unremarkable with no evidence of fractures. The facet joints are grossly normal. There is a soft tissue density extending the length of the cervical and thoracic cord which may represent the remains of the spinal cord. The shoulders, clavicles,

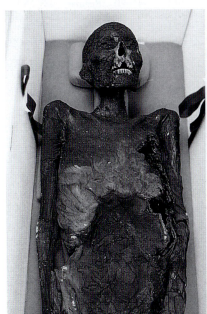
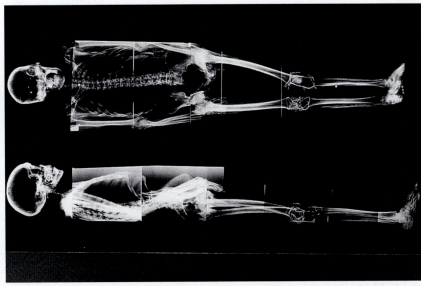

Fig. 14 (left). Mummy NMNH-385664. Fig. 15 (top). Composite anterior-posterior and lateral plain film radiographs of NMNH-385664. Note wire used to reattach legs at the knees

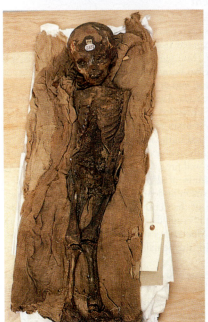
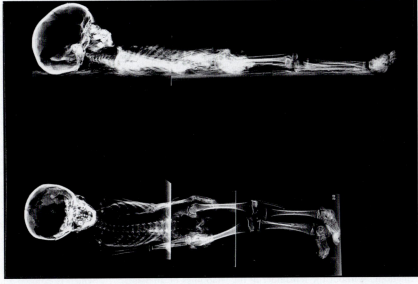

Fig. 16 (left). Mummy NMNH-381235. Fig. 17 (top). Composite anterior-posterior and lateral plain film radiographs of NMNH-381235

left and right humerus, sternum and ribs are all unremarkable. The intrathoracic organs have been removed. There is some soft tissue density seen anterior to the thoracic spine in the upper chest. These densities signify the subcutaneous packing of the mummification process. There are radio-opaque pins along the anterior surface of the chest in front of the mid-section of the sternum. The anterior aspect of the abdomen and pelvis are covered with a dense overlying material. There is also a round (0.5 cm) soft tissue density seen running along the anterior abdominal wall beneath several layers of wrapping extending down below the pelvis and ending just above the ankles. The lumbar and sacral spine are unremarkable. The intra-abdominal contents have been removed. The wrappings from the anterior abdominal wall push into the abdominal cavity and are situated anterior to the vertebral column, down to the pelvis which has a large amount of wrapping within it. The forearm, elbow, wrist and fingers are all present and appear normal. The pelvic organs have also been removed along with the abdominal organs. The majority of the pelvis is filled in with rag. The femurs are unremarkable with a small amount of residual muscle soft tissue identified surrounding. The patellae are both present, the knees appear to be unremarkable with no evidence of degenerative change. The tibiae and fibulae bilaterally are normal with some residual soft tissue and muscle identified around them. The ankles are normal with no evidence of degenerative change. The feet are normal with all phalanges and metatarsals present. There is soft tissue identified around the bones of the feet.

Case study #2 – Mummy NMNH-381234
(Figs. 10 & 11)

LOCATION: Unknown
COLLECTOR: WISTAR INSTITUTE, (Philadelphia, PA.) (# 2396)
ACCESSION NO.: 219346
ACCESSION DATE: 04/17/1958
RELATIVE DATE: New Kingdom (1550–1070 B.C.)
RADIOCARBON DATE: 2433 ± 80 BP (764–398 B.C.)
2935 ± 60 BP (910–486 B.C.) (AMS technique)

Catalog comments

This mummy was donated by the Wistar Institute, Philadelphia, PA (Catalog No. 2396) to the Department of Anthropology, National Museum of Natural History.

Specimen information

AGE: Late 20 years (from skeletal aging)
LENGTH: 161 cm
SEX: Female (from soft tissue and skeletal features)
CONDITION: Fair–partially unwrapped

The color of the mummification process, the state of preservation of the mummy, the method of wrapping and the positioning of the upper extremities across the chest are suggestive of an earlier form of mummification practiced during the early New Kingdom. Without the remainder of the wrappings or an associating sarcophagus, interpretation of cultural time period is tentative.

Radiologic studies completed: Fluoroscopy, plain film radiographs, CAT scans (Tables 2a & 2b).

Table 2a. Exposure guide for plain film radiographs for NMNH-381234

Region	Proj.	Decub. Filter	mAs	kVp	Screen	FFD	Grid
Skull	AP	NO	16	70	Detail	54″	Bucky
Skull	LAT	NO	5	60	Regular	48″	8:1
Chest	AP	NO	16	70	Detail	54″	Bucky
Chest	LAT	NO	4	70	Detail	48″	NO
Abdomen	AP	NO	16	66	Detail	54″	Bucky
Abdomen	LAT	NO	4	70	Regular	48″	NO
Pelvis	AP	NO	16	66	Detail	54″	Bucky
Pelvis	LAT	NO	4	70	Regular	48″	NO
Femur	AP	NO	16	62	Detail	54″	Bucky
Femur	LAT	NO	4	70	Detail	48″	NO
Knee (Bilat)	AP	NO	16	64	Regular	54″	Bucky
Knee Superimposed	LAT	NO	4	70	Detail	48″	Bucky
Tibia/Fibula (Bilat)	AP	NO	16	64	Regular	54″	Bucky
Tibia/Fibula Superimposed	LAT	NO	4	70	Detail	48″	NO
FEET (Bilat)	AP	NO	16	64	Regular	54″	Bucky
FEET Superimposed	LAT	NO	4	50	Detail	48″	NO

Table 2b. CT scan transaxial image measurements for NMNH-381234

Region	Slice (mm)
Head	10 mm
Chest	20 mm
Abdomen	10 mm
Lower extremity	10 mm

Physical assessment

This mummy has been partially unwrapped, only the last 3–7 layers remain. The exposed tissue is hard with a resinous appearance and quite dark (Munsell 5YR 2/1–2/2). The head is still wrapped, but the face has been exposed by the wrapping being cut away along the forehead and down along each cheek. The eyes are absent, the nostrils are packed with a granular looking material and the nose is somewhat distended from the packing. The packing in the nose would suggest brain removal from this orifice. The sella does not appear damaged in the radiographs or the CT images.

The mouth is slightly open exposing the front maxillary dentition. The dentition is well preserved and shows slight to moderate wear. Moderate amounts of dental wear are observable radiographically on the permanent dentition, the occlusal surface is essentially flat on most of the dentition except for the 3rd molars. The 3rd molars are fully erupted and Root Complete indicating an age over 18–20 yrs (Moorrees, Fanning and Hunt (1963a & b)). This feature along with the dental wear would suggest and age in the later 20's.

The arms and legs are still wrapped, except for the right hand which is exposed and is missing a number of phalanges. The left upper leg and patella are also exposed. The arms are crossed over the chest, the right arm over the left and the hands outstretched.

There is a large concretion of black resin under the left hand in the right armpit, a portion of the resin broken away having an obsidian-like appearance. The presence of the abdominal packing precludes observation for an evisceration cut. No indication for a cut was seen in the CT images. Plain film x-rays and in the CT scans both exhibit two rolls of packing inside the chest/abdominal cavity (Fig. 12). The only way this packing could have been inserted would be through an incision, so an evisceration cut must have been made.

In the groin region, wrappings and tissue have been cut away to expose the pubic symphyses (Fig. 13). The symphyses have ventral crests and the subpubic angle is oblique, features of the female pelvis (Suchy and Sutherland 1987; Stewart, 1979: 102–110). The pelvic anatomy seen radiographically is also female. The subpubic angle is wide, the sciatic notch is wide and the pelvic inlet is relatively large and oval (Bass, 1987: 201–206). The sex criterion of the pelvis and the overall bone size indicate sex as female despite a noticeable more male-like supraorbital eminence seen in the lateral projection.

All material has been removed from the dorsal side of the mummy exposing the back and buttocks. There is linen material packed into the parineal area, probably a plug for the anus and vagina. The lower legs have been packed to fill out the calves. If the upper legs had been similarly padded, the padding has been removed. The feet are detached at the ankles, the right foot has been lost at some point in the past. The left foot is damaged, missing the fifth toe phalanges. Where the wrappings are removed, the foot appears to have little tissue present, suggesting that the foot desiccated during mummification, thus the loss of the toes.

Looking at the radiographs for age criterion, epiphyseal union is complete on the proximal femur, tibia and humerus putting an age of over 20 yrs. There is a suggestion of fusion or nearly fused medial clavicles indicating the age of this individual at approximately 26–29 yrs (Webb and Suchy, 1985, Bass, 1987: 129–131). There are no clear indications of degenerative conditions such as articular joint wear or arthritis. The cranial sutures appear to be open and present, not to any point of fusion. The lack of skeletal degeneration and the level of dental wear (for this population) is consistent with an individual in her late 20's.

Physicians' report

Plain film study

The bony skull is remarkable with no evidence of fractures. The sella is normal in size. All cranial sinuses are well aerated. The bony mandible, teeth, and nasal bones are normal and show no evidence of fracture. The orbits are unremarkable. Cervical vertebral bodies are all well aligned with no evidence of subluxation and no joint sclerosis. The intervertebral disc spaces are slightly narrowed. The thoracic spine appears to be normal and in alignment with no compression of the vertebral bodies. Intervertebral disc spaces are narrowed. The ribs and clavicles are normal and show no evidence of fracture. The AC joints are normal. There is minimal narrowing of the gleno-humeral joints bilaterally. The humeri show no evidence of previous trauma or bony abnormalities. The radii, ulnae and hands are difficult to evaluate because of their positioning across the chest. There are no gross abnormalities or degenerative changes identified within the radius or ulna or in the hands. The second phalanx on the left hand is absent. The thoracic organs have been removed and replaced with rags which contain high density punctuate material throughout. There is rag identified in the right lower chest as well as in the left mid- and lower chest.

The lumbar and sacral spine show good vertebral body alignment with no spondylolisthesis evident. The disc spaces are well maintained, with some narrowing at the L5-S1 level. No fractures are identified. The right S1 joint is narrowed greater than the left side and the symphysis pubis is also narrowed. The hip joints also show evidence of narrowing. No other bony lesions are identified in the pelvis. The abdominal organs have been removed and rags with large amount of punctate calcific density are identified in the upper abdomen. The pelvis has some high density material identified in the region of the bladder. The etiology of this high density is not known. The femoral are both normal with no evidence of degenerative changes at the heads or evidence of previous trauma or bone lesions. The knees are remarkable only for slight narrowing of the joint spaces. The tibiae and fibulae show no evidence of previous trauma or bony lesions. The right foot is removed. The left ankle is dislocated with the talus anterior to the distal tibia. The left foot tarsals, metatarsals and phalanges are normal with no evidence of trauma. The fifth phalanges are not present.

CT examination

The soft tissues of the head are remarkable for the loss of muscles particularly in the occipital region and temporal regions with the fibrous fascia still in place. The bony skull shows no evidence of fractures. The sinuses are all well aerated. There is a high density material identified in the posterior dependent aspect of the skull. The thoracic and lumbosacral spine show a dense material layering posteriorly down to the level of T2 and T3. The vertebral bodies are all intact with no evidence of fracture or facet sclerosis. There is a higher density material identified at the L2–3, L5–S1 disc space. This material has a swirled appearance and most likely represents disc. There are no other regions that appear to be disc in this study. The S1 joints are normal. The hip joints show slight narrowing, but there is no evidence of degenerative changes or evidence of fracture. There is narrowing of the symphysis pubis. Examination of the soft tissues of the thorax, abdomen, pelvis, reveal loss of muscle which has been replaced with air. The fascial planes of fibrous tissue are still present. The loss of muscle is identifiable within all major muscle groups. The dependent portions of the mummy have a high density material layering within them. The significance of this is unclear. Within the chest cavity there is a small amount of high density material seen layering within the dependent portion of the left chest. There are also rolled pieces of rag identified in the right chest. There is a serpentine line of calcification identified within the region of the high density material previously described in the left chest. The abdominal cavity reveals removal of all abdominal organs. There is a rolled rag identified within the left abdominal cavity and three rolled rags within the right abdominal cavity. There is a small amount of high density material identified layering in the dependent portions of the left and right abdomen. The pelvic organs have been removed. There is a high density material identified layering out within the dependent portions of the pelvis. The fascial planes from the pelvic muscles are still present.

Case study #3 – Mummy NMNH-385664
(Figs. 14 & 15)

LOCATION: Unknown
COLLECTOR: Unknown
ACCESSION NO.: 348298
ACCESSION DATE: 05/14/1986

RELATIVE DATE: 3rd Intermediate (1070–945 B.C.) – Late Dynastic (712–332 B.C.)
RADIOCARBON DATE: 2451 ± 60 BP (764–405 B.C.)

Catalog comments

Found in wooden box behind Exhibition Hall 25. No further information associated.

Specimen information

AGE: Mid 30 years (from skeletal aging)
LENGTH: 169.5 cm
SEX: Male (from skeletal features)
CONDITION: Poor–unwrapped with damage

Aleš Hrdlička acquired human remains from Lisht, Egypt in 1909 coming from the 1906–8 excavations by the Metropolitan Museum of Art, New York (Lythgoe, 1907). Many of the less well preserved mummified specimens were macerated at the museum, some specimens were left in their mummified state, primarily heads. The state of preservation, color, and mummification practice and positioning of this mummy are consistent with other mummified specimens collected by Hrdlička from Lisht (identified as 20th–25th Dynasty), but the true provenance of this mummy is unknown.

Radiologic studies completed: Plain film radiographs (Table 3).

Table 3. Exposure guide for plain film radiographs for NMNH-385664

Region	Proj.	Decub. Filter	mAs	kVp	Screen	FFD	Grid
Skull	AP	NO	12	70	Detail	54″	Bucky
Skull	LAT	NO	4	60	Detail	44″	NO
Chest	AP	NO	10	66	Detail	54″	Bucky
Chest	LAT	YES	8	60	Detail	44″	NO
Abdomen	AP	NO	10	66	Detail	54″	Bucky
Abdomen	LAT	YES	8	60	Detail	44″	NO
Pelvis	AP	NO	10	66	Detail	54″	Bucky
Pelvis	LAT	YES	8	66	Detail	44″	NO
Femur	AP	NO	10	64	Detail	54″	Bucky
Femur	LAT	NO	8	60	Detail	44″	NO
Knee (Bilat)	AP	NO	10	64	Detail	54″	Bucky
Knee Superimposed	LAT	NO	8	60	Detail	44″	NO
Tibia/Fibula (Bilat)	AP	NO	10	64	Detail	54″	Bucky
Tibia/Fibula Superimposed	LAT	NO	8	60	Detail	44″	NO
FEET (Bilat)	AP	NO	10	60	Detail	54″	Bucky
FEET Superimposed	LAT	NO	8	54	Detail	44″	NO

Physical assessment

Mummy 385664 is essentially unwrapped except for a few layers of material remaining on the left arm, left leg and both feet. The exposed skin tissue is very dark, almost black (Munsell 5YR 1/1) and has a very resinous appearance. The head, neck and upper chest region has a crackled surface and it looks as though bitumen tar may have been applied in thin coats, especially on the head. Much of the tissue is absent from the nasal and upper mouth regions. The internal nasal region has suffered damage, and the endocranium can be seen through the nose. The sella and cribriform plate are broken and missing. This damage probably occurred during the removal of the brain. The front maxillary incisors have been broken out post-mortemly, the roots of the lateral incisors and canines are still in the alveolus. The mouth is slightly open and the dental arcade can be seen. The teeth show moderate to heavy wear on the occlusal surface. The tooth wear will be more extreme in these populations because of the higher amounts of grit in the diet (Harris and Pontz, 1988), but in using Brothwell's (1965: 69) dental wear chart, the age falls in the 30–40 yr range. No ante-mortem tooth loss is indicated. Radiographically, the dental wear is noticeable with significant loss of the crown. The maxillary third molars are present and Root Complete putting the age of the individual well above 20 yrs (Moorrees, Fanning and Hunt, 1963a & b). Agenesis of the mandibular third molars is observed.

The left chest has been broken open at the third rib and a 65 mm segment of the chest tissue and ribs is absent down to the bottom of the rib cage. This damage clearly occurred after mummification but appears to be old since the edges at the break are dark, not fresh and light in color. The chest and abdominal cavities can be seen through this fissure. Pleural lining, diaphragm and some indistinguishable dehydrated remnants of internal viscera are present. The abdominal concavity is covered by the lowest layers of resin soaked packing. A remnant scrotum is observed but no penis is present. Radiographically, the pelvic anatomy is characteristically male with a relatively narrow subpubic angle, narrow sciatic notch and more triangular and narrower pelvic inlet (Bass 1987: 201–206). No packing is observable in the anus.

The arms are situated along the side of the body, the hands laying outstretched and on top of the legs. The forearms and hands are raised off of the legs, rather than resting on them. This positioning is probably from tissue shrinkage and positioning for wrapping thickness. The hands have suffered damage, the right hand missing all phalanges from the second – fourth fingers. The left hand has lost all metacarpals and phalanges except for the first finger metacarpal.

Damage has also occurred to the legs. The lower legs are separated from the body at the knees. In the past, metal wire was used to re-attach the legs. This wire 'reconstruction' is clearly observable on the radiographs. There is a post-mortem fracture to the right proximal fibula, probably occurring at the time the legs separated at the knees. Much of the skin and tissue from the lower legs is damaged and conservation measures have been employed to contain the remaining tissues from further damage or loss.

Radiographs indicate completed growth and fusion of all the epiphyses. The cranial sutures are present, but some obliteration of the midcoronal and midsagittal sutures can be seen and increase age assessment of the individual to the mid 30's (Meindel and Lovejoy, 1985). There is observable vertebral osteoarthritis present on the corpus rim of the third – fifth lumbar. The lower thoracic spine shows very slight osteophyte activity, consistent with an athletic individual in their late 20's, or a sedentary individual in their mid to later 30's (Stewart, 1958). No other arthritic activity is seen on the joints. Taking into account all the above features and the dentition, the age of this individual is within the mid 30's but probably not older than the latter 30's.

Physicians' report

Plain film study

The bony calvarium is intact, without evidence of fracture. All sinuses are visualized and well aerated. The orbits are intact. The teeth are all present and in good condition with no evidence of dental disease or fractures. The soft tissues surrounding the skull are unremarkable. The cervical vertebrae and intervertebral spaces are well maintained and aligned with no evidence of subluxation. The facet joints are all normal with no degenerative changes. No fractures are identifiable. The bony structures of the gleno-humeral joint of the AC joints are not clearly visible but appear well maintained. There is a calcific density seen in the soft tissues lateral to the humeral head on the left; this is probably an artifact and does not present calcification within the rotator cuff tendons. Humeral heads are unremarkable with no evidence of degenerative changes; the humeri are intact and no evidence of old trauma or other pathologic processes. Elbow joints are well maintained. The radii and ulnae are unremarkable and show no pathologic changes. The radial-carpal joint and lunate carpal joint are unremarkable. The left hand is not present and the right hand is difficult to evaluate since it is crossed and superimposed on the hip and thigh. No evidence of arthritis or post dramatic changes are present on the remaining bones. The phalanges are not present. The clavicles are normal and without trauma. The ribs are remarkable for several fractures including the postoral lateral aspect of the right second and third ribs. Fractures are identified in the posterior aspect of the eights, ninth and tenth ribs. In addition to these fractures on the right, there are fractures on the left, identified at the posterior aspects of the sixth and seventh ribs. The rib fractures do not have surrounding callus but have sharp edges which would suggest this represents a peri- or post-mortem event. The thoracic vertebral bodies all have maintained their height, without evidence of compression. The intervertebral disc spaces are also all well maintained. There is a large amount of artifact projected over the thoracic cavity from the mummification process. The soft tissues are unremarkable other than the artifacts which are projected over them. The organs of the abdomen have been removed. The pelvis is rotated in the AP projection. The lumbar vertebral bodies are well maintained. The right SI joint is obscured, the left is well maintained. The bony pelvis is unremarkable with no evidence of fracture. There is a two centimeter well circumscribed density projected over the left iliac crest of unknown significance. The hip joints are normal with no degenerative changes. The femurs show no evidence of old fractures or bony abnormalities. The distal femurs have cerclage wires in place connecting them to the proximal tibiae bilaterally. The right knee is well maintained while the left knee joint appears unstable and there is narrowing on the lateral aspect and widening on the lateral aspect. This is most likely post mortem shifting since there is no evidence of degenerative change in the knee. The patellae are not clearly identified. The right tibia is remarkable for an avulsion fracture of the lateral aspect of the knee. There is a fragment located superiorly and slightly laterally. The fragment has sharp edges and probably represents a post mortem change. There are no other fractures identified. The left tibia and fibula are rotated at the knee joint, but show no degenerative changes or trauma. The ankle joints are well maintained with no degenerative changes. The feet show no evidence of fracture or degenerative change; and some bones of the right foot appear to have been lost. The fifth metatarsal looks somewhat shortened, the significance unknown.

CT examination

Due to time and machine use limitations, Mummy 385665 was not scanned since it was least well preserved of the group.

Case study #4 – Mummy NMNH-381235
(Figs. 16 & 17)

LOCATION: THEBES, EGYPT
COLLECTOR: DR. J. H. SLACK, to WISTAR INSTITUTE (# 2397)
ACCESSION NO.: 219346
ACCESSION DATE: 04/17/1958
RELATIVE DATE: Late Dynastic Periods (712–332 B.C.)
RADIOCARBON DATE: 1800q 70 BP (135–335 A.D.)

Catalog comments

This mummy was donated by the Wistar Institute, Philadelphia, PA (Catalog No. 2397) to the Department of Anthropology, National Museum of Natural History.

Specimen information

AGE: 2½–3 years (from dental and epiphyseal growth)
LENGTH: 86 cm
SEX: Male (from soft tissue & pelvic design)
CONDITION: Poor–Unwrapped; post-mortem damage to cranium and extremities.

The color of the mummification process and the position of the arms and hands along the side of the body and the forward inclination of the head suggests a later date to this mummy. This associated with the knowledge that Thebes was at its greatest prominence as a city in the 17th Dynasty (1640–1550) B.C. or in the 25th Dynasty (770–712 B.C.) would be plausible that this child was mummified in the Late Dynastic periods.

Radiologic studies completed: Plain film radiographs, CAT scans MRI was performed. After several attempts, the MRI study was aborted due to the inability to receive a signal. Active hydrogen must be present in the specimen in order to produce an image (Tables 4a & 4b).

Table 4a. Exposure guide for plain film radiographs for NMNH-381235

Region	Proj.	Decub. Filter	mAs	kVp	Screen	FFD	Grid
Skull-Chest	AP	NO	4	58	Detail	54″	NO
Skull-Chest	LAT	NO	4	58	Detail	44″	NO
Abdomen-Femur	AP	NO	4	56	Detail	54″	NO
Abdomen-Femur	LAT	NO	4	58	Detail	44″	NO
Femur-Feet	AP	NO	4	50	Detail	54″	NO
Femur-Feet	LAT	NO	4	58	Detail	44″	NO

Table 4b. CT scan transaxial imange measurements for NMNH-381235

Region	Slice (mm)
Head	10 mm
Chest	10 mm
Abdomen	10 mm
Lower extremity	10 mm

Physical assessment

This mummy is completely unwrapped with no remnant packing or wrapping materials present. It is resting on a coarse woven linen sheet to which the mummy now adheres. The linen is discolored but appears to be more recent in age. The coloration of the mummy tissue is dark brown mottled (Munsell – 5YR 3/4 to 5/6) with a slight resinous appearance, but no clear indications of bitumen use or tar. The head is inclined so that the chin is resting on the chest as if a prop was under the head. This feature is indicative of later mummification practices, possibly from the Ptolemaic or Greco-Roman Periods. The head is broken away from the neck at C1-C2 and a wooden hand-whittled dowel (14 mm dia.) has been inserted into the neural canal (Fig. 18). This dowel was probably placed there earlier in this century as a reconstruction measure since the neck muscles appear to be dry tissue breaks. There is no indication the brain was removed through the foramen magnum. There is a 63×65 cm segment of the left lateral parietal broken from the skull and the pieces resting inside the cranium. The jagged post-mortem type fractures do not radiate from the opening consistent with dry bone fractures, the skin and bone coloration at the fractures being light brown/tan, also indicating recent damage. Material inside the skull is granular and friable to the touch, but hardened enough with resins to remain solid. Some head hair is still present on the scalp but it is quite short, possibly cut or shaved. Head length is 166 mm and head breadth is 136 mm. The mouth is slightly open, exposing the incisors, the nose has collapsed with no packing in the nostrils. No indication of brain removal through the nose and the cribriform plate is present and complete radiographically and visually.

Radiographic evaluation of the dentition shows the deciduous incisors and canines and molars in place, the permanent first incisors are at Crown ¾ complete (Cr. ¾) to Crown Complete (Crc), the first deciduous molars are at Initial Root formation (Ri), and the second molars are not yet present indicating the child to be 2½–3 years old (Moorrees, Fanning and Hunt, 1963a & b) (Fig. 19). This estimate is not based on an ancient Egyptian sample. The child is too young to use El-Nofely, et al. (1982) dental eruption study of modern Egyptian children. However, dental growth rates vary only by a number of months between populations. For example, using Amerindian standards, this child's dental growth would be approximately the same age, 2–3 yrs (Ubelaker, 1989). Therefore, the dental standards for caucasians as stated above could be assumed to be a close approximation for this population.

The ribs have collapsed into the chest cavity and the body has shifted giving it a scoliotic appearance. No clear indications of scoliosis are present radiographically. The arms are situated along the side of the body, the right hand along the side of the leg while the left hand is resting on top. The abdomen appears intact, no evisceration cut noticed, but it might be hidden under one of the folds of the collapsed abdominal tissue. No evidence for evisceration cut in the CT images. Internal organs, however appear to be absent, so removal through the anus or by oil of cedar injection is probable. The pelvis is shifted from anatomical position by dehydration of the cartilage and tendons, but the pelvic shape radiographically looks male-like (Reynolds, 1945, 1947). A remnant penis is still present also identifying sex.

The legs have sustained damage with both lower legs being detached at the knee. The right leg has lost much of the overlying muscle and tissue and the fibula separated from the tibia and broken, the proximal end of the fibula found adhered to the right hip. The left foot is detached at the ankle. Tibial diaphyseal length is 137–139 mm, aging the child to approximately 2.5–3.5 yrs (Bass, 1987, p. 235) or 2–3 yrs (Ubelaker 1989, Tab. 14, p. 71). Epiphyseal ossification and growth are consistent with a child 2½–3 years old (Flecker, 1942, Fig. 20). Hand-wrist growth assessment was not possible from the images.

Physicians' report

Plain film study

The bony calvarium shows the removal of a fragment in the left occipital region. The fragment involves the portion of the left lambdoid suture. The sella is normal. The maxilla and mandible are normal with no bony lesions identified. The teeth are normal. The vertebral body heights are all well maintained. The disc spaces are severely narrowed. There is splaying of the posterior elements between C1 and C2. Significance of this splaying is unclear. The

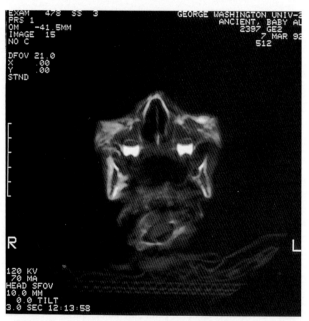

Fig. 18. Computer tomographic image of NMNH-381235 sectioned at the C3–C4 vertebral region illustrating a hand-whittled wooden dowel inserted in the neural canal

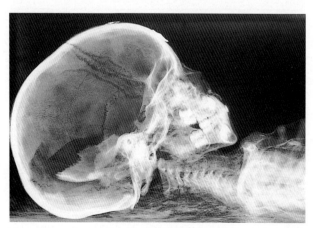

Fig. 19. Lateral cranial radiograph of NMNH-381235 indicating the dental maturation stage of a 2½–3 year old child

chest wall is sunken secondary to the removal of the organs. There are no organs identified within the chest cavity. The vertebral body heights are all well maintained. There is narrowing of the vertebral disc spaces. The ribs are unremarkable with no evidence of fracture. The shoulder shows normal clavicles and scapulae bilaterally. The epiphyseal plate has not yet fused. There is narrowing of the AC joint as well as the glenohumeral joints bilaterally. The upper extremities are all normal showing no bony fractures or abnormalities. The hands and wrists are not clear to evaluate in this study. The pelvis shows no evidence of fracture. There is slight narrowing of the right hip. The femoral head epiphysis and the iliac apophysis are not yet fused.

The SI joints are narrowed and not visualized. The pelvis has a male configuration. The femora show no evidence of fractures or abnormalities bilaterally. The proximal and distal epiphyses are not yet fused and there is cartilage seen surrounding the epiphyses. The cartilage is visualized on these studies because there is no joint fluid or other soft tissue which would silhouette-out the cartilage. Comparison of the skeleton's knees with those found in standard reference text are consistent with that of a comparable 30-month old male. The left tibia and fibula are normal and show no abnormalities. The cartilage is again visualized at both superior and inferior ends of the bones. The right tibia is unremarkable. The right fibula has been separated from the distal epiphysis and has its proximal end now at the mid-shaft of the tibia. There is a fracture just below the head of the fibula and this portion is found projected over the right ilium. The borders of the fractures are sharp indicating that this is a post-mortem fracture. The feet are normal and unremarkable.

CT examination

In the skull, there is a fragment removed from the left occipital region. This fragment is identified within the skull along the posterior lateral side of the right. There is a high density material identified layering out mainly over the right posterior aspect of the skull. The frontal sinuses are not visualized. The ethmoid and maxillary sinuses are not yet developed. There is no sphenoid sinus seen. The bony structures at the skull base are all intact. The teeth are present and normal. The spinal canal has high density material within it which is similar to the material identified in the skull. The vertebral bodies at all levels are intact without evidence of fracture or other bony abnormalities. The dense material identified in the cervical cord is identified down to approximately the level of the 2nd and 3rd thoracic body. The clavicles are normal. The organs of the chest and abdomen have been removed and the anterior abdominal wall is collapsed throughout most of the chest, abdomen, and pelvis. There is soft tissue density seen within the chest cavity which most likely represents the rag which was placed in the cavity. The same holds true in the abdominal cavity. Scans through the extremities reveal no abnormalities. The pelvis is normal without fractures and the SI joints are well maintained. The lower extremities are unremarkable except for the proximal displacement of the right fibula and fracture of the proximal head, re-oriented over the right iliac crest.

Discussion

From this study, it has again clear that radiography is one of the most diagnostic methods for studying internal structure of objects (and in this case human mummies)

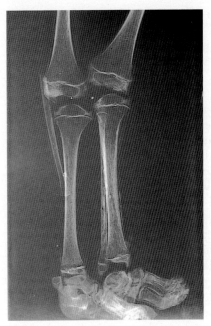

Fig. 20. Anterior-posterior plain film radiograph of NMNH-381235 lower extremities. Epipheseal ossification is consistent with 2½–3 year old child

without having to do dissection, cross-sectioning or do any other form of destructive investigation. The limitations of plain film radiography; distortion, photo-radiographic restrictions by film type and cassette type or speed, artifact, super-impositioning can be circumvented by the use of computerized tomography. But even CT is restricted by the abilities of the x-ray penetration and the pickup quality, the capabilities of the computer equipment and the attenuation of the grey scale or windows.

It is definitely an art to produce sharp, accurate radiographic images. The abilities of the equipment are in many cases only as good as the experience of the radiological technician. Techniques will change between subjects and may even change within a subject based on their densities, position (or restrictions made upon movement of the specimen) and associating artifact. All these variables must be taken into account when preparing an image. Although there was variation between an within each specimen, a consistent decreased shift of mAs and kVp from living subjects was made. Bucky systems or detail cassettes were required for most of the plain film imaging due to x-ray dispersion from densities. The use of the decubitus filter greatly aided in modifying the x-ray beam to reduce 'burn-out' of less dense areas while allowing penetration of more opaque areas in the same region. It is highly suggested that these filters be used in the radiography of mummified remains to produce uniform images. The CT imaging followed standard protocol for bone attenuation but constant care had to be made in the power adjustment at different regions of the body.

The results of the investigation indicate that each of these individuals were normally active and in relative health throughout their lives. None of them experienced trauma or debilitating disease to an extent which had left traces of these events in or on the bone. The causes of death were acute enough not to leave traces of their effects to be visible radiographically. This is not to say that the cause of death could not be determined through autopsy or by histological examination, but that type of invasive research was not the objective of this investigation.

The adults had accelerated dental wear with respect to today's standards, and probably experienced some level of tooth pain or loss through infection/disease or attrition. All skeletal damage is post-mortem, some of recent past while other may have occurred much earlier by ancient Egyptian looters, in particular Mummy 385664. The damage to the chest is quite similar to that seen in Queen Tiy and two other male mummies from the 1898 royal mummy cache (Metropolitan Museum of Art Neg. # TAA 1241B, in *Egypt:* 30). No clear features of nutritional or physical stress are present in any of the mummies. The child manifests no indications of growth stress in the bone, again suggesting an acute rather than chronic cause of death. Mummy 126790 lived to a relatively long life for Egyptian standards at that time but his cause of death was probably not from old age but some other complication.

The accuracy of the anthropological assessments of these mummies must be addressed. Physical anthropologists are normally trained to assess age, sex and cause of death from the bony skeleton. The images produced on radiography are in many cases visually different from the appearance of the hard tissue. Case in point is the amount and extent of osteophytes on the vertebrae. Small, minute osteophytic activity which would be an indication of physical activity or advancing age in physical specimen may not be manifested on the x-ray because of its low density. These subtle changes necessary for anthropological evaluation would be missed or mis-interpreted. Other progressive changes in bony structure such as cranial suture closure, metamorphosis of the pubic symphysis and auricular surface are also not as clearly defined in x-ray images making age assessment more difficult. Conversely, the remnant epiphyseal lines in the long bones are present radiographically for a slightly longer period of time than would be observed on the actual bone, so a slightly younger age may be attributed to an individual if this was not realized.

The observations of the radiologist and the anthropologist agree in most cases. Those cases of discrepancy are usually involving missing skeletal elements. The radiologist reported on the internal features and the elements associated in the subject but a observational examination of the subject had not been done. The anthropologist examined the external features of the subject and reported on the physical attributes of preservation and damage and then took into account the missing elements on the radiographs. This exemplifies the different perspectives which are usually taken by the two professions.

There were some discrepancies between the estimated age of the mummies by body positioning, mummification techniques and coloration to the radiocarbon dates from skin or wrappings next to the body. In the case of Mummy 381234, the lighter coloration of the tissues, the position of the arms across the chest and the general good preservation of the mummy would suggest a New Kingdom time frame. The radiocarbon dates for this mummy are approximately 700 years younger than that expectation. Therefore the embalming practice performed on this mummy seems culturally conservative to the absolute date if one were to consider mummification practices to modify homogeneously across Egypt through time.

This assumption is clearly not supportable when the mummification techniques of mummies of known ages are organized and evaluated as was done by Gray (1972). He found all Dynastic mummies had their arms along their sides after the 25th to 27th Dynasty, but in the Ptolemaic Period arms were sometimes folded on the chest. Hand position was not consistent for various time periods or geographical areas, so regionalization and time lag in mummification changes are most probable in technique. He also suggested that child mummies are more prevalent in the later periods, this could be a factor of preservation or burial practice. Based on this information along with the head position similar to that seen in Christian Period mummies (Prominska, 1986), the child mummy (381235) would be more consistent with the radiocarbon date of Greco-Roman Period in Thebes rather than the visual Late Period estimate. It should be noted again that cultural dating may be deceptive since the practices of mummification may vary regionally or between contemporaneous religious sects. Care should be taken in assessing ages to mummies by body positions and coloration only rather, especially when there is no other associating cultural artifacts with the specimen.

There should be an re-emphasis made to involve all museums and universities which house mummies of any sort in the International Mummy Data Base (Pettitt and Fildes, 1986). The utility of this data base for research, systematic study of mummification practices, positions and the past and ongoing research on them would be invaluable. I believe that more museums need to become aware of this facility and participate in its information gathering and reporting. A short report on the state of the data base and its utility should be sent to holding institutions with a sample data form, and instructions for reporting mummies, so that the institutions may begin to share information and be aware of the need for a central data base involving these invaluable specimens.

Acknowledgements

First and formost, we are endebted to Dr. David O. Davis, Chairman, Department of Radiology, George Washington, University Medical Center for providing access and availability of the department's radiological equipment and the highly qualified staff who gave their time for this research; Angie Dopkowski, Julie Crismond, Jennifer Ettienne, Anna Lawrence, Penny Bulter. We would also like to thank Carol Handrahan, Administrator, Department of Radiology, for allowing the research time in the radiology department and the use of her personnel. The physician reports were rewritten from the plain film and CT reports read by Dr. Peter Morrison who must be acknowledged for his important contribution to this study.

We would also like to acknowledge the participation of the National Museum of Natural History Anthropology Conservation Labortory staff; Natalie Firnhaber, Lynn Schneider, Margaret Schweitzer and Ken Foster for their help in preparing the mummies for transport from the museum to the Medical Center.

And finally, we would like to thank photographer Chip Clark for providing his expertise in recording this project.

Summary

As part of an exhibition renovation at the National Museum of Natural History, Smithsonian Institution, four Egyptian mummies from the Department of Anthropology were investigated at the George Washington Medical Center, Washington, D.C. using fluoroscopy, plain film radiography, and Computerized Tomography (CT). The objectives of this study were two fold: anthropologically we were interested in identifying the age, sex, cause of death (if possible) and individual characteristics without destroying the mummies by autopsy. Radiologically we were interested in evaluating the methodological radiographic techniques required for this analysis in order to use the results in future anthropological and forensic research.

Special techniques were required to provide clear and uniform imaging because of radio-opaque resins, thick wrappings and the inability to remove one mummy from its sarcophagus. Sophisticated techniques and filtering were employed to produce images with minimal fogging, distortion and high contrast definition in plain film radiography. CT imaging procedures were adjusted to minimize shadow and ghosting from opaque masses.

From the results there are two adult males, one in his mid 30's and the other in his early 40's; one adult female in her late 20's; and one 2½–3 year old male child. Modifications in radiation, time, filtering techniques and film screens produced clear definition of skeletal structures and some remnant internal viscera. CT scanning clarified areas too opaque for plain film imaging, superimposed features, some muscle groups, mummification techniques and even woody growth ring structure in the sarcophagus. None of the individuals exhibited evidence as to the cause of death such as trauma or gross pathology in the plain film or CT images.

Zusammenfassung

Im Rahmen einer Ausstellungserneuerung im Naturgeschichtlichen Museum des Smithsonian Institute wurden vier ägyptische Mumien aus der Abteilung für Anthropologie am George Washington Medical Center, Washington D.C. mit Hilfe von Fluoroskopie, Radiographie und Computertomographie (CT) untersucht. Mit der Studie verfolgten wir ein zweifaches Ziel: eine anthropologische Untersuchung auf Alter, Geschlecht, Todesursache (soweit möglich) und persönliche Merkmale, ohne die Mumie durch eine Autopsie zu zerstören, sowie eine methodische Bewertung der dafür erforderlichen Röntgentechniken, um die Ergebnisse in künftigen anthropologischen bzw. gerichtsmedizinischen Untersuchungen anwenden zu können.

Spezielle Techniken waren erforderlich, um trotz der röntgenundurchlässigen Harze, dicken Umhüllungen und in einem Fall der Unmöglichkeit, die Mumie aus dem Sarkophag zu entfernen, klare und gleichmäßige Darstellungen zu erzielen. Hochentwickelte Techniken und Filter wurden eingesetzt, um Bilder mit minimaler Schleierbildung und Verzerrung und hoher Kontrastauflösung in den Röntgenbildern zu erhalten. Die CT-Verfahren wurden so angepaßt, daß Schatteneffekte und Doppelkonturen von opaken Massen so gering wie möglich waren.

Nach den Ergebnissen handelt es sich um zwei erwachsene Männer, einer Mitte 30 und der andere Anfang 40, eine erwachsene Frau um Ende 20, und einen zweieinhalb bis drei Jahre alten Buben. Entsprechend modifizierte Bestrahlung und Belichtungszeit, Filtertechniken und Folien resultierten in klaren Konturen der Skelettstrukturen und Reste von inneren Organen. Mit CT-Scans konnten Gebiete dargestellt werden, die für die Radiographie zu undurchlässig waren, übereinanderliegende Strukturen, einige Muskelgruppen, Mumifizierungstechniken und sogar die Wachstumsringe im Holz der Sarkophage. Bei keiner der Mumien waren im Röntgenbild oder im CT-Scan Hinweise auf die Todesursache, wie Traumata oder auffallende Pathologien festzustellen.

Résumé

Au cours des travaux de rénovation de l'exposition dans le musée d'histoire naturelle du Smithsonian Institut, quatre momies égyptiennes provenant du département d'antropologie du George Washington Medical Center, Washington D.C., ont été examinées à l'aide de fluoroscopies, de radiographies et de topographies par ordinateur. Nous avons poursuivi avec ces examens un double objectif: d'un côté celui d'une étude antropologique permettant de déterminer l'âge, le sexe et (dans la mesure du possible) la cause de la mort, ainsi que les caractéristiques personnelles de la momie sans pour autant la détruire par une autopsie, et de l'autre celui d'une évaluation méthodique des techniques radiologiques employées, dans le but d'en pouvoir appliquer les résultats dans des futurs examens antropologiques ou médicaux-légaux.

Des techniques spéciales étaient nécessaires afin d'obtenir des images nettes et régulières en dépit de la présence de résine résistante aux rayons X et d'épaisses enveloppes, et malgré l'impossibilité d'enlever la momie du sarcophage. Des technologies hautement développées ainsi que des filtres ont été utilisés afin de réduire au maximum et la formation de voiles et la distorsion des images radiologiques tout en assurant une acutance maximale.

Les tomographies par ordinateur ont été faites en sorte d'éviter dans la mesure du possible les effets d'ombre et les doubles contours des masses opaques.

Les résultats des examens nous indiquent qu'il s'agit de deux adultes de secx masculin, dont l'âge estimé est d'environ 35 et 40 ans respectivement, d'une femme âgée d'à peu près 28 ans ainsi que d'un garçon de 2 ans et demi – trois ans. Une adaptation adéquate de la lumination énergétique et du temps de pose, ainsi que les techniques de filtrage et l'emploi de feuilles ont abouti à des contours précis des structures squelettiques et des restes de viscères. Les tomographies par ordinateur ont permis de montrer les parties trop opaques sur les radiographies, ainsi que des structures superposées, certains groupes musculaires, les techniques de momification et même les anneaux annuels dans le bois des sarcophages. Dans aucune des radiographies ou des tomographies des momies il n'a été possible de trouver des indices indiquant la cause de la mort, comme par exemple des traumatismes ou des pathologies majeures.

Riassunto

Nell'ambito di lavori di ripristino nel Museo di Storia Naturale del Smithsonian Institute sono state analizzate quattro mummie egiziane, provenienti dal reparto di antropologia presso il George Washington Medical Center, Washington D.C., con l'aiuto della fluoroscopia, radiografia e tomografia computerizzata. Con tale studio abbiamo voluto perseguire un duplice obbiettivo, da un lato uno studio antropologico riguardante l'età, il sesso e la causa di morte, nonché caratteristiche individuali della mummia, senza tuttavia distruggere la mummia con l'autopsia, nonché une valutazione metodologica delle tecniche radiologiche indispensabili per poter applicare i sisultati di tali indagini anche in analisi antropologiche o di medicina legale future. Per ottenere risultati nonostante la presenza di resine resistenti al raggi Röntgen, i grossi involucri e in un caso anche l'impossibilità di estrarre la mummia dal sarcofago, risultati chiari ed omogenei, é stato necessario ricorrere a tecniche speciali. Sono state applicate tecniche altamente sviluppate e filtri per ottenere radiografie con un minimo di imprecisioni, velature, distorsioni di immagine, ed una massima nitidezza di contrasto. I procedimenti di tomografia computerizzata furono adattati in maniera tale da ridurre al massimo effetti di ombra, linee doppie di masse opache. In base ai risultati si tratta di due uomini adulti, uno intorno ai trentacinque anni, l'altro circa quarantenne, una donna adulta intorno ai ventinove anni ed un bambino di due anni e mezzo – tre anni. In base alle varie tecniche di irradiazione modificata, al tempo di esposizione, a tecniche di filtraggio, si sono ottenute strutture scheletriche e resti di organi interni dai contorni molto netti. Con gli scanner di tomografia computerizzata é stato possibile rappresentare aree radiologicamente impermeabili, strutture sovrapposte, alcuni gruppi muscolari nonché tecniche di mummificazione e persino gli anelli annuali nel legno dei sarcofaghi.

In nessuna radiografia ed in nessuno scanner di tomografia computerizzata si sono potute riscontrare indicazioni sulle origini di morte, traumi o patologie di rilievo.

References

Bass, W. M. (1987) *Human Osteology. A Laboratory and Field Manual.* Missouri Archaeological Society, Special Publication No. 2.

Braunstein, E. and D. Contes (1992) The Wenuhotep Project. *Images.* Indiana University School of Medicine, Department of Radiology Newsletter, Fall., p. 4.

Brothwell, D. R. (1965) *Digging up Bones.* Cornell University Press, Ithaca, NY.

Budge, E. A. W. (1925) *The Mummy.* Reprinted 1987 by Kegan Paul International, London.

Bushong Stewart C. (1993) Anode Heel Effect. *Radiographic Imaging.* St. Louis, MO. Mosby-Yearbook, Inc., p. 124.

Cahoon, J. (1974). Techniques. *Formulating X-ray Techniques.* Duke University Press, Tennessee. pp. 133–134.

Černý, M. (1983) Our experience with estimation of an individual's age from skeletal remains of the degree of thyroid ossification. *Acta Universitatis Palackianae Olomucensis.* 3: 121–144.

David A. R. (1979) *The Manchester Mummy Project.* Manchester University Press.

David, A. R. (ed) (1988) *Science in Egyptology.* Manchester University Press.

D'Auria, S., P. Lacovara and C. Roehrig (1988). *Mummies and Magic, The Funerary Arts of Ancient Egypt.* Museum of Fine Arts, Boston.

Egypt: Land of the Pharaohs. (1992) Time-Life Books. Alexandria, VA.

El Mahdy, C. (1989) *Mummies, Myth and Magic in Ancient Egypt.* Thames and Hudson.

El-Nofeley, A., A. W. Soliman and A. W. Abuzeid (1982) Interrelation of physical growth and dental maturation in Egyptian Nubians. *Anthropos* 22: 249–256.

Flecker, H. M. (1942) Time of appearance and fusion of ossification centers as observed by roentgenographic methods. *American Journal of Roentgenology and Radium Therapy.* 47: 97–159.

Fleming, S. (1980). *The Egyptian Mummy Secrets and Science.* University Museum Handbook 1. The University Museum, University of Pennsylvania: Philadelphia.

Goyon, J.-C. and P. Josset (1988) *Un Corps Pour L'Eternité.* Le Leopold D'Or, Paris.

Gray, P. H. K. (1972) Notes concerning the position of arms and hands with a view to possible dating of the specimen. *Journal of Egyptian Archaeology.* 58: 200–204.

Gray, P. H. K. and E. Slow (1968) *Egyptian Mummies in the City of Liverpool Museums.*

Harris, J. E. and P. V. Pontz (1988) Dental Health in Egypt. In A. and E. Cockburn (ed) *Mummies. Disease and Ancient Cultures.* Cambridge University Press, New York, pp. 46–55.

Harris and Weeks (1973) *X-raying the Pharaohs.* Scribner's, New York.

Harris, J. and E. Wente (1980) *X-ray Atlas of the Royal Mummies.* University of Chicago Press.

Jonckheere, F. and P. H. K. Gray (1951) Civico Museo di Storia ed Arte, Trieste.

Konig, W. (1896). 14 Photographien mit Roentgen-Strahlen, aufgenommen. In *Physikalischen Verein,* Frankfurt a. M., J. Barth, Leipzig.

Krogman, W. M. and M. İşcan (1986) *The Human Skeleton in Forensic Medicine.* Charles C. Thomas, Springfield, IL.

Lythgoe, A. M. (1907) The Egyptian Expedition. *Bulletin of the Metropolitan Museum of Art.* 2: 169.

Meindel, R. S. and C. O. Lovejoy (1985) Ectocranial suture closure: A revised method for determination of skeletal age at death based on the lateral-anterior sutures. *American Journal of Physical Anthropology* 68: 57–66.

Moodie, R. L. (1931) *Roentgenologic Studies of Egyptian and Peruvian Mummies.* Field Museum of Natural History, Chicago.

Moorrees, C. F. A., E. A. Fanning and E. E. Hunt (1963a) Formation and resorption of three deciduous teeth in children. *American Journal of Physical Anthropology* 21: 205–213.

Moorrees, C. F. A., E. A. Fanning and E. E. Hunt (1963b) Age variation of formation stages for ten permanent teeth. *Journal of Dental Research* 42/6: 1490–1502.

zur Nedden, D. and K. Wicke (1992) Der Eismann aus der Sicht der radiologischen und computertomographischen Daten. In Höpfel, F., W. Platzer and K. Spindler (ed), *Der Mann im Eis*. Band 1. Bericht über das Internationale Symposium 1992, Veröffentlichungen der Universität Innsbruck, Bd 187. Universität Innsbruck.

Notman, D. N. H. (1986) Ancient scannings: Computed tomography of Egyptian mummies. In R. A. David (ed) *Science in Egyptology*. Manchester University Press, pp. 251–320.

Pahl, W. M. (1986) Possibilities, limitations and prospects of computed tomography as a non-invasive method of mummy studies. In R. A. David (ed) *Science in Egyptology*. Manchester University Press, pp. 13–24.

Pettitt, C. and G. Fildes (1986). The international Egyptian mummy data base. In A. R. David (ed), *Science in Egyptology*, Manchester University Press, pp. 175–181.

Prominska, E. (1986) Ancient Egyptian traditions of artificial mummification in the Christian Period in Egypt. In A. R. David (ed), *Science in Egyptology*, Manchester University Press, pp. 113–121.

Reynolds, E. L. (1945) The bony pelvic girdle in early infancy. A roentgenometric study. *American Journal of Physical Anthropology* 3: 321–354.

Reynolds, E. L. (1947) The bony pelvis in prepuberal childhood. *American Journal of Physical Anthropology* 5: 165–200.

Smith, G. E. (1914) Egyptian Mummies. *Journal of Egyptian Archeology*. 1:189–196.

United States National Museum, Report for 1886, Smithsonian Institution Press, Pt. II, p. 50.

Stewart, T. D. (1958) The rate of development of vertebral osteoarthritis in American Whites and its significance in skeletal age identification. *The Leech*. 28/3, 4, 5: 144–151.

Stewart, T. D. (1979) *Essentials in Forensic Medicine*. C. C. Thomas, Springfield, IL.

Suchey, J. and L. D. Sutherland (1987) Use of the ventral arc in sex determination of the Os Pubis. *Proceedings of the 39th Annual meeting of the American Academy of Forensic Sciences*. (Abstract)

Ubelaker, D. (1989) *Human Skeletal Remains: Excavation, Analysis, Interpretation*. Taraxacum, Washington, D. C.

Webb, P. A. O. and J. M. Suchey (1985) Epiphyseal union of the anterior iliac crest and medial clavicle in a modern multiracial sample of American males and females. *American Journal of Physical Anthropology*, 68: 457–466.

Correspondence: Dr. David R. Hunt, Department of Anthropology/MRC 112, National Museum of Natural History, Washington, D.C. 20560, U.S.A.

Dry human and animal remains – their treatment at the British Museum

A. Rae

Department of Conservation, The British Museum, London, U.K.

Human and animal remains in the museum

The richest sources of human and animal material in the British Museum are from Ancient Egypt and a variety of ethnographic cultures.

In recent years the Department of Conservation has been involved with the care and treatment of pre-dynastic human bodies from Ancient Egypt, dried llama foetuses from Bolivia, and preserved human trophy heads from South America, which will form the main case studies of this paper. The conservation of the waterlogged body, known as Lindow Man, will be discussed by Dr. Daniels (see in this volume, pp. 173–181).

The very extensive collections of mummified, rather than desiccated, humans and animals in the Department of Egyptian Antiquities cannot be considered. Similarly, the more recent Ethnographic collections, including Peruvian mummies, New Guinea head trophies and North American Indian material, will be excluded.

The great wealth of untanned and semi-tanned skin products, including Inuit fur clothing and equipment, such as skin covered Kayaks (1), seal floats (2) and gut parkas (3), as well as raw hide artefacts, West African tandu (4) for example, will not be considered here, although aspects of their deterioration and conservation treatment have similarities to those used with human and animal remains. The collections also contain skeletal material from many parts of the world.

Causes of deterioration

Dry human and animal remains are susceptible to damage from many environmental factors once removed from the conditions which resulted in their preservation. In particular the combination of skin and bone with their differential responses to changes in relative humidity (RH) and temperature can lead to the splitting and lifting of skin layers. High visible and ultra-violet light levels can also lead to the fading and embrittlement of collagenous materials.

Human and animal material has little or no resistance to biological attack. High RH is likely to lead to the development of mould growth and may initiate renewed putrefaction. The remaining proteins form a rich food source for insect pests, such as clothes moth (*Tineola*), carpet beetle (*Dermestidae*) and boring beetles, including biscuit beetle (*Stegobium*) which can severely damage specimens, as can rodents.

As with so many delicate artifacts, thoughtful handling, storage and display are essential. Whilst bodies may appear robust, the desiccation which had led to their preservation inevitably results in a loss of the natural flexibility of the epidermis and subcutaneous tissues which may become friable and detached. Bones also become increasingly brittle so that rough or careless handling can easily lead to fractures.

Equally dust, if allowed to accumulate, is disfiguring, and may lead to additional handling in order to remove it. Dust also behaves as a hygroscopic layer which may hold moisture in close proximity to the skin leading to mould growth, and it also provides an additional food source for insect pests.

The long term preservation of abody is frequently best achieved by controlling these environmental factors in order to provide stable conditions (5). Recommended conditions for organic materials in the British Museum are an RH of 50 %±5 %, temperature of 18–25 °C and light levels of below 200 lux and below 70 microwatts per lumen ultra-violet light. In certain instances where remains have stabilized at a lower RH this is maintained as is the case with the pre-dynastic Egyptian bodies.

However, many objects are received into the collections having already been exposed to adverse conditions. Conservators are therefore frequently faced with artifacts which are suffering from varying degrees of deterioration which require an interventive approach.

Case studies

Bolivian llama foetuses

The smell of decaying animal matter is unmistakable, and when staff in the Department of Ethnography opened some boxes of recent acquisitions from Bolivia it was particularly pungent. Llama foetuses (see Figure 1) are used as religious offerings in Bolivia and are prepared by drying in the sun. They may then be decorated with cotton wool, or coloured wools, or left in a more or less natural state, after which they are offered for sale in local markets. The British Museum acquired a number in 1985, in preparation for an exhibition on Bolivia the following year.

The smell from the foetuses was so strong that their condition was investigated by conservation staff. The bodies appeared to be completely dry and there was no evidence of mould. However it was suspected that the interiors of the bodies might have residual moisture and after some experimentation a number of the foetuses were freeze dried. Before, during and after the process the weight of the foetuses was monitored and although the weight dropped following treatment this was regained with the return to ambient conditions. The appearance of the bodies was completely unchanged by the freeze drying process. It was concluded that whilst the smell was extremely unpleasant it did not indicate continuing putrefaction, and it was reduced by thoroughly airing the foetuses in a well ventilated area.

Jivaro shrunken heads

Another biological problem encountered in the Museum more recently has been a moth infestation which threatened some of the llama foetuses and some Jivaro shrunken heads. Unlike the llama foetuses, the trophy heads (or tsantsa) of the Jivaro Indians of Ecuador are prepared in a complex and highly ritualized way (see Figure 2). Information on the preparation of tsantsa varies somewhat (6, 7) but essentially the decapitated head is cut up the back of the neck and head and the skin of the face and scalp eased off the skull, which is then discarded. The mask is boiled in water, after which the lips and eye lids are sewn closed and the head filled with hot sand, which is constantly replenished over a period of days. During this time the outside of the face is rubbed with hot stones and shaped with the fingers to maintain the features. The end result is about a third the size of a normal head, thoroughly dry and hard to the touch. The hair is intact, and the trophies are often decorated with various materials including feathers and beetle wing cases.

The llama foetuses and shrunken heads were sterilized by freezing to ensure that no live moth larvae or eggs were present. The previous freeze-drying of some of the foetuses, of course, offered no protection against an infestation. The objects were enclosed in polythene bags from which the majority of the air was excluded before being sealed. They were then frozen at –30 °C for 72 hours, after which they were removed from the freezer chest and allowed to gradually return to ambient temperatures. The freezing process was successful and no further insect activity has been apparent, although they remain susceptible. The heads have required very little other conservation treatment.

Mundurucu trophy heads

Preserved human heads from the Mundurucu Indians in Brazil exhibit different characteristics (see Figure 3). The heads are preserved with the skull intact, the brains having been removed. The skin is then preserved over a fire, so that as well as drying it is likely that there is also a degree of smoke tannage. The hair is natural and the ears are often decorated with feather ornaments. The eyes of many are sealed with a black resinous gum and rodent teeth.

One such head (BM Registration number 1854.5-3.1) required cleaning and repair for exhibition. Initially superficial dust was removed with a soft brush and low powered vacuum cleaner. The hair was then cleaned using saw dust soaked in white spirit. The saw dust was worked into the hair, and then removed once the solvent had evaporated, with a vacuum cleaner and soft brush. This method was considered the safest way of cleaning the hair without allowing solvent or water to penetrate to the scalp. The feather ornaments were cleaned with industrial methylated spirits (ethanol and methanol) brushing in the direction of the feathers onto absorbent tissue. The skin on the face was extremely taught and in one or two areas, particularly around the jaw, had split, owing in part to the weight of the jaw and previous insect damage. In the past the largest split had been filled rather crudely with a mixture of shellac and calcium carbonate which was then heavily over painted. The earlier filler and pigment were removed, and replaced with polyfilla (calcium carbonate and polyvinyl acetate) to protect the edges of the split and to improve the appearance of the face. It was painted to match the surrounding skin using Cryla acrylic paint. The underside of the chin had also been damaged and in order to protect this area Japanese Kozo tissue paper was selected as a repair material. It was coloured using Solophenyl cellulose dyes and adhered over the torn and fragile skin with a 50 % solution of Mowilith DMC2 (polyvinyl acetate/dibutyl maleate copolymer and cellulose ether) in distilled water. This proved extremely effective in securing the damaged area.

Mounting the head was also a concern. In storage a thick wooden down had been inserted through the neck hole into the hollow head. This meant that all the weight was supported by a small area of the inside of the skull, and the lower jaw was not supported at all. For exhibition it was desirable to show the head on a tall stake as it would have been in ritual use. A new mount was therefore devised which would support the head more fully and be suitable for both storage and display.

Perspex was selected to form a support under the head and chin. In order to match the support to the contours of the base of the head, a mould was taken using sheet dental wax which was warmed and pressed into place. Once cool it was removed, and a plaster cast made from it. Perspex sheet was heated and pressed into the contours of the plaster cast giving an accurate reproduction of the base of the head. A hole was cut through the perspex corresponding to the neck hole and the edges trimmed so that it would not show when the head was positioned on it. A thick piece of wooden dowelling was fitted through the perspex support, the upper end being padded with Plasterzote polyethylene foam to rest against the inside of the skull. In this way the weight of the head was securely supported by the contoured perspex and the mount could be left in situ at all times. The dowel could be set into a long stake for display or a secure base board for storage (see Figure 4).

Pre-dynastic Egyptian bodies

At the end of the nineteenth century the British Museum acquired six human burials which were discovered at Gebelein, one hundred miles south of Luxor in Egypt. Dating from before 3,100 BC, the bodies were found in shallow sand graves where the combination of easy drainage of body fluids and the intimate contact with warm sand led to rapid desiccation resulting in a remarkable level of preservation. A male corpse (BM Registration number EA 32751) (see Figure 5), which became affectionately known as "Ginger" due to remains of red hair, was quickly placed on exhibition in the First Egyptian Room where he has remained ever since.

In 1985 Ginger was examined by conservation staff and found to require cleaning and repair. This was also considered a useful opportunity to improve the support of the body in its reconstructed grave. As Ginger has always been an extremely popular subject with the public and conservation would take some months, one of the other bodies from the same grave site was selected as a temporary replacement. The body of a young adult female was chosen (BM Registration Number EA 32751) which also required conservation before display (8).

The female body (see Figure 6) was in a tightly flexed position, and exhibited a number of major fractures to the brittle bones. The left leg was broken at the femur and knee, and the right wrist and right shoulder were also broken. The lower jaw and skull had been crushed in antiquity. In many areas the skin had contracted, splitting and lifting away from the underlying tissue, which was very brittle and fragile.

A soft sable brush and low powered vacuum cleaner, fitted with a flexible plastic hose, were used to remove superficial dust. Further cleaning was desirable, but the very degraded nature of the skin ruled out the use of water which had an adverse effect on the collagen. IMS was found to remove dirt safely. Cotton wool swabs were therefore moistened in IMS, and gently rolled over the surface.

The fragility of much of the subcutaneous tissue necessitated intensive consolidation in order to provide a surface strong enough to repair the broken limbs and lay down the brittle skin. This was achieved with solutions of Paraloid B72 (ethyl methacrylate copolymer) in xylene using a sequence of 3 %, 5 %, 10 % and 15 % until sufficient strength was achieved. Where possible the breaks to the limbs were repaired with 20 % Mowilith 50 (Polyvinyl acetate resin) in IMS. This adhesive, whilst providing a strong bond, retains a degree of flexibility and will fail under stress rather than allowing a fresh break to occur to the bone and was used to repair the right shoulder and right wrist.

However, the fractures to the left leg supported more weight when the body was moved and therefore stronger joins were required. The broken femur was secured with a wooden dowel, initially held in position in the bone cavity in one side of the break with 20 % Mowilith 50. The join was then reinforced with Devcon 5-minute epoxy adhesive bulked with fumed silica to strengthen the degraded tissue around the break before bringing the broken surfaces together. This worked well, and the same adhesive was used to repair the break at the knee without a dowel.

Great care was taken in devising a method of relaying the skin. In order to achieve this it was necessary to induce some flexibility in the brittle epidermal layer, or it would simply crack when pressure was applied. After much experimentation the following procedure was found to be successful. White spirit was applied by brush to the area to be repaired, followed by 10 % Mowilith 50 in IMS which was introduced beneath the cracked skin from a fine sable brush. The white spirit had a dual role, firstly rendering any excess adhesive easily removable from the surface of the skin with absorbent tissues, and secondly acting as a carrier, or wetting agent, drawing the resin into the repair. The skin was then slightly softened with a heated spatula set at 70 °C, which was applied gently to the area using Melinex polyester sheet, or silicone release paper between the skin and spatula head. The warmth had the effect of temporarily making the skin more flexible so that it could be eased back into position with finger pressure. The heat speeded the evaporation of the solvent and therefore the setting of the adhesive.

The techniques developed during the treatment of the female body were now used to stabilize the male body. Ginger was structurally in a fairly strong condition with no major fractures, but the skin, particularly on the head, face, neck, knees, ankles, toes and right hip was badly cracked and lifting away from the underlying tissue. At its worst the skin stood one centimetre away from the skull at the forehead. The remaining hair was also very brittle and poorly attached. The same cleaning, consolidation and re-laying methods were applied with great success. The hair was then consolidated with two applications of 5 % v/v Plaintex (polyethyl acrylate) in ace-

Fig. 1. Bolivian llama foetuses

Fig. 2. Jivaro shrunken head, decorated with feathers, vegetable fibre thread and bone beads

Fig. 3. Mundurucu trophy heads, after conservation supported on storage/display mounts in their storage box

Fig. 5. A desiccated Egyptian body, dating from 3,100 BC, before conservation

Fig. 4. Mundurucu head, after conservation, on its new storage mount

Fig. 6. Female Egyptian body, before conservation

Fig. 7. Mr Colin Johnson finishing the new grave and support for 'Ginger'

tone, resulting in a considerable increase in strength. Loosely attached tufts could then be secured with 10 % Mowilith 50 in IMS.

Both bodies required new mounts to provide complete support, and simplify future handling. The male body was relatively flat on the underside, but the head had become set at a steep angle from the body during burial and required support. A wooden headrest was attached to a support board matching the angle of the head to the body. The boards were then padded with layers of polyester wadding, and covered with display grade textile (see Figure 7).

The underside of the female body, in contrast, was very uneven and it was desirable to create a contoured support. A plywood box slightly larger than the body was filled with polystyrene flakes. The body was laid on top and gently settled in to form a negative impression of the underside. The body was then removed, and the mount set with one and a half litres of 40 % Mowilith DMC2 applied with a pressurized garden spray.

After two weeks the adhesive had dried thoroughly and a rigid contoured mount was produced. The surface was covered with three layers of display grade textile, and the body settled into place.

Conclusion

The varied collection of human and animal remains in the BM collections have provided some interesting challenges from the conservation perspective. Where possible the control of environmental conditions is strongly recommended to increase the preservation of this valuable material. When interventive treatments are necessary a variety of cleaning, repair and stabilization processes have been used successfully, drawing on the broad experience of conservators working with a wide range of organic materials. The application of techniques developed for work with leather, ethnographic artifacts and waterlogged materials have been particularly valuable in dealing with the complex problems posed by the combination of skin, bone and hair.

Great care is needed when selecting adhesives, consolidants and repair materials, as in any conservation treatment, to ensure that they have good aging properties, remain reversible and are as compatible as possible with the material of which the object is made. This is particularly true with bodies where it is likely that the materials of which they are composed will continue to age and respond differently to each other.

The importance of providing effective support and mounting systems which minimize handling is also fully recognised.

Acknowledgements

My thanks are due to my colleagues in the Organic Materials Conservation Section at the British Museum, who carried out many of the treatments I have described. I am particularly grateful to Colin Johnson and Barbara Wills for their work on the pre-dynastic Egyptian bodies and Andrew Calver, now Conservation Officer at Nottingham Museum, for his work with the llama foetuses and Mundurucu heads. I would also like to thank Andrew Oddy, Keeper of the Department of Conservation, for his encouragement.

Summary

The collection of the British Museum incorporates a wide range of human and animal remains from a variety of periods and cultures. These include human bodies or parts of bodies, which have been preserved naturally by desiccation or by burial in waterlogged environments, and those where conscious intervention has inhibited the natural processes of decay. This paper outlines the types of preserved human and animal material in the Museum, before considering the causes of deterioration which have been observed. The conservation treatments used are described in some details focusing on case studies of dry, as opposed to waterlogged, remains.

Zusammenfassung

Die Sammlungen des British Museum beinhalten eine breite Palette von menschlichen und tierischen Resten aus verschiedenen Epochen und Kulturen. Im Inventar befinden sich menschliche Körper und Körperteile, die entweder durch natürliche Austrocknung oder durch Lagerung in feuchten Milieus konserviert worden sind, wie auch solche, bei denen durch bewußte Intervention der Fäulnisprozeß unterbrochen wurde. Dieser Artikel beschreibt die verschiedenen Beispiele konservierten und tierischen Materials des Museums, die durch die Anwendung neuerlicher Konservierungstechniken vor einem weiteren Verfallen geschützt werden sollen. Weiters werden die angewandten Konservierungsmethoden in einigen Details beschrieben, wobei wir uns auf Fallstudien konzentrieren, in denen trockene und wassergelagerte Reste einander gegenübergestellt werden.

Résumé

Les collections du British Museum inclurent une grande gamme de restes humains et animaux de différentes époques et cultures. Celles-ci comprennent des corps humains et des parties des corps, qui ont été préservés naturellement par dessèchement ou par logement dans des lieux humides et des corps, lors desquelles une intervention conscieuse a bloqué le processus naturel de pourriture. Cet article veut décrire les différents types de matériel humain et animal préservé dans le musée, avant de considérer les raisons pour la détérioration qui a été observée. Ensuite suivra une description détaillée des méthodes de conservation appliqués, qui se concentrera sur des études de cas pratiques des restes secs en opposition aux restes humides.

Riassunto

Le collezioni del British Museum comprendono una vasta gamma di resti umani ed animali provenienti da diverse epoche è culture. Contengono anche corpi è parti di corpi umani conservatisi sia per essicazione naturale sia per deposizione in ambienti umidi, oppure anche in base ad interventi tesi ad interrompere il naturale processo di putrefazione. Il presente articolo descrive i vari tipi di materiale umano ed animale conservato nel museo è considera le ragioni per il deterioramento del loro stato di conservazione. In seguito vengono descritti i metodi di conservazione applicati concentrandosi su casi di studio in cui vengono confrontati i resti essicati con resti posti in ambienti umidi.

References

1. Stone, T.: The conservation of skin and semi-tanned leather at the Canadian Conservation Institute: three cases studies. *International Leather and Parchment Symposium* ICOM (1989), pp. 228–242.
2. Calver, A., Wills, B. and Cruickshank, P.: The Freeze-drying of Ethnographic Skins and Gut. *ICOM 8th Triennial Meeting* Sydney (1987), pp. 225–230.
3. Hill, L.: The Conservation of Eskimo seal-gut Kagools. *Scottish Society for Conservation and Restoration Bulletin* No. 7 (1986), pp. 17–19.
4. Vandykee-Lee, D. J.: The Conservation of Tandu. *Studies in Conservation* 21 (1976), pp. 74–78.
5. Bradley, S (ed).: A Guide to the Storage Exhibition and Handling of Antiquities, Ethnographic and Pictorial Art. *British Museum Occasional Paper* No. 66 (1990), pp. 53–55.
6. Flornoy, B.: *Jivaro*. Norwich (1953).
7. Karsten, R.: *The head hunters of the Western Amazon* Helsingfors (1935).
8. Johnson, C., Wills, B.: The Conservation of Two Pre-dynastic Egyptian Bodies. *Conservation of Ancient Egyptian Materials* UKIC (1988), pp. 79–84.

Correspondence: Ms. Allyson Rae, Department of Conservation, The British Museum, Franks House, 38–56 Orsman Road, London, N1 5QJ, U.K.

Egyptian mummies

Research on mummies in Egyptology. An overview

W. Seipel

Kunsthistorisches Museum, Vienna, Austria

The publication of Thomas Joseph's book "History of Egyptian Mummies" in 1834 in London marks the beginning of scholarly research on mummies. Doctor Pettigrew, who was in the habit of unwrapping mummies in front of large paying crowds, was a fashionable doctor at the time, he even inoculated Queen Victoria against smallpox. His voluminous oeuvre can still serve as a source of interesting detailed observations today to help illustrate the amount of scientific knowledge available at the time.

The transfer of the Egyptian royal mummies to the Egyptian Museum in Cairo in 1881 awakened the interests of Egyptology in mummy research. The French Egyptologist Gaston Maspero, then director-general of the Administration of Egyptian Antiquities and of the Egyptian Museum, started the systematic examination of the royal mummies in 1889. This work was continued by Elliot Smith, who was already able to take the new finds of royal mummies, which had been transfered to the museum in 1902, into account. In 1912 he published his extensive survey "The Royal Mummies" as part of the Complete Catalogue of the Egyptian Museum, which still remains the only comprehensive publication on the collection of royal mummies in Cairo. Smith, a doctor, surgeon and historian (of cultural phenomena), was always interested in demonstrating connections and interrelations, including neighbouring sciences in his approach. His work, be it his description of mummification-techniques in the light of contemporary scientific knowledge or his pathological findings, must be considered a milestone in the history of mummy research. "Migrations of Early Culture", published in 1915, in which he shows the dissemination of the technique of mummification all over the world, is characteristic of his methods. He gave public lectures on "the history of mummification with an account of the ancient Egyptians and the evidence of disease in their remains" at the Royal College of Surgeons of England. In his introduction he speaks of "the influence of the Egyptian custom of embalming on the history of anatomy and medical science", considering in particular the Greek anatomists of the Ptolemean era. After a historical introduction, documenting contemporary scientific knowledge of the mummification-techniques of ancient Egypt, he gives a description of the "physical characters and affinities of the earliest Egyptians". He analyses "stature, skull, face, hair and skin colour" as well as the different mummification-techniques, from the earliest time up to the Christian era. Particularly interesting is "his attempt to restore a lifelike appearance to the mummy so that it becomes the statue of the deceased".

Smith's last lecture dealt with questions concerning pathology, such as osteoarthritis, tuberculosis, "the question of syphillis, mastoid diseases and malignant diseases, leprosy, injuries and their treatment". Insofar as scientific methods of the time allowed, Smith's lectures already covered the whole spectrum of modern academic research on mummies. Together with Warren Dawson he published "Egyptian Mummies", a comprehensive survey of his research on Egyptian mummies, in 1924.

Although the German physicist Wilhelm Konrad Röntgen had discovered X-rays in 1895, their use in mummy-research only became prevalent several decades later – with one exception. Already in 1896 a little book was published in Frankfurt containing 14 X-rays of an Egyptian mummy. At the time the equipment used was extremely expensive and cumbersome; this made the use of X-rays almost impossible. Only the mummy of Thutmosis IV had already been X-rayed by Smith, but without any particularly significant results. The mummy of Thutmosis IV was brought to a hospital in Cairo by taxi, where the X-rays were then taken. The examination of mummies with the help of X-rays was continued only in 1913 by the Italian Bertolotti, who was, however, mainly looking for amulets and jewellery in the mummy-wrappings. He was able to document for the first time an anomaly of the spine of a mummy dating from the 11th Dynasty. This was the beginning of radiopathology in mummy research.

The examination of mummies using X-rays was only continued by Douglas E. Derry, an anatomist, surgeon and professor at the Faculty of Medicine at the Egyptian University in Cairo. Apart from different examinations of mummies, such as those of Djoser, Setka and Tutanchamun, Derry X-rayed the mummy of Amenophis I. The first extensive use of X-rays in connection with the examinations of mummies was made by members of the University of Chicago, which published "Roentgenologic Studies of Egyptian and Peruvian Mummies" in 1931. The work of the British doctor and radiologist P. H. K. Gray, who in the 1960's X-rayed 133 different mummies in English and European collections, must be considered a milestone in X-ray assisted mummy-research. Together with Dawson he was instrumental in compiling the "Catalogue of Egyptian Antiquities in the British Museum, I. Mummies and Human Remains" of 1968. Another research paper on Egyptian mummies in the City of Liverpool Museum was also published in 1968.

Continuing in the chronological order of X-ray research, we must now discuss the X-ray assisted examination of royal mummies in the Egyptian Museum in Cairo carried out between 1966 and 1971 under the supervision of the University of Michigan School of Dentistry, and published in conjunction with the University of Chicago. Two publications originated from this research; the first is "X-raying the Pharaohs", a popular version for laymen, written by Harris and Weeks and published in London in 1973. The second is Harris and Wente's scholarly report "An X-ray Atlas of the Royal Mummies" published by the University of Chicago Press in 1980. This important research-project started as part of the Unesco-sponsored project to rescue the Nubian antiquities threatened by submersion. It was started in the spring of 1965 when the University of Michigan was invited, in collaboration with Alexandria University in Egypt, to examine the skeletal remains of the early Nubian inhabitants in the area of the second cateract. The research was carried out under the supervision of the Egyptologist Dr. Kent Weeks and the dental surgeon Dr. Nicholas Millet, and studied the etiology of malocclusion and the changes in man's face through history. 750 examinations of skulls were carried out with the help of a specially designed Ytterbium-Cephalometer. The particular interest in their findings, especially in the X-ray examinations, shown by the Egyptian Administration of Antiquities led them to invite the American-Egyptian research-team to carry out a complete X-ray survey of the Royal Mummies Collection in the Egyptian Museum. This resulted in a complete X-ray examination of all the royal mummies in the years between 1967 and 1971. The mummies were kept in show-cases covered with a type of glass containing lead, from which they could not be moved. As Elliot G. Smith had already noted earlier, some substances used in the mummification-process, especially particular types of resins, impaired the quality of the X-rays. These problems made the development of new techniques of X-raying necessary. Instead of the Ytterbium-Cephalometer used in Nubia, they used a conventional electric medical 90 KV X-ray head as well as an optical cephalometer, which enabled the Michigan expedition to take excellent cephalometrical X-rays in lateral view of the skulls. These were of particular importance in efforts to compare the different mummies. In this way it had become possible to compare the individuals within the Collection of Royal Mummies with any population, living or dead, Egyptian or non-Egyptian. In the following years they X-rayed not only all the royal mummies, but they gradually covered the whole mummy collection of the Egyptian Museum, which includes mummies from the Middle Kingdom up to the Graeco-Roman period. The resulting "X-ray Atlas of the Royal Mummies" gives not only a more or less comprehensive survey of the X-rays, whole-body as well as zoom-X-rays, of all the royal mummies, it contains also a wealth of scholarly contributions ranging over the wide spectrum of medical-historical, pathological and Egyptological research carried out within the framework of scientific mummy research. The second part of this paper will look at the scientific importance of the research on mummies in more detail.

Apart from the examination of the royal mummies which I have just discussed, the Manchester Mummy Project, started in 1973 and directed by Rosalie David, must also be considered a milestone in the history of the research on mummies. This multidisciplinary research-project was carried out by Manchester University and the Museum belonging to this University. It set itself two scientific aims: "first, to discover evidence of disease and causes of death, and to gain further information relating to funerary practices and living conditions in ancient Egypt; and, secondly, to establish a methodology for the examination of mummified remains which other institutions could adopt and adept for the study of their own collections" (R. David). In 1979, Rosalie David published "Manchester Museum Mummy Project: Multidisciplinary Research on Ancient Egyptian Mummified Remains", a survey of the findings. In the first phase of the project, between 1973 and 1979, an extensive X-ray examination of all human and animal mummies was carried out. In addition to this, a number of mummies were also examined using an electronic microscope. The Greater Manchester Police Force Division developed a technique enabling them to take fingerprints of mummified tissues. Attempts were also made to create different scientific reconstructions, using clay and wax, of the heads and facial appearances of the mummies.

In June 1975, the first scientific dissection of a mummy carried out in England arroused widespread interest. The findings, though, were only available two years later. It established the ante mortem amputation of the lower half of the legs, as well as an infection with worms. The radio-carbon dating of the bones and of the wrappings indicated a substantial discrepancy between the age of the body and its wrappings, and suggested that this mummy may have been rewrapped in the Graeco-Roman period. The collected findings of the Manchester Mummy Project were presented to the public in June 1979 during the first international symposium on the subject of "Science in Egyptology". The International Mummy Data Base, founded at the same time, was the first of its kind and was intended to serve as a basis for comparative studies.

In the second phase of this research-project, carried out between 1979 and 1984, other non-destructive methods – such as endoscopy and serological studies – were used besides the X-ray examinations. The findings of both phases were published by Rosalie David as "Science in Egyptology" in Manchester in 1986.

Probably the most important single project ever undertaken, reaching far beyond the mere X-raying of mummies, was the examination of the mummy of Ramses II, which was started in 1977, and published comprehensively in 1985.

This project, initiatived by the then director of the Egyptian Collections of the Louvre, Mme Desroches-Noblecourt, to try to conserve the mummy of Ramses II and thereby to stop its further deterioration, led to the stay of Ramses II in Paris between 26. 9. 1976 and 10. 5. 1977. The Egyptian pharaoh was received at Orly airport in the style of a visiting head of state – he was met by the Minister for Research, a representative of the President of the Republic, high-ranking members of the Armed Forces, the Egyptian ambassador to France and a deputation of the Republican Guard – and became the subject of extensive scientific examination- and conservation-projects.

A team of over 100 scientists, headed by Lionel Balout of the Musée de l'Homme in Paris, was formed: anatomists, anthropologists, botanists, endoscopists, entomologists, microbiologists, museologists and Egyptologists, cooperating with numerous technical assistants, worked on the examination and conservation of the mummy of Ramses II for over a year. This is not the place to elaborate on the scientific findings of this examination – which, at any rate, most experts will be familiar with – suffice it to say that the use of dental evidence to determine the age of the pharaoh showed him to have reached the age of 80 – a result which corresponds with our historical knowledge of the reign of this most important pharaoh of the New Kingdom. I only want to mention the surprising fact that Ramses II apparently had auburn-coloured hair because this led the Egyptologist Desroches-Noblecourt to far-reaching conclusions regarding questions of religious history.

Let me also cite from among the many research-projects on mummies, the X-ray research carried out by the Pennsylvania University Museum which led to the publication of "Mummies, Diseases and Ancient Cultures"
by Cockburn in 1983. In the same year, the University of Melbourne started the examination of a mummy dating from the Middle Kingdom and found in a tomb near Schech Farag. The head of this mummy was reconstructed in wax in a technique similar to the one used for the head of Ramses II, and a report on this was published in the collected papers of the symposium on "Science in Egyptology", held at the University of Manchester in 1984. This collection of papers is the most comprehensive ever published and deals with a substantial part of the research on mummies carried out until 1985. It also mentions other research-projects, for example the interdisciplinary project carried out by the University of Munich, still the most important scientific undertaking carried out in German-speaking countries.

No doubt the research on mummies has attained an important place in academic interdisciplinary research today. Though mainly concentrating on mummies from ancient Egypt, due to their large numbers an almost inexhaustible source, research on mummies today also covers studies on human remains from all the other old civilisations.

Within the space of a few years, research on mummies has become accepted as an independent academic discipline – the growing number of symposiums and the World Congress on Mummy-Research taking place in the near future emphasise this assertion. Even though a large degree of interdisciplinary exchange in mummy-research is noticable, it is still often difficult – at least at this stage of scientific development – to make its findings and methods available to those areas of academic research most interested in them. Looking through the countless academic papers published in the last few years in, for example, "Science in Egyptology" published by Rosalie David or Harris and Wente's "X-ray Atlas of the Royal Mummies", one notices the fact that interdisciplinary exchanges are usually restricted to "related" areas of academic research. The cooperation of a specialist in Egyptology with a scientist is the exception, not the rule. Symtomatically, the papers of Krogman/Baer and Wente in the "X-ray Atlas" both deal independently with questions of age at death of pharaohs of the New Kingdom, the first relying solely on results of the X-ray examinations, the second solely on historical sources. Quite apart from the fact that neither paper refers to the other one – although each must surely have been available to the other author before publication – the enormous differences in the findings on the ages of the most important pharaohs of the New Kingdom cannot be explained by referring to methodologically-induced variations alone. For example, according to the X-ray examination an age at death of between 18 and 22 was calculated for Thutmosis I, while Wente finds historical evidence to assume that this pharaoh reached a minimum age of 27 (+x) years. The differences in the findings on the age of one of the most important kings of the 18th Dynasty, of Amenophis III, are particularly striking. According to the historical evidence, he must have reached at least the age of 50; judged by the X-ray evidence of his skull he died aged 30 or 35. An opposite example are the estimates on the age at death of Ramses VI; X-ray examinations give his age at death as 30 or 35, while according to the historical evidence he had died already aged 23 or 26. Just as striking are the diverging results on king Merenptah, who was pronounced to have died aged 45 or 50 and, according to historical sources, to have lived until the age of 70. The extensive examination of the mummy of Ramses II, to which I have already referred, gave his age as about 80 ± 5 years. This already differs greatly from Wente's historical findings, which give the age at death of the pharaoh as between 87 and 93 years, and makes the findings of Krogman and Baer, published in their paper, and giving his age at death as between 50 and 55 (+) somewhat difficult to understand. This is not to say that scientific examinations are always unreliable, but these examples should be seen as warnings not to give preference to scientific findings in each and every case. Egyptology in particular is an academic discipline based on a large body of evidence, numerous written sources and extensive archaeological excavation-work and is therefore able to provide a comprehensive survey as well as a detailed picture of ancient Egypt and its culture. Egyptologists still regard research on mummies as an auxiliary science, troubled in many areas with its own methodological problems and of limited use in expanding existing knowledge. As an example let me cite again the extensive examination of the mummy of Ramses II, carried out by over 100 scientists and published on over 450 pages – only three of which contain findings directly connected to Egyptology proper. And even these are not of great importance, apart from the surprising statement that Ramses II seems to have had auburn-coloured hair, a colour which might throw new light on him and the religious policies emphasising the importance of the god Seth carried out during his reign. This interpretation of a scientific result – itself not completely without questionmarks – by Mme Desroches-Noblecourt seems to me, though, not convincing enough to allow such conclusions.

Without doubt, mummyresearch extending beyond pathological and radiological questions, mainly concerned, as they are, with anatomy, size, possible illnesses etc., makes results available that justify extensive scientific efforts. Nonetheless, it should be the aim of mummyresearch – especially the work studying Egyptian material – to try to forge closer connections with the different branches of scientific research than have been achieved so far, in order to avoid fruitless parallel work. The danger of specialisation already threatens to cut off the necessary interdisciplinary exchanges, and to lead only to unconnected individual results and findings that have lost their place of reference in the context of history. May this congress help to reduce these increasingly apparent divisions, so that a comprehensive conclusion incorporating all areas of the different cultures concerned may be reached. Perhaps a definition of the aims of research and of its role in the body of science would lead to a new positioning and a new, and in my opinion very important, conception of itself.

Summary

Research on mummies, which started with a scholarly publication in 1834, is nowadays, due to its scope and interdisciplinary quality, one of the fastest growing areas of research in the last few years. From initial research carried out in 1889 on Ancient Egyptian royal mummies to the first examinations using newly discovered X-rays carried out in 1896 and to the X-ray Atlas of the royal mummies published by Wente and Harries in 1980, research on mummies has developed rapidly. Among the numerous extremely important individual examinations, the examination of the mummy of Ramses II, carried out in Paris in 1970, the results of which were published in 1985, should be mentioned. In spite of the numerous international cooperations on most of the current mummy research projects, parallel research work does occur and overspecialization does seem a danger. Therefore, a definition of the current position of mummy research and a clearer definition of its scholarly aims seems imperative.

Zusammenfassung

Die Mumienforschung, die 1834 mit einer wissenschaftlichen Publikation begann, ist heute dermaßen komplex und interdisziplinär, daß sie zu einem der am schnellsten wachsenden Forschungsbereiche der letzten Jahre geworden ist. Von den ersten Untersuchungen an altägyptischen königlichen Mumien im Jahr 1889, über den Einsatz der eben erst entdeckten Röntgenstrahlen im Jahr 1896 bis zu dem von Wente und Harris 1980 herausgegebenen Röntgenbildatlas der königlichen Mumien hat sich die Mumienforschung rapide entwickelt. Unter den zahlreichen bedeutenden Einzelprojekten ist auch die Untersuchung von Ramses II. zu erwähnen, die 1970 in Paris durchgeführt wurde, und deren Ergebnisse 1985 veröffentlicht wurden. Trotz zahlreicher internationaler Kooperationen bei den meisten laufenden Mumienprojekten kommt es zu Doppelgleisigkeit in der Forschung und auch Überspezialisierung scheint durchaus eine Gefahr darzustellen. Eine Standortbestimmung der derzeitigen Mumienforschung und eine klarere Definition der wissenschaftlichen Ziele scheinen daher dringend erforderlich.

Résumé

Les recherches sur les momies débutèrent en 1834 avec une première publication spécialisée et représentent aujourd'hui un des domaines de recherche avec un devéloppement récent parmi les plus rapides. Depuis les premières analyses faites sur les momies royales d'Égypte en 1889, suivies, en 1896 par des examens aux rayons X, qui venaient d'être découverts, et la publication de l'atlas radiographie que des momies royales, publié par Harris et Wente en 1980, la recherche sur les momies a connu jusqu'à nos jours un développement très rapide. Parmi les très nombreuses analyses de grande importance figure celle de la momie de Ramses II, debutée en 1970 à Paris et dans les résultats fûrent publiés en 1985. Bienque la plus grande partie des projets de recherche sur les momies connaise une bonne participation internationale. Il est toutefois inévitable que des recherches soient faites parallèlement et qu'il y ait parfois trop de spécialisation dans ce domaine. C'est pourquoi il serait opportun de définir nettement la situation actuelle de la recherche sur les momies ainsi que ses butes scientifiques.

Riassunto

Le ricerche sulle mummie ebbero inizio nell'anno 1834 con la prima pubblicazione specializzata ed appartengono oggi, in base alla loro complessità e diversificazione delle discipline, ad uno dei campi di ricerca più rapidamente sviluppatosi negli ultimi tempi. Partendo dalle prime analisi effettuate sulle mummie reali dell'antico Egitto nell'anno 1889 a cui seguirono, nell'anno 1896, gli esamir radiografici con i raggi x, che allora erano stati appena scoperti, e la pubblicazione dell'atlante radiografico delle mummie reali di Harris e Wente nel 1980, la ricerca sulle mummie ha subito fino ad oggi uno sviluppo rapidissimo. Tra le numerose analisi di particolare importanza bisogna annoverare quella della mummia di Ramsete II, iniziata nel 1970 a Parigi e i cui risultati furono pubblicati nel 1985. Benche la maggior parte dei progetti di ricerche sulle mummie goda di una notevole partecipazione internazionale, ciò non ostante non è evitabile una certa contemporaneità nel lavoro di ricerca scientifica; e tol volta il rischio di una eccessiva specializzazione. Sembra quindi opportuno perseguire una più esplicita determinazione delle ricerche sulle mummie ed una più chiara definizione degli scopi scientifici.

Correspondence: Dr. Wilfried Seipel, Kunsthistorisches Museum, Burgring 5, 1010 Vienna, Austria.

Development of a prototype storage and display case for the Royal Mummies of the Egyptian Museum in Cairo

S. Maekawa[1] and **N. Valentin**[2]

[1] The Getty Conservation Institute, Marina del Rey, CA
[2] Instituto de Consevacion y Restauracion de Bienes Culturales, Madrid, Spain

1. Introduction

The goal of conservation is to preserve objects of art and of culture for future generations. To achieve this goal, the factors causing deterioration must be countered or, at the very least, abated. Depending upon the particular art or cultural treasure, the process of preventing deterioration may be straightforward or nearly impossible.

A sculpture elegantly crafted from camphor-loaded nitrocellulose is doomed to a life-span of less than a century because of the inherently unstable nature of this plastic. A mummy, properly protected from biological attack and other external hazards both in ancient times and modern, can be preserved for scores of centuries. First and determinative are the internal chemical and physical composition and the construction of the object.

Second come the influences of the environment, its relative humidity – extremes as well as their duration and rates of change, temperature changes (usually more important in their rapidity and frequency than the annual extremes of temperature themselves), the light intensity and wavelength, and biological concomitants from fungi to insects. For outdoor art objects or monuments, other factors become important: the strength, direction and extent of impinging wind, the quantity and type of wind-blown grit or sand, the influx of moisture from the air and from the ground, and most reprehensible, wear and vandalism caused by individuals.

Organic materials in museums and collections are especially susceptible to deterioration due to changes in humidity and temperature, to attack by fungi or bacteria or insects, to photo-oxidation, and to attack by gaseous and particulate air pollutants commonly found in urban and industrialized areas.

Obviously, the deleterious effects of oxygen-dependent organisms such as insects, fungi, or aerobic bacteria upon organic art objects [1, 2, and 3] (especially, but also upon stone etc.) would be eliminated if the objects could somehow be maintained in a nitrogen or other oxygen-free atmosphere. The catastrophic effects of insect attack on cultural treasures are well known: on delicate feather objects themselves or ornamentation consisting of feathers, on historically unique parchment documents, on mummified human remains which are national or ethnographic exemplars, and even on stones where blackening or discoloration occurs. Just as well known is the fact that all such biological attack is accelerated by high humidity, a constant threat in an uncontrolled environment.

For some 3000 years, the remarkable skill of the ancient Egyptians in preserving the bodies of their Pharaohs and others was coupled with equally remarkable environmental conditions of extreme dryness and stability. This combination resulted in superb conservation of an unparalleled bridge to the past, the actual life forms of ancient rulers of Egypt and some of their subjects. Relatively recently the Royal Mummies were moved to simple museum cases in the Egyptian Museum in Cairo.

Unfortunately, Cairo is an exceptionally busy metropolis with very few environmental regulations and all of the disadvantages of a commercial city; the air heavily polluted with oxidants as well as with particulate matter, both of which are particularly deleterious to organic materials. The Museum, because it is so close to the Nile, is more humid in the night and early hours than the Valley of the Kings and the Valley of the Queens where the mummies were kept for over 3000 years. Moreover, the Egyptian Museum does not have either air conditioning to control the humidity and temperature nor air filters to remove the particles of diesel soot or other deleterious matter and dust. The foregoing environmental conditions, together with the large numbers of visitors to the world-famous Museum which affect the microclimate within it, constitute a serious hazard to the continued preservation of the Royal Mummies of Egypt.

Thus it was in 1985 that the Getty Conservation Institute (GCI) was first approached by the Egyptian Antiquities Organization (EAO) with their concerns about alleviating the threats to the present and future state of the royal mummies. It was the specific request of officials of the EAO that the mummies not only be protected from harmful environmental effects by a newly designed case but that such a case be both technically effective and visually superior for the display of mummies. In other words, the case should have dual utility, both safe for storage and desirable for viewing by visitors.

In 1987, an agreement was signed by the EAO and GCI for developing a long-term display and storage case for the Royal Mummies in the Egyptian Museum in Cairo. GCI was to be primarily responsible for design and construction. EAO was to document and restore the mummies to exhibition state as well as to investigate the microbiology of the mummies, their wrappings, cases, and exhibition room environment. The case should maintain an inert gas atmosphere as well as a stable relative humidity microenvironment.

A number of requirements for the hermetically sealed case were defined at the beginning of the project, requirements which would make it have an even more important scope than that of protecting only the Egyptian mummies: the preservation of any kind of organic artifact by means of a controlled micro environment in any museum in the world, including those with minimal funds for complex cases.

However, past attempts to develop display cases or storage units which had a nitrogen atmosphere whose relative humidity was controlled were either extremely expensive or required considerable maintenance or monitoring of gas flow equipment. Additional problems always arose if mechanical or electrical devices were attached to a case in order to control humidity. Accordingly, we undertook the design and construction of hermetically sealed cases which could be filled with nitrogen or another inert gas and left unattended for periods of a decade or more. An important constraint on the research and development was cost; we believed that the final product must be as inexpensive as possible within the parameters of the goals. Thereby, the advantage of freedom from the possibility of aerobic biological harm to an art object could be achieved by a relatively low cost microenvironment with equally low maintenance.

To achieve this major goal, (1) the cases should not be dependent on any mechanical or electrical systems, (2) they should require as little maintenance as possible, arbitrarily a minimum of two years, (3) the cases should be capable of being manufactured in developing countries, and (4) the cost per case should be kept as low as possible.

2. Project development

The work plan for the project included two major areas of research; one, to establish ideal storage conditions; and two, to develop hardware that could maintain such conditions. The parameters studied were selection of inert gas, allowable level of oxygen, and optimal levels of relative humidity, temperature and illumination. The design of hardware proceeded parallel with research on storage conditions.

Once the conceptual design was completed, a prototype was fabricated. Detailed mechanical design and fabrication of a case (1,000 liter volume) for the Egyptian Royal Mummy were contracted to Lightsense Corp., Laguna Hill, California. The case was tested and various improvements of its design were implemented at the GCI.

The presentation of the project report and the installation of the prototype, an hermetically sealed display and storage case, to the EAO were made at the Egyptian Museum, Cairo, Egypt in May 1989. The performance of the case was evaluated in-situ in a proposed room for the permanent exhibition of the Royal Mummies in the Museum. The oxygen leak rate of the case was better than 20 ppm/day and was acceptable to both the EAO and the GCI.

Additional design improvements for both improved performance and simpler fabrication were made at the GCI, and three new half-size (500 liter volume) cases were fabricated with these improvements. The cases had oxygen leak rates which were better than 10 ppm/day. The design improvements were forwarded to the EAO for their production. The case's components, such as laminated aluminum sheets for bellows, O-ring cord for its hermetic seal, and high quality valves and tube fittings, which the EAO was unable to obtain in Egypt were identified, and the necessary quantity of the components to build 27 cases were donated to the EAO. In addition to above items, a set of aluminum extrusions for fabricating one complete case was donated to the EAO for their first case.

Production of the design has now been carried out by the EAO for their Royal Mummy collections in the Egyptian Museum in Cairo under the supervision of the GCI. The EAO has completed the assembly of 10 cases as of December 1993.

3. Proposed storage condition

The research on effects of reduced oxygen atmospheres and variation in relative humidity on biological activities was conducted [4, 5, and 6]. Research on the chemical and physical stability of proteinaceous materials such as mummified tissues and parchment paper also was carried out and the following recommendations are based on it and literature reviews by above investigators at the GCI.

An oxygen-reduced inert atmosphere of nitrogen at low relative humidity suppresses both biological activity and oxidation reactions in the proteinaceous material. The recent study [7] has shown that the complete eradication of all stages of major museum pests, except cigarette beetles, was observed in a nitrogen gas environment at room temperature, where the oxygen concentration was maintained at less than 0.1 %, after 10 days. While ideally the oxygen level should be as close to zero as possible, in reality this is impractical. Repurging of the case is therefore recommended once the oxygen concentration reaches 2 %, better at 1 %.

An ideal temperature level is 10 °C to 15 °C for the preservation of organic materials. A lower temperature reduces reaction rates in general denaturation in collogen. Due to the fact that humidity fluctuations are a result of temperature fluctuations, lower temperature levels are only desirable if the amount of variation is not increased by the requirement of this lower temperature. Furthermore, other museological considerations such as human comfort sometimes make low temperatures undesirable in display and storage room. However, it is recommended to keep the temperature as constant and as low as possible.

An ideal relative humidity level exists between 30 % and 50 % for organic objects. Values below 30 % RH should be avoided due to the possibility of permanent deformation of proteinaceous material. The microbiological activities on mummified tissue and parchment samples were significantly reduced for the microenvironment with relative humidity below 50 % RH. Most oxygen scavengers require moderate levels of moisture to react with oxygen. (Reaction rates decreased considerably after a scavenger had been in a 33 % RH atmosphere for a year.) Therefore, the higher level of about 45 % RH is advisable if an oxygen scavenger is used to maintain the inert gas atmosphere.

The spectral distribution of the light reaching an object should be restricted to the longest wavelengths that result in its acceptable appearance. Restricting light to as close as 400 nm cut-on should be accomplished through ultraviolet and possibly violet filters, proper choice of an illuminant or a combination of the two. The intensity of the lighting should be limited to less than 100 lux.

4. Design concepts of the case

The basic concept of the display and storage system was to provide and maintain a reduced oxygen environment for stored or displayed objects. Requirements were established such that the system should not rely on a mechanical or electrical sub-system for maintaining its performance; should require minimum maintenance over an assigned period, which was arbitrarily set to be two years; should be easily manufactured, even in developing countries; and the cost per unit should be contained within a reasonable range. Reasonable range meant that the price should be comparable to that of high quality cases.

4.1 The display and storage case

A passive system for maintaining the inert gas atmosphere was selected over an active system, because its performance is independent of mechanical or electrical accessories. It would result in low maintenance requirement of the system over an extended period, since it does

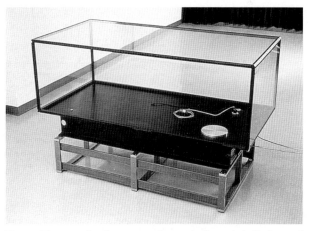

Fig. 1. Photograph of prototype hermetically sealed display and storage case built for the Egyptian Royal Mummy

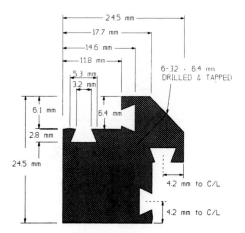

Fig. 2. Mechanical drawing of extrusion profile

not require power or a specialized electrical/mechanical service crew.

The case consists of a display section, a pillow shaped bellows and a steel base assembly. The case is filled with humidity-conditioned nitrogen, hermetically sealed from the surrounding atmosphere, and monitored for oxygen content and relative humidity without opening (see Figure 1). Because of the extremely low leak rate of oxygen into the case, when initially filled with nitrogen, less than two percent of oxygen will be present before ten years. If packets of oxygen scavenger are placed in the case, an additional ten to twenty years may elapse before the two percent oxygen level is exceeded. The relative humidity buffering material that has been conditioned at the selected RH is put in the case to provide a constant relative humidity level despite any moisture from the displayed object or ingress over time. The case can be equipped with a septum port for removing a sample for gas chromatographic analysis (for the analysis of potential volatiles emitted from displayed objects). As an option, activated carbon packets can be placed in the case to act as a sorbent for pollutants and volatiles. "Pollutants" may be generated by the displayed objects, or less probably, from construction materials, or infiltration via the minute leaks.

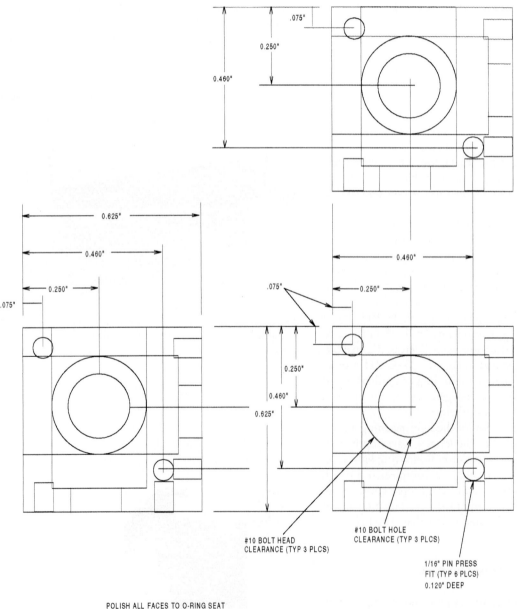

Fig. 3. Mechanical drawing of corner element

Pressure equilibration of the case with the outside atmosphere is achieved by inflation and deflation of a bellows attached to the display case. This pressure-free condition of the case allows light construction of its hermetic seal. The temperature induced volume changes of the case's gas are similarly automatically adjusted by the inflation and deflation of the bellows.

During installation of object in the case the nitrogen gas supply system is connected to provide humidity-controlled nitrogen for simultaneously purging oxygen from the case and conditioning of its contents to the selected relative humidity. The oxygen measurement system is also attached to the case for determination of the oxygen level during the purging and the leak rate of oxygen following the purging.

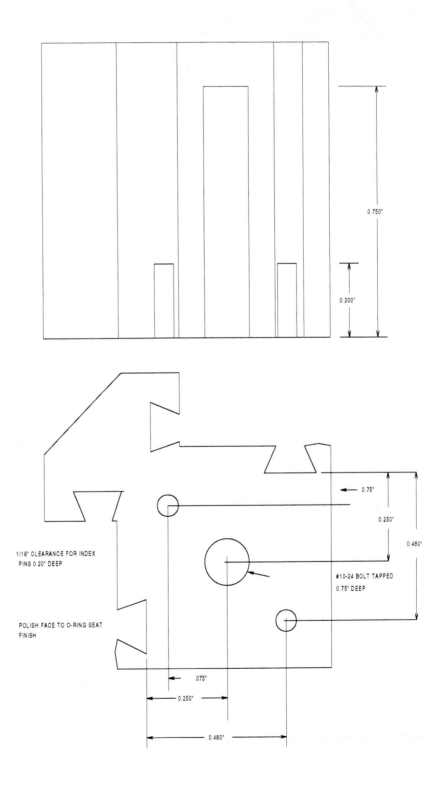

Fig. 4. Mechanical drawing of drilled extrusion end for corner assembly

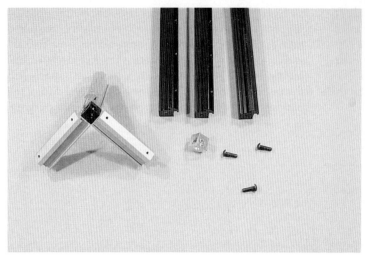

Fig. 5. Photograph of corner assembly

Fig. 6. Schematic of hermetic seal

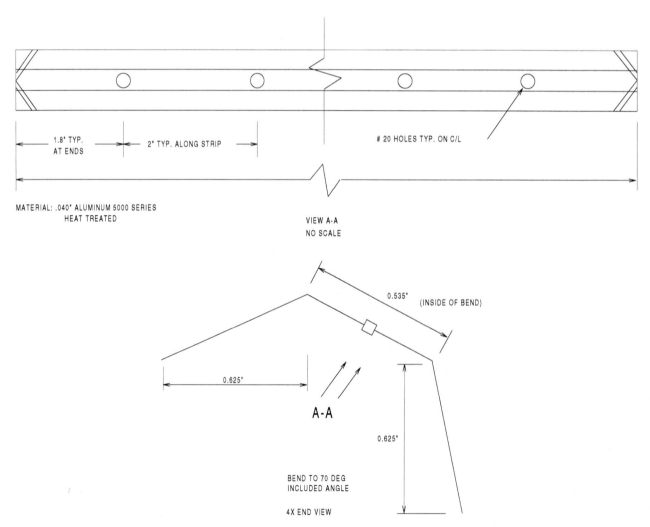

Fig. 7. Mechanical drawing of retention strip

4.1.1 Display section

The display section consists of five panels of 6.4 mm thick laminated glass plate and an aluminum base plate which are mounted to the aluminum extrusion frame. The base plate is fabricated from 6.4 mm thick 2024 aluminum alloy with two pairs of aluminum channel supports beneath it for providing clearance for a bellows. The frame is constructed by bolting the specially designed extrusion frame members (see Figure 2) to the machined corner elements (see Figures 3, 4, and 5) which are also made of the aluminum alloy of the same composition, with a thin film of neutrally curing silicone filler. (This design simplified the fabrication and reduced the cost as compared to helium-arc welding.) Seals between the glass or aluminum plates and the frame (see Figure 6) are effected by the Viton® O-rings under the pressure exerted by the retaining strips, precisely bent and heat treated aluminum sheet (see Figure 7). All the aluminum components are black anodized for the surface stability against moisture and polluted atmosphere and reduced reflection of the surfaces.

Six ports of 0.64 cm diameter and one port of 20.32 cm diameter are drilled or machined on the aluminum base plate (see Figure 8). These small ports are designed for nitrogen purging, recirculating, sample taking, attachment of a pressure relief valve, and connection to the bellows. The large port is equipped with an O-ring sealed cover and designed as an access for insertion and replacement of the oxygen scavenger, the relative humidity buffer, the pollution sorbent and the sensors. The case can be equipped with a septum port which allows taking air samples from the case with a syringe. The air samples can then be analyzed for volatiles potentially emitted from the exhibit. Should this analytical work be planned, one should not place activated carbon in the case.

An oxygen sensor and a relative humidity transmitter are positioned inside the display section and their electrical leads soldered to the hermetically sealed electrical connector mounted on the aluminum base plate. Excitation voltage is applied to the RH sensor from outside the case by its monitor; similarly, the voltage developed by the oxygen sensor is recorded as oxygen percent on a monitor.

The case is equipped with a removable U-tube manometer for static pressure monitoring of the case during purging or conditioning of the contents, replacement of the oxygen scavenger and RH buffer packets and servicing of sensors. A pressure relief valve is mounted on the aluminum base plate to provide additional safety against pressure build-up in the case during purging of oxygen and conditioning of objects.

4.1.2 Bellows

A pillow-shaped bellows (see Figure 9) is connected to the aluminum base plate with an O-ring sealed fitting. The bellows is made by heat-sealing two oxygen barrier films, such as aluminized films, approximately 0.13 mm thick at all four edges. Its function is to expand and contract during any external temperature and barometric pressure fluctuations, thereby preventing any pressure on

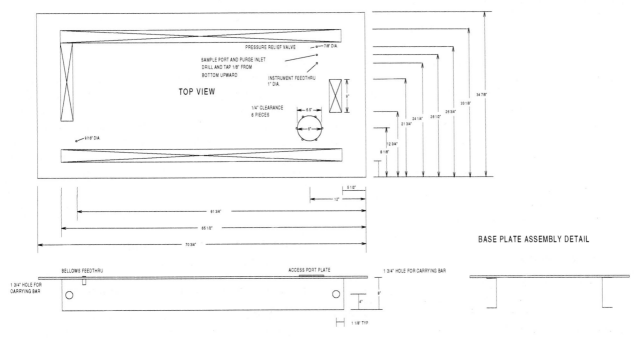

Fig. 8. Mechanical drawing of base plate

Fig. 9. Photograph of aluminum bellows

Fig. 10. Photograph of oxygen scavenger, Ageless®

the display section. The bellows, whose volume is recommended to be at least one-tenth of the display section's volume for a 10 °C temperature variation in the environment and rests directly below the aluminum base plate with enough vertical clearance for its expansion. A larger bellows is recommended if larger temperature variations in the environment are expected.

5. Passive control agents

5.1 Relative humidity buffer

The case does not require protection against infiltration of moisture to or from the surroundings, because the leak rate of the case is minimal. However, moisture could enter or leave the case during the replacement of the passive control agents or servicing of the sensors. A typical oxygen scavenger, such as Ageless®, itself contains a small amount of water, and it releases the moisture in time. Temperature changes in the case will also result in relative humidity fluctuations, therefore, the placement of the relative humidity buffering material is considered advisable.

The selection of relative humidity buffering material, such as silica gel, artsorb and molecular sieves for the case, has to be made on the effectiveness of the material in the relative humidity range at which the objects will be stored, the costs, and its availability. The amount of the material depends on the volume of the case and hygroscopic capacity of the object. Thinly spreading of the buffer, to produce the largest surface area, is recommended for the most rapid response of the buffer.

5.2 Oxygen scavenger

A commonly used oxygen scavenger, such as Ageless® (see Figure 10), has been developed for stored food industries. It consists of a finely divided iron powder, potassium or sodium chloride, and a zeolite containing water, packaged in oxygen-permeable plastic packets, and has been successful in extending the shelf-life of packaged processed foods. The reaction is most rapid at 75 % RH, but it is significant even at 45 % RH. The packets are marked with their oxygen absorbing capacities. Therefore, the number of packets that must be placed in the case can be determined from its oxygen leak rate and the time required before maintenance. The ability of oxygen absorption is affected by the moisture retention of the packets. The kinetics of the reaction of Ageless with oxygen in sealed cases are reported by F. Lambert et al. [8].

5.3 Pollution sorbent

The case is designed with the packets of pollution sorbent for protection of the contents from internal pollutants and from ingress of any indoor/outdoor pollutants. The pollutants generated inside the case are the volatile gases emitted from the object and minute out-gassing of the case's components, such as O-ring material and oxygen scavenger. Even though the composition and quantity of these internal pollutants have not been thoroughly investigated, the composition of outdoor and indoor pollutants and their levels have been well studied. Therefore, the selection of the pollution sorbent material has been based on the indoor pollutants in a typical museum; the quantity is calculated from the leak rate of the case and its required service-free duration. Activated carbon was selected to absorb typical pollutants in a museum atmosphere that may contain HC, NO_x, SO_2, CO, CO_2, and O_3 at a level of approximately 50 ppb.

6. Evaluation of performance and start-up of operation

6.1 Leak test and performance evaluation

First, the case and bellows were tested individually for major leaks using freon or helium gas in the case and measuring levels of transmission through all the hermetically sealed locations from the outside. Then the case and bellows were purged with dry nitrogen. The oxygen concentrations in the case and the bellows were then equilibrated and the initial oxygen concentration was measured by using a trace oxygen analyzer.

During the testing period, the temperature of the case and the room were periodically raised as much as 10 °C above normal room temperature to simulate variations of the temperature and the barometric pressure changes that might occur under actual extreme operating conditions. Oxygen concentrations in the case were measured daily for 7 to 10 days to determine the average oxygen (and thereby air) leak rate of the case.

The leakage rate of atmospheric oxygen into the case ranged approximately 20 ppm/day for the prototype, which was 1,000,000 cc volume, and it improved to less than 3 to 8 ppm/day for 5 new cases (500,000 cc and 96,000 cc volumes) produced with improvements in the design. If the case's oxygen concentration is brought down to 0 % at the time of installation and this leak rate is maintained, it will take approximately eleven years for the oxygen concentration inside the case to reach 2 % (even without the presence of an oxygen scavenger).

6.2 Purging of the case with objects

It is essential to control relative humidity (RH) in the display case. Therefore, nitrogen gas used in the case must be carefully humidified. Figure 11 is the simplified schematic of the nitrogen gas supply system. RH-conditioned nitrogen is produced by controlling the mixing ratio of dry nitrogen from the tank to the humidified nitrogen which is produced by the dry nitrogen bubbling through water at room temperature.

The case can be equipped with a micro-filter to prevent bacterial and fungal spores on the object from being dispersed out to the museum atmosphere during the purging. The pleated polytetrafluoroethylene (PTFE) membrane filters with a retention rating 0.5 µm are recommended. Pleated type micro-filters with a large filtration area should be selected for the application in order to prevent pressure build-up in the case.

The appropriate amount of the passive control agents, relative humidity buffer, oxygen scavenger, and pollution sorbent, are placed in the case through the access hole at the end of the purging process. Once the above procedures are completed, and the access door tightly bolted, the nitrogen supply system and micro-filter are removed from case with the valves closed. Re-purging of the case is necessary when the oxygen in the case reaches 2 %.

7. Conclusions

An hermetically sealed display and storage case was successfully designed, fabricated and tested at the Getty

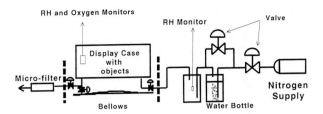

Fig. 11. Schematic of purging setup

Conservation Institute. The case requires no electrical or mechanical system to maintain its performance. It is relatively easy to be build without special tools and can be fabricated in developing countries. Humidity-conditioned nitrogen was used as an inert atmosphere in the display case, but of course other inert gases such as argon could be used. It can be built so that less than 10 ppm/day of the atmospheric oxygen will leak into the case. The maintenance-free period of the case is approximately 11 years when the case is purged to 1000 ppm oxygen level without the oxygen scavenger, and its performance can be extended to longer than 20 years by introducing the oxygen scavenger after purging. The display cases are now being produced by the Egyptian Antiquities Organization for display and storage of 27 Pharaonic Royal Mummies. Because construction of the case is labor intensive, its cost varies, depending on the country where it is produced. In the U.S. it costs approximately $ 0.01/cc in 1993 to build, but this could be reduced drastically if multiple orders are placed at the same time.

Acknowledgements

The authors wish to thank Dr. Neville Agnew and Dr. Frank Preusser for their corporation and technical supports throughout the project.

Summary

An hermetically sealed display and storage case was designed, built, and tested by the Getty Conservation Institute for long term preservation of objects and artifacts that are prone to biodeterioration and/or are oxygen sensitive. The case is designed to prevent deterioration of organic and inorganic objects in museums resulting from changes in humidity, biodeterioration, thermally and photolytically induced oxidation, and degradation by gaseous and particulate air pollutants.

The hermetically sealed, nitrogen filled case maintains its pressure equilibrium with the surrounding atmosphere by a compensating bellows for applications in temperature-varying non-air-conditioned environments. The case does not use mechanical or electrical control for maintaining the microenvironment. It contains sensors for oxygen and humidity, a relative humidity buffer, oxygen scavenger and pollution sorbent. Under normal conditions the case can be maintained at less than two percent oxygen concentration without major maintenance for about ten years, and probably twenty years with the oxygen scavenger in the case. After that period it can be flushed with humidity conditioned nitrogen, and the buffers and sorbents replaced. The case can be built in a standard machine shop and does not require special tools. Production of the case is underway by the Egyptian Antiquities Organization for the display and storage of the Royal Mummy collections of the Egyptian Museum, Cairo.

Zusammenfassung

Das Getty Foundation Institute hat für die langfristige Konservierung von Objekten und Artefakten, die für biologischen Abbau anfällig sind oder empfindlich auf Sauerstoff reagieren, einen luftdichten Schau- und Aufbewahrungskasten entwickelt, gebaut und erprobt. Der Kasten soll Schäden an organischen und anorganischen Objekten in Museen durch Feuchtigkeitsschwankungen, biologischen Abbau, thermisch und photolytisch ausgelöste Oxidation und gas- und teilchenförmige Luftschadstoffe verhindern.

In dem luftdicht verschlossenen, stickstoffgefüllten Kasten wird durch einen für Räume mit Temperaturschwankungen und ohne Klimaanlage geeigneten Kompensationsblasebalg ein Druckausgleich mit der Umgebung hergestellt. Es kommen keine mechanischen oder elektronischen Regler zur Aufrechterhaltung des Mikromilieus zum Einsatz. Der Kasten enthält Meßfühler für Sauerstoff und Feuchtigkeit, einen Puffer für die relative Feuchte, einen Sauerstoffaufnehmer und ein Schadstoffsorptionsmittel. Unter normalen Bedingungen kann in dem Kasten ohne größere Wartungsarbeiten etwa 10 Jahre lang eine Konzentration von weniger als 2 Prozent Sauerstoff aufrechterhalten werden; mit dem Sauerstoffaufnehmer im Kasten wahrscheinlich 20 Jahre. Nach dieser Zeit kann er mit feuchtigkeitsreguliertem Stickstoff gespült und Puffer und Sorptionsmittel ausgetauscht werden. Der Kasten kann in einer normalen Werkstätte, ohne irgendwelche Spezialwerkzeuge, hergestellt werden. Er wird zur Zeit von der ägyptischen Organisation für Altertümer für die Lagerung und Ausstellung der Königlichen Mumiensammlung des Ägyptischen Museums von Kairo gebaut.

Résumé

Le Getty Foundation Institute a conçu, construit et testé une caisse hermétique pour la conservation à long terme et la présentation d'objets et d'artefacts menacés par la désintégration biologique ou sensibles à l'oxigène. Cette caisse a la fonction d'éviter l'endommagement des objets organiques et anorganiques exposés dans les musées, dû aux variations d'humidité, à la décomposition biologique, à l'oxidation d'origine thermique ou photolytique ainsi qu'aux polluants atmosphériques sous forme de gaz ou de particules.

Dans la caisse hermétique remplie d'azote, un soufflet de compensation adapté aux espaces à forte variation d'humidité et dépourvus d'une climatisation assure une compensation de pression avec l'extérieur. Aucun régulateur mécanique ou électronique pour la conservation du microenvironnement n'est employé. La caisse est munie de capteurs d'oxigène et d'humidité, ainsi que d'un tampon pour l'humidité relative, un absorbeur d'oxigène et un sorbant de substances nuisibles. Dans des condition normales il est possible de maintenir dans la ciasse pour une période de dix ans – et sans de majeurs travaux d'entretien – une concentration d'oxigène de moins de 2 %, la présence d'un absorbeur d'oxigène dans la caisse pouvant doubler cette période. Au terme de celle-ci, on peut laver la caisse avec de l'azote régulant l'humidité et remplacer le tampon et le sorbant. La caisse peut se construire dans un simple atelier dépourvu de tout outillage spécial. L'organisation égyptienne pour les antiquités est en train de la construire pour la conservation et l'exposition de la collection de momies royales au Musée Égyptien du Caire.

Riassunto

Il Getty Foundation Institute ha sviluppato, costruito e sperimentato, per la conservazione a lungo termine di oggetti ed artefatti suscettibili di un degrado biologico oppure che reagiscono sensibilmente all'ossigeno, una cassa ermetica di presentazione e conservazione. La cassa dovrebbe ovviare a danni arrecabili ad oggetti organici e anorganici in musei, dovuti a variazioni dell'umidità, al degrado biologico, all'ossidazione termica o fotolitica e all'azione degradante di fattori inquinanti sotto forma di gas e particelle. Nella cassa chiusa ermeticamente e riempita di azoto viene raggiunta un compensazione con la pressione esterna, attraverso un soffietto di compensazione adatto ad ambienti con variazioni di temperatura e privi di impianti di condizionamento dell'aria. Non vengono utilizzati regolatori meccanici ed elettronici per la conservazione del micro-ambiente. La cassa contiene sensori per ossigeno e umidità, un rilevatore di ossigeno e un prodotto di assorbimento di fattori inquinanti. A condizioni normali si possono raggiungere nella cassa concentrazioni di meno del 2 % di ossigeno per un periodo di dieci anni, senza maggiori lavori di manutenzione. Con il rilevatore di ossigeno nella cassa ció é possibile per vent'anni. Dopo questo periodo la cassa puó essere lavata con azoto regolante l'umidità, e possono essere sostituiti il rilevatore di ossigeno ed i mezzi di assorbimento. La cassa puó essere prodotta in qualsiasi officina, senza ricorso a strumenti speciali di sorta. Attualmente viene costruita dall'organizzazione egiziana per antichità, per la conservazione e l'esposizione della raccolta di mummie reali del Museo Egiziano del Cairo.

References

1. Balout, L., ‚La Momie de Ramses II (1976–1977) Contribution Scientifique a l'Egyptologie'. Ed. Recherche sur les Civilisations. Paris: Museum National d'Histoire Naturelle, Musee de l'Homme (1985).
2. Bucaille, M., ‚Les Momies des Pharaons et la Medicine' Paris: Librarie Seguier (1987).
3. Pääbo, S., ‚The Mummy of Ramses II Reconsidered' Österreichische Leder-Zeitung (OLZ) [Austrian Leather Newsletter]. **83** (1988) 389–392.
4. Valentin, N. and Preusser, F., ‚Nitrogen for Biodeterioration Control on Museum Collections', *The Third Pan-American Biodeterioration Society Meeting*, Washington, D.C. (1989).
5. Valentin, N., Lindstrom, M. and Preusser, F., ‚Microbial Control by Low Oxygen and Low Relative Humidity Environment' *Studies In Conservation*. **35** (1990) 222–230.
6. Valentin, N. and Preusser, F. ‚Insect Control by Inert Gas in Museums, Archives and Libraries' *Restaurator* **11** (1990) 22–33.
7. Rust, M. and Kennedy, J., ‚The Feasibility of Using Modified Atmospheres to Control Insect Pests in Museums' (Final Report, prepared for The Getty Conservation Institute) Dept. of Entomology, University of California, Riverside (1991).
8. Lambert, F., Daniel, V., and Preusser, F., ‚The rate of absorption of oxygen by Ageless®; The utility of an oxygen scavenger in sealed cases'. *Studies in Conservation* **37** (1993) 267–273.

Correspondence: Dr. Shin Maekawa, The Getty Conservation Institute, 4503 Glencoe Avenue, Marina del Rey, CA 90292–7913, U.S.A.

Asiatic mummies

Excavation and preliminary studies of the ancient mummies of Xinjiang in China

B.-H. Wang

China Xinjiang Institute of Relics and Archaeology, Xinjiang, Republic of China

Introduction

China has a vast territory and the natural geographical conditions differ a lot between its regions. The ancient Chinese believed honestly that the spirit of those who died would not perish. Therefore it became a custom among the relatives of the diseased to attach particular significance to the preservation of the corpses so that they should not decay. In order to achieve that, it was not sufficient to ensure a rich burial, with precious gifts including objects of jade and other valuables. They even developed a set of procedures and techniques to treat the bodily remains in order to prevent decaying for a long period. The upper class favoured different kinds of chemical treatment, for instance involving mercury as a preserving agent. Even the graves were at some occasions dug very deep and were densely sealed. This has contributed to the conservation of some of the bodies during several thousand years.

A large number of such "humid" (or wet) mummies dating from the Han dynasty has been discovered in the Hu-bei province in Middle China and in Mawangduei at Changsa city in the Hunan province, dating from the Ming dynasty. These provide examples of locations within the Yangzi jiang river valley where well preserved mummies have been found. All these mummies provide valuable cultural and historical information.

Since the climate of Xinjiang is extremely arid, a large number of graves with mummies have been discovered, particularly during the last two three decades. In contrast to the above mentioned wet mummies, these are generally natural mummies, having rapidly dessicated after burial because of the extremely arid environment. When these ancient corpses were excavated, they were still well preserved, with all their clothing, hats and decorations and the detailed characteristics of the dead bodies to a large extent retaining the conditions just as they were buried. In this way, a rich and varied information about history and culture is preserved. Information may be obtained about racial peculiarities, characteristics of the different nationalities and national traditions, clothing, food, lodgement, communication, religion, morphology and appreciation of beauty, views on art, as well as about nutritional conditions given by the diseased, causes of disease, presence of parasites, and even about the ecological environment and its change. In this way, scientists of different research areas can get correct and reliable information based on the bodies and corresponding archaeological data.

The present article gives a survey of the finds of ancient mummies from Xinjiang.

1. Excavations of ancient mummies in Xinjiang

Excavations providing mummies in Xinjiang are known back to the turn of the century, before the new China was established. About ninety years ago, several excavations were carried out in the area, notably by F. Bergmann, A. Stein, Whuang Wenbi and the Japanese Ju-Rui-chao. They all both discovered and reported finds of mummies. Archaeologists from Xinjiang have also excavated several mummies and conducted a fair amount of research. However, it is not before the end of the 1950s and onwards, and especially within the last 20 years, that excavations of mummies have been carried out in a proper, archaeologically scientific manner. Therefore, particular attention was paid to this kind of finds. In the course of ten excavation periods, approximately 400 mummies have been discovered to the south of the Tien-shan mountains during the last decennia.

The results of the ten excavation periods, some of them extending several years, whereas other are more limited in time, may be summarized as follows:

1.1 During the longest excavation period, conducted by the Xinjiang Institute of Archaeology, Xinjiang Museum, extending from 1959–1975, 456 graves were excavated, 305 containing mummies, dating the Gin South-North Dynasty until the Sui-Tang Dynasty, and altogether 11 tombs in Turpan Astana, Karahoja. The statistics is not exact because a certain amount of mummies had been disturbed and remained incomplete.

1.2 In October, 1959, the Xinjiang Museum had discovered and excavated a wooden coffin containing the mummies of two adults, one male and one female. The discovery was made about 3 km northwest of the ancient site of Niro in the Hetian region. The skin of these mummies was well preserved.

1.3 During excavation campaigns in 1978, 1986, and 1991, the Xinjiang Institute of Relics and Archaeology excavated a total of 114 graves, among which 11 contained well preserved mummies. Though 70 individuals displayed partial mummification, only the 11 were well preserved.

1.4 In February, 1970 one mummy was excavated in one of the valleys of Tien-Shan, at the southern shore of a salt lake at the southern suburb of Urumqi. It is dated to the Yuan Dynasty. Unfortunately, it has later been destroyed.

1.5 October, 1979. Forty-two graves were excavated by the Xinjiang Institute of Archaeology in the cemetery Gu-mu-gon (Valley of ancient tombs). Among these, one contained a well preserved mummy of a young female and another a well preserved mummy of a baby.

1.6 April, 1980. The Xinjiang Institute of Archaeology excavated the mummy of an adult female at the northwest part of Lop-nor, where the delta of Tieban river enters the lake.

1.7 September, 1985. In an ancient cemetery of the Jahongluk county of the Chercheng district, the Xinjiang Museum excavated 5 tombs, of which 3 had already suffered destruction. Within the better preserved 2 tombs, three well preserved mummies were excavated, one mummy of an adult male, one adult female and one baby.

1.8 A particular discovery of mummies was made in the area of the Beita-Shan in the Qi-tai district, to the *north* of the Tien-Shan mountain ridge. In an abandoned mining shaft of a gold mine, the Relics Administrative Bureau of the Qi-tai district discovered several well preserved, dessicated bodies of gold panners. Some of them were tied up with ropes. The state of preservation of the clothes on the bodies as well as of their belongings was also good. The permission to pan gold and their tools were still beside their bodies. It looked as if there had been a quarrel about money during their panning labour which had led to the killing. The reason they were preserved the way they were was that the wet soil in the mining shaft contained a high percentage of salt. To the north of the Tien-Shan mountains, the conditions for natural mummification is otherwise not particularly suitable because of higher precipitation than to the south.

1.9 October, 1989. In the course of the excavation of 82 graves in the cemetery Yang-he in Luk-qin county of the Shaan-shan district, carried out by the Xinjiang Institute of Relics and Archaeology, the mummy of a baby was discovered.

1.10 April 1991, less than six months before the discovery of the Iceman at Hauslabjoch on September 19, the Xinjiang Institute of Relics and Archaeology was excavating the ancient sites and one cemetery at the Subash valley side within the Flaming mountain in the Shaan-shan district in the Turpan Prefecture. In the course of this, 23 mummies were discovered in 34 graves in the Warring-Countries cemetery. Among these mummies, 7 were perfectly preserved, and completely preserved clothing with decorations were also recovered.

In addition to the excavated mummies mentioned above, there are also records about excavated mummies from the Ba-chu district, from Lianmuquin within the Shaan-shan district and east Gobi of the Hami Oases. In general, the locations of these excavated ancient mummies indicate that the cemeteries were basically located at the southern parts of Tien-Shan, all sourroundings of the Tarim Basin or deep in the Taklamakan. The Turpan Basin, the Lop-Nor desert and the Hami Oasis are all very dry regions. But at the corresponding northern regions, to the north of the Tien-Shan mountains, no ancient mummies have been found, although a lot of archaeological excavations have been carried out and a several human skeletons have been recovered. The well preserved mummies of the gold-panning workers discovered in a mining shaft in Bieta-Shan by the Qi-tai Relics Administration mentioned above (1.9), appears to be an exception caused by the high content of salt in the soil of the shaft, and cannot be recognized as a general case of preservation in that area.

2. Conditions for mummification of ancient bodies in Xinjiang

When an individual dies, the cells in the body normally begin to self-dissolve. By the process of hydrolysis of the body different proteins and acids are formed, and the hydrolysis successively causes breakdown of larger molecules into smaller and smaller molecules. Beside this self-dissolving process, natural invading processes by various putrifying bacteria occur, which cause the dead body to corrupt and decompose fast. However, in Xinjiang in many places the ancient buried human bodies betrayed this natural process: the bodies did not corrupt but turned into dessicated, dried corpses. The conditions that formed the ancient Xinjiang mummies can be summarised by the following factors:
– The climate was very dry.
– The soil had a high content of salt.
– The human bodies were shallowly buried or their tomb-caves were not tightly sealed.

– The burials took place during winter, that is, during the cold season when bacterial activity is low.

A brief explanation is given as follows.

2.1 Dry climate

As described above the Xinjiang dried mummies were mostly found in the Tarim Basin and its neighboring places that had a generally dry climate. Nevertheless some also neighboured the Lop desert, the Turpan Basin, and the Hami Oasis, but as a macro-geographical climate criterion, it can be said that the locations share the same characteristics: a very enclosed relief, far from the ocean, very dry climate and nearly no hydro-atmospheric content.

Looking at the map, it is realized that the Tarim Basin's southern border is limited by the Altyn-tagh, the Kun-lun mountains, and the Kara-kun-lun mountains. At the western border, one finds "the roof of the world", the Pamir Plateau, and at the northern border there is the high Tien-Shan mountain ridge that extends several hundred kilometers east-westward. These high mountains prevent moist air currents from the Indian Ocean, the Atlantic Ocean and the Arctic Ocean. In the eastern part where they neighbour the Gan-su province, there is a small break that forms a natural "communication tunnel", but due to the distance of 2000 km from the Pacific Ocean, no south-east wind can enter. Therefore, the natural geographical conditions tend to prevent clouds to enter into the Tarim Basin, and this explains why the above mentioned regions have rare precipitation.

When compared with the Zhungarian Basin of north Xinjiang, which is belongs to the temperate zone, the Tarim and Turpan basins are extremely arid deserts belonging to the warm-temperate zone, whereas the Hami basin is a temperate arid desert. From Miocene and onwards, and especially since the early Pleistocene, this region changed into an arid and extremely arid desert zone. The analysis of Quaternary sediments indicates that after the last glaciation (7000 years ago), and especially within the last 4500 years, this region was getting continuously more and more arid, whereas during the same period the human population was increasing.

2.2 High salt accumulation in the soil

Arid regions normally form salt accumulation zones, and the rate of salt accumulation is generally highly correlated with the aridity. In the process of and because of basin formation, rivers move the salts from the surroundings mountains together with fluvial debris into the basin where it is sedimented. When the soil dries out, evaporation tends to accumulate the salts on the ground surface. Additional salt-absorbing of plants and accumulated leaves and branches on the ground tend to increase the accumulation of salt in the shallower layers of the soil. Each of the Tarim, Turpan, and Hami Basins has its own enclosed hydrological unit, and the drainage of the water in the basin is limited to the evaporation from ground surface and plant transpiration. The mineralization of the ground water generally reaches 10 g/l, and the upper one seventh reaches several tens g per liter. Usually the cemeteries were preferably selected at the higher places near or at the outskirt of the human dwelling places. The salt accumulating process had been strong. The salts deposited are predominantely chlorides. Except for some few salt-addicted bacteria, other bacteria cannot exist and live in the salt environment. In the human daily activities, salt is frequently applied to the food for preventing corruption, preventing from bacterial decay being one of the reasons.

Take the cemetery of Jahongluk in the Chercheng district as an example: large amounts of salt blocks were contained in the sand dunes at the surroundings of the tombs. The salt layer was one metre thick. At Wu-bao of Hami city there occured gypsum layers (which are frequently found in connection with salt layers) and in Turpan there occurred salt layers. These high-concentrated salt layers which obviously restrained the activities and growth of the corruption-causing micro-organisms thus favoured the protection of the dead bodies.

2.3 Shallow burials, not well sealed

Analysing the excavation conditions of the ancient Xinjiang mummies, a very distinct peculiarity is noted, because common for them was a shallow burial and a cover which permitted evaporation of body moisture and a quick dessication.

The Niya mummy, for instance, was put in box-shaped wooden coffin, and was placed in a shallow grave in the sandy soil. The female mummy at the Tieban River was excavated on a high, flat terrace in the Tieban River Valley, at a depth only 1 m. On top it was a cover of dried branches, reeds, and sand, providing a good ventilation. In the Gu-mu-gou cemetery, where altogether 42 tombs were excavated, all the females and babies mummified had been buried less than 1 m deep. On top of these graves there was a coverage with wood, furs, then sand, loosely piled. From the corpses buried in deeper graves only skeletons could be found. The ancient baby grave of Jahongluk in the Chercheng district had been dug only 30 cm deep in the sand, and the baby was therefore mummified. A second tomb showed another extreme case of mummification of an adult individual: the grave itself was 2.4 m deep, but not sealed off. The body had been covered with wood, and on the wood there was a coverage of blocks of salt, then reeds, furs, tamarix, and mats. Strangely, on the top of the tomb, an aperture measuring 0.3–0.6 m exsited, covered with a wool overcoat and a bag made of animal furs.

In Turpan Astana and Karahoja, cemeteries from the Jin-Tang Dynasty exist. From here a large number of mummies have been recovered, the dead having been placed on mats. If layers of gravel existed, the ventilation of the tomb was improved. The tombs of the Hami mummies were recovered from the most arid-hot, salt containing desert apart from the Wubao Oasis. These tombs were shallow. The bodies were covered with wood followed by earth, but the thickness of the cover did only amount to some tens of cm, and was not compact. The water could therefore evaporate easily. A shallow burial covered by loose soil could therefore let the tomb be directly influenced by the extreme arid air and soil conditions. Then the corpse rapidly got dehydrated. The microbes very soon soon lost their living conditions and could not develop an effective decomposition process.

2.4 Burying in the cold season

Within the same cemetery many dead were buried. However, sometimes, even within the same grave, what remained of some of the bodies were only skeletons, whereas there were some, whose clothing, hats, shoes were not corrupted, and the skin perfectly preserved. Since the differences are so great, an explanation has to be sought.

Several phenomena indicate that the main reason for this difference was the season of burial. The conclusion which can be drawn is that all the well preserved ancient mummies were buried during the cold season. The testimony is not only found in Hami Wu-bao, but even in the Lop-nor desert, Tieban River (lower reach of Kunque daria), Subashi of Shaan-shan district, and at Jahongluk of Chercheng district. All the mummies had their fur overcoat, fur trouser, two legs wrapped with felt, fur hat or felt hat. One some dead bodies there was a cover consisting of a large piece of fur (at the Tie-ban River), whereas some were covered with wool cloth or furs on their coffins. This care was obviously intended to protect the diseased from the cold, that is the equipment to keep warm clearly indicated that these funerals had taken place during the cold season.

The microorganisms that existed in the body, such as saecharomycetae, and bacteria which produces actinomyces could still exist and act in the organs of the dead corpse after the individual had died. To stop the body decomposition would require the elimination of the activity of the microorganisms. In a bitter cold environment, the microorganisms would cease their metabolism, but the exchange of the air in the grave and uptake of water from the corpse in the dry soil soil would still continue. When the cold period had passed, the corpse would already have lost large amounts of water, and with increasing temperature, it would dry out quickly, the microorganisms would again loose their acting environment, and would not effectively contribute to the decomposition of the dead bodies.

Beside the above mentioned geographical and climatic conditions, we have also noticed that on some dead bodies it seemed that animal protein had been painted as some kind of antiseptic. Tian-lin of the Xinjiang Museum has observed that large amounts of animal protein were stained on the gunny clothing of the female mummy from the Tie-ban River. At a later occasion, some antiseptic residues have been found and collected from the armpit of the mummy (though most of it might have disappeared in the process of handling the mummy in connection with prior analysis and dissection). On the ancient mummy which was found in Jahongluk a similar phenomenon was observed during the preliminary analysis, though already decomposed into amino acids. The mummies which have painted animal protein differ somewhat in their exterior appearance compared with the Turpan mummies that have not painted animal protein. The possible role played by addition of animal protein to the process of the mummification is currently under study.

To sum up, the basic reasons for the good preservation of mummies in Xinjiang are the extremely arid climate, the high salt content of the soil, the loose and shallow burying, burial during the cold winter season, and, as a consequence of this, that the activities of the microorganisms which normally cause a corpse to decompose, tended to cease or even terminate.

3. Research on the ancient Xinjiang mummies

The large amount of Xinjiang mummies are considered as precious samples that are important for different kinds of scientific research. At present, our research work on this theme is insufficient. After the Hami and the Lou-lan mummies were excavated, Xinjiang Institute of Archaeology has cooperated with Shanghai Museum of Natural History, Xinjiang Institute of Medicine, Shanghai Medical University, Shanghai Institute of Biochemistry of Academia Sinica, Shanghai Institute of Physiology, and Shanghai Institute of Labor-hygiene and Occupational Disease, which have carried out various analyses and research projects, and provided lots of precious scientific data, for instance based on anthropological and morphological observations, blood typing, studies of the lungs, chemical composition, trace elements of the hair, and the state of preservation of carbohydrates. Here, a brief introduction is given as follows:

3.1 The Lou-lan mummies include a mummy usually called "the beautiful Lou-lan woman" by Japanese researchers, excavated from Tie-ban River, and another excavated in the prehistoric Gu-mu-gou cemetery in the lower region of Kunque Daria denoted "the beautiful Lou-lan girl".

3.1.1 *"The beautiful Lou-lan girl"*: When excavated she was still having her hat on the head, shoes on her feet, and her whole corpse was wrapped in a wool blanket. The skin of her face was perfectly preserved; her soft organs and others part of her body had carbonized, she was mainly supported by her skeleton system, so that the contour of her figure was preserved. Further anatomical analysis has not yet been carried out.

3.1.2 *"The beautiful Lou-lan woman"*: Her head, neck, torso, arms and legs were all perfectly preserved, her facial features were regular, her forehead was narrow and slightly raised, eyes closed, cheekbone raised, eyebrow and eyelashes remained, and she possessed a high, narrow nose-bridge with a tilted nose-tip. Her hair, combed from the middle, drooped to both sides, and fell naturally to her shoulders. Her skin was slippery, dry, hard, reddish-brown, the dermayoglyphic pattern of her fingers was clear, ten finger nails were still well preserved, the cuticle of the skin and the deeper layers of the skin were all well preserved and distinct. After autopsy, her viscera was found to remain in a perfectly normal position, but had undergone a certain degree of shrinkage and become thinner. The length of her corpse was 152 cm. So, by inference, she should have had a body height 155.8 cm while she was alive and well propotioned. The weight of her dried dead body was 10.7 kg. According to an X-ray analysis of her skeletal system she died at an age of 40–45 years.

In the pulmonary alveolus cavity of her respiratory system, large amounts of black granulated powder were sedimented, concentrated especially at the surroundings of the blood vessels. Polariscope examination showed that crystals grains with double refraction did also exist, in addition to the black particles. Further analysis of the powder showed it to consist of predominantly silicate and coal dust. The coal dust made up 44 % of the total powdered dust and the silicate dust 38 %. This indicates that at that time eolian sand in the outside air and carbonized smoke particles from the fireplaces inside the dwellings were frequently inhaled and greatly influenced the living condition of the ancient Lou-lan people to the worse. The content of iron, aluminum, magnesium and titanium in the dust was rather high, whereas the copper content was low.

After electron microscope examination an analysis of protein, fat and carbohydrate content of the mummy was carried out. Most of the muscle texture of the organs had been atrophied, but the collagen fibers were still well preserved. Phosphides existed. Moreover, the cholesterol and the nerval amino acid content remained high. Her fatty tissues had mostly oxidized and hydrolyzed into free fatty acids and saturated fatty acids.

After the analysis and observations of the mummy's muscles, ribs and hairs, her blood type was determined as O according to the ABO system. Her hair was observed and measured morphologically. After applying the pull-test, it was shown that the pattern of small cuticles had remained perfect and distinct. The cortex and the cord marrow layers were slightly curved but perfectly retained. The mechanical rigidity of her hair had been greatly reduced compared with that of a living woman. From X-ray inspection it was demonstrated that the protein pattern of her hair remained unchanged. Protein X-fluorence (PIXE) analysis of the mummy's hair revealed the trace elements sulfur, calcium, titanium, manganese, iron, copper, zinc, lead, and strontium. Spectral analysis showed the chromium content to be particularly high.

Near the root of her hair, large amounts of well preserved head louse and louse eggs were found. Some of them had migrated to her eyebrows, eyelashes and pudenda. Their density was so high that it would not be tolerable for an ordinary human being. Besides, one bed bug (*Cimex centicularis*) was discovered, but no body ice were found.

Special attention may be focused on the following fact: By means of electron microscope observations large amounts of bacteria or bacterial gemma textures were observed on the skin, in the cartilage, on the striated muscles, on the kidney and other intestinal organs. This indicates that after the woman died, there was a process of bacterial activity, bacterial growth and interactions between bacteria and the process of self-dissolving of the muscles and organs on the dead body. Due to the external environmental factors, however, the water content in the corpse decreased. The body was rapidly dessicating, the growth of the bacteria and the corruption of the organs quickly restrained, and the corruption of the organs brought to an end. This explains why the body could be preserved.

3.1.3 *About the Hami mummy:* The three excavation campaigns at the Wu-bao cemetery in Hami have brought several well preserved or partly preserved mummies (see 1.3). Not enough research has so far been carried out on these mummies. The young female mummy excavated from the 24th cemetery in 1978, has, however, been studied in connection with a joint research project between the Xinjiang Institute of Archaeology and the Shanghai Museum of Natural History, the Shanghai Medical University, the Shanghai Institute of Biochemistry and the Shanghai Institute of Physiology. This cooperation has resulted in a rather thorough analysis.

The female mummy had a length of 156.2 cm, her subcutaneous tissues were luxuriant, and she was well grown and nourished. Beneath her knee joint, her lower limbs had been teared off and were not preserved.

Her face was narrow and rather flat, the cheekbones lightly raised, the root of her nose high, and the bridge of the nasal bones straight. The hair was brownish to yellowish in colour, straight and combed in a queue, tightly grown on her head, the eyebrows, the eyelashes, the body hair and even the vibrissae of the nose were still retained. The teeth were yellowish-white, neat and regular. Her heart, lungs, abdominal cavity and all her viscera, such as liver, spleen, kidney, intestines and stomach were all kept in normal position, but dried out, thinned and shrunk. In the cells, bacterial gemma was found. This means that the process of corruption of the corpse once started, but ceased because of the arid circumstances, and did not develop further.

X-ray examination showed her skeleton to be well preserved, and the density of her bones to be basically normal. The coronal and sagittal sutures were obliterated. Her first rib had been calcified. She died at an age of approximately 35 years. The blood type was found to be O according to the ABO system.

Histological observations demonstrated that the epidermis of the skin had been lost, as well as the reticular fibres and the elastic fibres of the corium. The collagenous fibres were however retained. According to the analysis on the lungs, its trypsin restrainer haemochrome in the heart had lost its activity, and the haemochrome substances had already been destroyed. Trace element analysis of her hair revealed aluminum, nitrate, titanium, and manganese, whereas the spectrographic analysis demonstrated an augmentation of chromium.

4. Archaeological and cultural importance

The above-mentioned dessicated mummies excavated in different regions of Xinjiang derive from different time periods, and belonged to different nationalities and different social strata. Because they have retained their original conditions while they were buried they provide plenty of historical and cultural information. This inevitably arose the interests of various scientists from archaeology, history, folk costume, religion and arts.

Here the author would like to state some of the main content which will help the readers to a better understanding of their importance:

4.1 Racial affinities

The mummies from Xinjiang have offered a reliable raw material for understanding the racial affinities of the ancient Xinjiang residents. During a long historical period, Xinjiang had been a region inhabited by several nationalities. This is according to the general impression, but if we want to draw definite and rather accurate conclusions, we must rely on archaeological research. Normally, the anthropologists mainly have to rely on large amounts of human skeletons and bone samples, carry out anthropological analyses by means of research work on the skeletal material, and seek conclusions about race peculiarities of the residents on that basis. Due to the scarce number of the mummies in general, only occasionally they will be able to carry out research based on well preserved mummies. In Xinjiang more exact conclusions may be reached.

The mummies excavated from the ancient cemetery valley of Gu-mu-gou and along the Tieban River had narrow faces, a high nose bridge, and light brown colored hair. On the other hand, the mummies which were excavated from Astan and Karahoja of the Turpan region had wide faces, high cheekbones and black hair. The first mentioned mummies clearly pertain to the Caucasian race, and the later group to the typical Mongolian race. This shows that the residents of ancient Xinjiang had their pluralistic peculiarities of racial components. Until now, we have carried out only very few studies and accumulated very little data. Therefore it is at the moment not possible to present a distinct theory about the different races which have lived in the area nor has a clear methodology been developed.

Research on nationality recognition and racial differentiation may be carried out under profitable conditions in Xinjiang. But because so few cases until now have been available for research, the anthropologists have not yet had the chance to use these materials for study, so the recognitions about the nature are certainly restrained. Until now, it seems that there is still not a proper theory or an analytical method which could be generally recognized and accepted. This is an aspect of research which has to be developed.

4.2 Research on folk-custom

Each nationality generally has its own peculiar, traditional customs. This indicates that the different nationalities possess different psychological characters. The study and the analysis of the accompanying clothing and artefacts of the excavated dead bodies lead to an understanding of their everyday life and customs. Abundance of such materials may substitute the lack of written records, and contribute to the research on nationalities and folk customs.

4.2.1 The Gu-mu-gou mummies from the Tieban River represent the Lop-nor residents who lived about 4000 years ago. The people at that time wore a felt hat with a conical top. Their hair fell loosely on their shoulders and their back. The mummies wore no clothing or trousers, but were decorated with pearls of jade around neck and wrists. The whole corpse was usually wrapped with a blanket made of wool and with fur for preventing cold. On their feet shoes of leather were worn.

4.2.2 The Jahongluk people lived in the Chercheng Oasis about 3000 years ago. They also wore a felt hat or a hat of woven wool. On their bodies, they wore a woolcloth overcoat and long trousers. The female mummies were dressed in long skirts or long trousers, stockings of wool and long boots. They were often wrapped with a brilliant wool belt of wool around the waist. In some cases, face and corpse were painted.

4.2.3 During the same epoch, the residents of the Hami Oasis wore also felt hats or hats of woven wool and the women combed their hair into a queue. They were clothed in a woolcloth overcoat, long trousers, long felt stocking, and a belt of wool wrapped around to protect their legs, they wore leather shoes, and sometimes they put small copper bells on. Their overcoat was made of a kind of wool cloth, a long gown without collar, rather wide, and a belt was tied in their waist for keeping warm. In winter, they had long, sleeved fur overcoats. Their cloth and gown were quite different from that of the Jahongluk people.

4.2.4 The Subashi mummies from the Shaan-shan district in the Turpan region, during the Warring Countries period, assumed the living condition of the ancient residents of Turpan (see 1.10). The males wore helmet-formed felt hats, the females had no hat but showed great variations in hair style, often in a dragon-shape, covered by a very fine net of wool. Both the males and females were dressed in wool cloth and trousers. The males were dressed in a fur gown, and had trousers connected with the boots. The females were dressed in wool trousers, and had a long skirt outside their trousers. They were carrying tools for preparing fur at their waist. The face of some of the males was painted with geometric figures.

4.2.5 The Jing-jue people of the Han dynasty that lived in Niya Oasis wore long trousers of cotton, silk clothing and skirts. The male mummies were dressed in silk gowns and lying with the head on pillows decorated with a crying cock, with cloths covering the face; whereas the female mummies were equipped with mirrors of the Han dynasty, with their heads supported by a rattan case to save the elegant, elaborated hair style.

4.2.6 The people who lived in the Turpan Basin during the Jin-Tang Dynasties wore cotton or hemp cloth clothing and trousers. The faces of the mummies were covered with cloth and their closed eyes were also covered. They kept coins in the mouth, and both of their hands were grasping on wood for praying the blessings of the dead. They wore paper hats, paper belts, and paper shoes, showing superstitious believes. The method of treatment to the dead, which reflects the living people's view, indicated the practical living standards and living conditions. From the above mentioned clothing and decorations, as well as the treatment of the burials, not only we can see the differences between cultural epochs and cultural ideas, but also the differences between nationalities which have existed or coexisted within the region.

4.3 The cultural peculiarities of the western territory

Scholars from various countries have different views on the cultural peculiarities of the western territory of Xinjiang. With a conscientious analysis, the most popular hypothesis that absorb attentions from all scientists concerns higly developed, advanced cultures of the neighboring regions which have had a great influence to Xinjiang (such as the ancient civilization of Yangzi jiang river valley, India, Persia, Greek and Rome). But this popular hypothesis did not pay enough attention to the development of local traditional culture, which peculiarities were ignored. Large amounts of these local peculiarities are represented by artefacts recovered in connection with the excavation of the Xinjiang mummies, fully showing the characteristics of the ancient Xinjiang culture. In summary, they make it possible to study the development of the local culture of the region. From the research done so far we feel assured to conclude that within Xinjiang, there was a capability to develop a brilliant and colorful local culture, demonstrating the intelligence of the ancient Xinjiang people.

This culture has its own merits and peculiarities. Yet they were not completely isolated, but showed various interactions and communications with the neighboring regions, absorbed influence from these regions, but at the same time also developed their own peculiar characteristics. The ancient Xinjiang people were dependent on the characteristic properties of their own local resources. All their tools for everyday life were made from various kind of bones, horn and wood, but techniques for making nets and other equipment made of grasses as well as ceramics were comparatively dragged behind. On the other hand, their techniques to produce felt, to weave wool to cloth and to treat leather rather early reached a high, technical level. All these production techniques accompanied with the development of the society, continuously progressing, resulted in an enormous success. The Hami Wu-bao and Chercheng Jahongluk cemeteries, dated to some 3000 years, have provided evidence that the techniques producing wool textile had already reached a rather high level. They had developed techniques for grading the wool, and for spinning the wool to a fine and uniform thread. They were also experienced with dyeing techniques, and applied different colors such as red, green, blue, yellow and brown, colors which have retained their brilliance until today. The colors were simply, but elegantly displayed in various different geometrical patterns such as squares, stripes or spirals, but also in animal patterns. Two thousand years ago, they were already well experienced with the technique of weaving wool by the tapestry method to produce textile. Soft leather was also tanned, and both boots and clothing have all their special peculiarities and characteristics. These achievements resulted from an economy based almost exclusively on animal husbandry, which differs from the production developed within the settled, agricultural cultures.

4.4 Medicine and sorcery

Though the ancient human beings were as intelligent as people are today, their state knowledge was of another kind and on another level, so they were very superstitious. The changing life-cycle from birth, becoming old, sick and death made the human beings perplexed. They struggled their minds with the profound mystery of life and death and believed that when people died, everybody had a soul which passed on to another world that just resembled the practical world they used to live in. Therefore, they all paid the outmost attention to the burial of the dead. They made perfect arrangements of clothing, and provided food to accompany the dead. In the Gu-mu-gou cemetery, we usually found a grass basket beside the head of the dead, containing several or several dozens of wheat grains. That means that the dead body should not suffer from hunger. About 1000 years later, i.e. during the Jin-Tang Dynasty, those caring for the Astana cemetery had put an abundance of various sorts of food, sweet cakes, fruits and meat beside the dead body. This meant that they had the same understandings as 1000 years before.

At the state of ignorance, our ancestor did not completely understand the cause of illnesses and how to properly cure them, but they tried to observe attentively, analyze, conclude and seek the remedies to conquer the illness. The Gu-mu-gou people, without any exception, carried with their bodies a bundle of ephedra branches. This is growing scattered almost everywhere in the Lop desert, and contains ephedrine and pseudoephedrine. The bio-alkalis that existed in the ephedra has the function of inducing perspiration and bring down the fever and stopping cough in connection with asthma, and it is also effective against inflammation of the respiratory tract. Ephedra was therefore regarded as possessing a magic power for easing and solving the pain and illness. The Gu-mu-gou people understood that clearly, and that is why they had a special concern about the ephedra. This, at the same time, uncovered the process of developing formal medicial treatment. Because of superstitions of the ancient people of ghosts and gods, the cemetery for burying their dead naturally had a holy position. At the Hami Wubao cemetery, we have clearly seen that their tombs had widely been destroyed, and the corpses of the dead had been humiliated and distrusted, so that the partly well preserved dessicated mummies even suffered a rough, cruel tearing. This destroying action to the dead body's tomb, obviously implied a serious curse. It is regarded as an attempt to destroy the blessings of the living, but gave also evidence about the existence and activities of sorcery.

4.5 The change and development of eco-environment

Generally speaking, dissection of the mummies provides information about the food that was eaten, but even information about the environment in which they were living and where the food was produced can be also be obtained. Because of this, the sample of dessicated mummies provides a first hand material for studying the environment during their lifetime.

Take the Lop Desert environment as an example. In the female mummy from the Tieban River, large amounts of black granular dust were sedimented in her lung. Beside the carbon dust, there was also silicon dust. This means that the woman had lived in an environment containing smoke, dust and eolean sand for a long time. We can imagine the residents there used lots of plants as fuel at that epoch. This kind of fuel can usually not be completely burned, but produces a large amount of smoke. Furthermore, the area had been already become very arid, windy, and sandy, which means that the environment was not at all healthy.

The Gu-mu-gou people which lived nearly in the same epoch as the Tieban River people were cutting enormous trees for constructing the tombs of their cemetery. At least one of the tombs, which was circular, with a radius corresponding to the length of the stem of the tree, covered with tree stems, required, according to uncompleted statistics, nearly 600 trees to be cut for its construction. What a disaster to the environmental protection in such an arid fragile environment. This information about the cemetery clearly showed the related local environment of that time, which was a relatively more humid and better environment than today. The nowaday barren land of Gu-mu-gou and the Tieban River had a rather dense forest and abundance of water and reed enabling the people to carry out both animal husbandry and simple agriculture. The Lop-nor of that time, even though at that period the region had started to get arid, provided much better living conditions than today. The main factor that caused the circumstances to change to the worse was the activities of the humans.

Through studies of the ancient mummies and the preserved artefacts of the dead by means of modern scientific methods we can obtain a large amount of irreplacable, important scientific data. For instance, we are now cooperating with scientists from American and Italy to analyze the DNA of the mummies, and carry out comparisons of the DNA of the contemporary Xinjiang nationalities with the neighboring nationalities, hoping to find out some convincing scientific basis about the relations of race peculiarities between ancient Central Asia residents and the contemporaty nationalities. In addition, study of the stomach and gut content of the residents of previous epochs, their health conditions, the causes of their getting ill, and the origins of their sicknesses provide valuable information, whereas the "blood type" analysis leads to an understanding of the marriage pattern and life within a certain residential community, and to an understanding of the influences of human communities and their connections to each other.

5. Conclusions

The Xinjiang Institute of Relics and Archaeology bears an important duty in collecting, preserving and further researching the mummies. As the excavations are carried out in an arid region, mummies will frequently be encountered. In order to do better and more perfect excavations, it is important to secure that the mummies and cultural materials connected with them are completely preserved. This is an important point that should not be neglected.

While mummies are excavated, all the related materials should be carefully collected and recorded. The mummies and their equipment should be kept for long-term preservation, and we strive our best to create the same kind of micro-environment which as closely as possible corresponds to the conditions under which the mummy has been lying buried, so that it is not being destroyed because of being unearthed and exposed to changes in temperature and humidity and possible attacks of modern micro-organisms.

The mummies, being very precious samples of ancient human corpses, need a multidisciplinary scientific collaboration for their study and analysis. The Xinjiang archaeologists wish to make a vivid cooperation with all scolars from related disciplines in order to make this research as complete and successful as possible.

Reviewing the past 30 years, the Xinjiang prehistorians have done lots of important work in excavating, preserving and protecting the mummies, and were concerned by many administrations. Anyhow, still many shortcomings remain. We cannot help to say that our efforts to strive better has not paid off, but different conditions have restrained us, so a lot of duties have not yet been done perfectly. Because of this, some of the mummies have unfortunately suffered various degrees of destruction. Some of the expected research results until now have not been realized, and this might restrain our further research.

In order to enforce the protection of the excavated mummies, we have tried to obtain the concerns and supports of China and from international collegues. Xinjiang is the most adequate, ideal place to preserve the ancient dead bodies and to construct an organization to protect and to carry out research on them, equipped with all necessary modern installations and devices, so that the ancient dead bodies can be preserved for a long period of time. Certainly, to preserve the mummies is not our terminal goal, the goal is to use the ancient mummies to carry out multi-disciplinary studies comprising fields such as archaeology, nationality, anatomy, pathology, biochemistry and environmental protection.

Various scholars from different subjects that may seem to have no close relations to each other can yet profit from all perfect devices for research here. Thus scholars from every discipline may not only obtain his analytical research results from his own professional domain, but even cooperate and help each other to deepen their meaningful complex conclusions. This kind of research projects will yield very meaningful results to the human's present, past and future.

This plan, which due to the restraints of finance and equipment has yet not gone into practice, we hope may be realized in cooperation with our international colleagues so that we may strive together.

Summary

The dessication of dead bodies in the province of Xinjiang was the result of a natural process, depending on an arid climate, salt-containing soil, tomb caves which were not completely sealed off, and burials carried out during cold winter periods. The dead bodies and their clothing and funerary articles are generally all well preserved. This is a help in researching and understanding the racial peculiarities of the ancient Xinjiang, their economic activities, clothing and equipment, as well as their views on religion their environment.

At present, from a medical point of view, the amount of research performed on these ancient bodies is not satisfactory. Until now, only the mummies of the Lou-lan and Hami women have been dissected. All their viscera were perfectly preserved, but shrunk. In their lungs, there were large amounts of dusts. Analysis by electron microscope permitted a number of conclusions to be drawn relating to the proteins, fats, and carbohydrates.

Zusammenfassung

Die Dehydratisierung der Mumien in Xinjiang erfolgte durch einen natürlichen Vorgang, der auf trockenes Klima, Salz in der Erde, unvollständig geschlossene Gräber und Bestattung während des kalten Winters zurückzuführen ist. Die Toten, ihre Kleidung und übrigen Grabbeigaben sind fast immer wohl erhalten. Dies fördert die Erforschung und das Verständnis für die ethnische Zusammensetzung und die physischen Eigenschaften der früheren Einwohner in Xinjiang, für ihre wirtschaftlichen Tätigkeiten, Bekleidung und Ausrüstung, religiöse Auffassung und ökologische Umwelt.

Zur Zeit ist die Zahl der Untersuchungen an den Mumien aus Xinjiang noch nicht befriedigend. Bis jetzt sind nur die weibli-

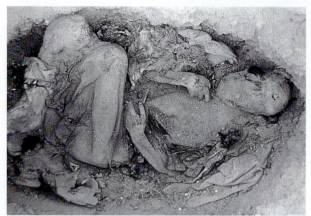
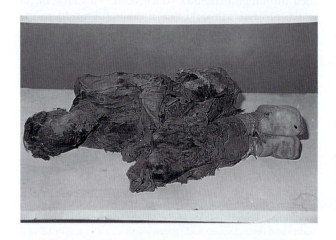

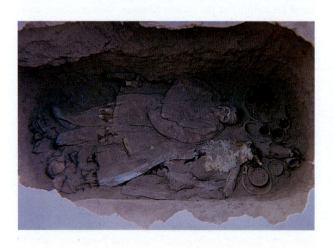

chen Mumien aus Lou-lan und Hami seziert worden. Ihre Eingeweide waren vollständig erhalten, aber geschrumpft. In ihren Lungen befanden sich Unmengen von Staub. Durch elektronenmikroskopische Analysen konnte eine Reihe von Schlußfolgerungen über Proteine, Körperfette und Carbohydrate gezogen werden.

Résumé

La déshydratation des momies de Xinjiang s'est faite par un processus naturel dû à un climat sec, à la présence de sel dans la terre, à une mauvaise étanchéité des tombeaux et à une inhumation réalisée pendant un hiver froid. Les morts, leurs vêtements, les objets funéraires sont pour la plupart bien conservés, ce qui facilite nos recherches sur la composition ethnique et les caractéristiques physiques, toutes les activités économiques, les vêtements, les outils, la religion et le contexte écologique des habitants préhistoriques de Xinjiang.

Actuellement, les résultalts des études réalisées sur les momies de Xinjiang ne sont pas encore satisfaisants. Jusqu'à présent l'autopsie a été éffectuée seulement sur des momies-femmes de Lou-lan et de Hami. Les intestins étaient parfaitement conservés mais atrophiés, leurs poumons étaient complètement envahis par la poussière. Grâce à l'analyse faite au microscope électronique, il a été possible de tirer des conclusions sur la présence de protéines, lipides et hydrates de carbone dans leur organisme.

Riassunto

La disidratazione delle mummie di Xinjiang è avvenuta tramite un processo naturale riconducibile al clima secco, alla presenza di sale nel terreno, alla permeabilità delle tombe ed all'inumazione durante un inverno rigido. I defunti, il loro abbigliamento e i doni funerari sono per lo più ben conservati, il che facilita lo studio e l'interpretazione della composizione etnica e delle caratteristiche fisiche dei primi abitanti, delle loro attività economiche, dell'abbigliamento e degli attrezzi di lavoro nonché della religione e dell'ambiente.

Attualmente il numero e gli esiti degli studi sulle mummie di Xinjiang non sono ancora soddisfacenti. Finora sono state esequite dissezioni solo su mummie femminili provenienti da Lou-lan e Hami, le cui interiora erano completamente conservate ma notevolmente atrofizzate e i cui polmoni presentavano notevoli quantità di polvere. Analisi al microscopio elettronico hanno permesso di trarre delle conclusioni sulla presenza di proteine, lipidi e carboidrati.

Collagen in 300 year-old tissue and a short introduction to the mummies in Japan

T. K. Yamada[1], T. Kudou[2], H. Takahashi-Iwanaga[3], T. Ozawa[4], R. Uchihi[5], and Y. Katsumata[5]

[1] Department of Zoology, National Science Museum Tokyo, Tokyo, Japan
[2] Department of Forensic Medicine, Tohoku University School of Medicine, Sendai, Japan
[3] Department of Anatomy, Niigata University School of Medicine, Niigata, Japan
[4] Department of Earth Science, Nagoya University, Nagoya, Japan
[5] Department of Legal Medicine, Nagoya University School of Medicine, Nagoya, Japan

Collagen in 300 year-old tissue

Introduction

Mummified human bodies or tissues have been reported from various parts of the World (Smith & Dawson, 1924; Brothwell et al., 1969; Sandison, 1969; Brothwell, 1986). Even cell structure has been examined by electron microscope (Lewin, 1967). It has now become possible to make DNA analyses with ancient specimens (Pääbo et al., 1988).

In Japan, on the other hand, preservation of human soft tissue has been regarded as very rare (Sakurai & Ogata, 1980; Brothwell, 1986; Morimoto, 1993). Several exceptional cases (most of them were those of Buddhist priests in certain regions) had been known, but no scientific examinations had been made until after the second World War (Suzuki, 1950). More then 20 mummies have been examined thereafter, but no histological study nor DNA examination has been made in this country.

This article deals with tissue pieces exceptionally preserved for more than 320 years. Examinations were made, to prove the presence of human tissue, to make identification of the tissue, to reveal the cause of preservation, and to obtain part of the genomic DNA sequence. For these purposes, immuno-electrophoresis, electron microscopy and PCR methods were carried out.

History

The Date clan was appointed by the Tokugawa Shogunate as a lord of "Sendai Han" in 1603 (Han is a feudal estate in Edo Era: 1603–1867, and the Sendai Han approximately corresponds to the present Miyagi Prefecture). The first lord of the Han was Masamune Date and the lordship had been hereditary until the "Meiji Restoration" in 1867. In the city of Sendai (Fig. 1), mausolea of the first three lords of the clan, namely those of Masamune, Tadamune and Tsunamune, were built, and were named "Zuihoden", "Kansenden" and "Zen-noden", respectively. All three mausolea were totally burned down by an air-raid in 1945. In the 1970's, the Date family, Sendai city municipal government and Miyagi prefectural government decided to rebuild them after relevant excavations of the underground structures of these mausolea. Several years after the excavation of "Zuihoden" (Suzuki, 1979), remaining two, "Kansenden" and "Zen-noden" were excavated in 1981. An anthropological report on the skeletal remains of these two lords was pub-

Fig. 1. Location of the Sendai city in Miyagi Prefecture

lished elsewhere (Suzuki and Yamada, 1985). During the investigation of Tadamune's skeletal remains, pieces of thoracic viscera were found. A considerable amount of mercury used for the burial was found around the chest region. It is suspected that the mercury had much to do with the preservation of the soft tissue.

In a stone chamber under the ground ...

Tadamune's skeleton was discovered in a heap of lime and pieces of decayed timber at the bottom of the underground stone chamber (Fig. 2). It is inferred as follows: Tadamune's body, dressed in special silk garments, had been set in a sitting posture, in a wooden tub (sarcophagus), which was traditionally used for burials in those days. The tub was then put in a lacquered palanquin prepared for his last journey.

The skeleton was preserved in a good condition presumably because of the lime. It was quite extraordinary that several pieces of soft tissue were also discovered from the right half of the chest region. Numerous droplets of mercury totaling more than 1,000 grams were found around the chest region (Fig. 3).

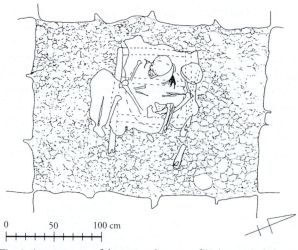

Fig. 2. Arrangement of the major elements of Tadamune's skeleton in the stone chamber. Note the spread of lime (white area) used at the burial

Fig. 3. Mercury droplets scattered around the thoracic region

Fig. 4. Major pieces of Tadamune's soft tissue preserved for more than 320 years. Impressions presumably caused by the ribs are clearly seen

Preliminary observations

Prior to the detailed analyses, the gross observation was made. The wafer-like materials were in several pieces. The tissue pieces had many small compartments that contained minute droplets of mercury. Three major pieces bore clear impressions of the ribs indicating that the materials had been in the thoracic cavity and in a close relationship with the rib cage, at least in the earlier stage of the preservation (Fig. 4). These impressions were found on the convex outer surfaces of the pieces secured from the inner cavity of the skeletal thorax. These materials were subjected to the following investigations.

Immunological analyses

Immuno-electrophoresis and the Ouchterlony method were carried out on the material to confirm the presence of human protein.

Preparation of the materials

After being homogenized, the material was filtrated by 0.5 mm mesh. A small amount of the specimen was stirred in Triton X-100 (1 % w/v), kept still overnight, and centrifuged at 3,000 rpm for 5 minutes. The supernatant was used for the analyses. Saline and diluted water extracts were prepared in a similar process.

The gel buffer was diluted barbiturate-Tris, at the pH level of 8.9 with 0.01 M glycine and 0.038 M Tris. 1 % agarose (Nakarai, Kyoto, Japan) dissolved in the gel buffer was poured onto glass plates to make gel layers of 1 mm thick.

Immuno-electrophoresis

5 µl each of the Triton extracts and human serum was applied to wells punched out on a gel plate. After electrophoresis at 10 mA for 60 minutes, a groove was made between the two wells and filled with anti-human serum. The plate was left overnight to let the anti-human serum diffuse. Finally, the plate was washed and stained with acid violet.

Compared with the arcs formed by the human serum, the single arc caused by the Triton extract was concluded to be that of human albumin.

Ouchterlony method

To confirm the presence of human albumin the Ouchterlony method was also carried out. 5 µl each of anti-human albumin serum, human serum, Triton extract and again human serum was applied into well Nos. 0, 1, 2, and 5 respectively. The gel plate was left two days and stained with acid violet.

The precipitation lines of well Nos. 1 and 2 were fused. The reaction bands of well Nos. 1 and 5 were those formed by the interaction between anti-human albumin serum and human albumin contained in human serum. This result specified that the arc obtained in immuno-electrophoresis corresponded to that of the human albumin.

Saline and diluted water extracts showed no significant results.

Electron microscopy

Scanning electron microscopy

The multi-layered tissue in question was teased out into small pieces. The pieces were fixed by OsO_4 vapour, sputter-coated with gold-palladium, and examined with a scanning electron microscope (Hitachi S-450 LB).

The tissue pieces contained a large number of fibres about 0.1 mm in diameter. No cellular elements were found. Some tissue pieces were composed of densely packed sheets of fibrous structures (Figure 5a & 5b). Fibres in one layer were parallel to each other and those in the next layer were oriented at almost right angles to the former. In other tissue pieces, the fibrous structures converged into thick twisted bundles with frequent branchings and anastomosings (Figure 6a). In the interstices of the meshwork of these bundles, the fibrous structures were extended separately from each other in random directions (Figure 6b).

The fibrous structures in the specimen appear similar to collagen fibrils in some dense connective tissue. It is well known that the aponeurosis is usually composed of tight sheets of straight collagen fibrils. A mixture of separated collagen fibrils and its bundles with twisted courses would suggest a tissue with some degree of plasticity, such as the pleura.

Transmission electron microscopy

Some pieces of the specimen were fixed with 1 % OsO_4 in phosphate buffer (pH 7.4). After fixation the tissue pieces were stained *en bloc* with uranyl acetate, dehydrated and embedded in Epon 812. Ultra-thin sections were examined with a transmission electron microscope (Hitachi H-2000) with Sato's lead staining (Sato, 1968).

The fibrous structures under the electron microscope showed, periodical cross-striations as shown in Fig. 7a. The striating pattern was characterized by a major period of 49 nm within which nine striations were resolved. Electron-dense deposits, which were irregularly ovoid in shape with various sizes, were frequently found among the fibrous structures.

It is generally accepted that collagen fibrils have their specific pattern of cross-striation with the major period of 67 nm. According to Bruns (1976), one major period of collagen fibrils shows twelve dark striations, I to XII, after staining with uranyl acetate. The present finding strongly suggests that the fibrous structures in the specimen would correspond to collagen fibrils. Although the striation period in the former is much shorter than that of the latter, their patterns of the striations found in the fibrous structures

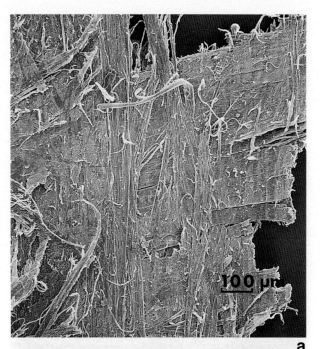

Fig. 5. (a) A scanning electron micrograph of the outermost surface of Tadamune's soft tissue. (b) The same layer at a higher magnification showing the details of the texture

are identical to the striations I, III, IV, VI, VII, VIII, IX, X and XI in collagen fibrils (Fig. 7b).

The electron-dense material between the fibrous structures probably contains mercury in a high concentration, because numerous droplets of metal mercury were found in the specimen by macroscopic observation.

DNA analysis

The HLA-DQA1, one of the HLA-classII genes, is highly polymorphic and is useful for individual identification and also for pa-

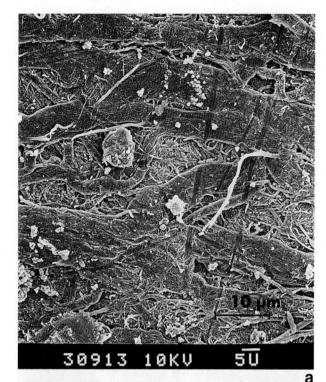

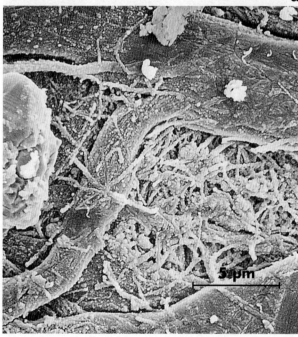

Fig. 6. (a) A deeper layer exhibiting a different scene. (b) With larger magnification, fibrils in the interstices are seen branching from the thick bundles

ternity disputes. Since hairs of his father and his son had been recovered from their mausolea, Tadamune's soft tissue together with those hairs underwent DNA analysis to ascertain their kinship. HLA-DQA1 genotypes were determined with genomic DNA extracted from the soft tissue and hairs.

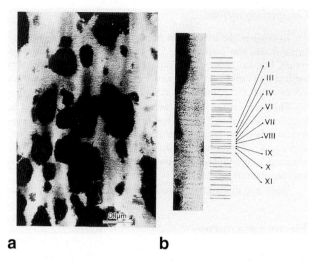

Fig. 7. (a) A transmission electron micrograph of densely arranged fibrils shown in Fig. 7 (×150,000). Electron-dense spots are surmised to contain mercury. (b) Nine striations out of the twelve described by Brun's are identifiable in the present material

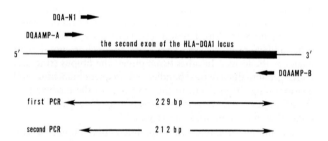

Fig. 8. Locations of the primers fro the semi-nested PCR and lengths of the PCR products

Preparations of the material

Some 2 g of Tadamune's soft tissue was used for DNA analysis. Several hairs, each 5 to 10 cm long, were used for DNA analysis with Masamune Date (Tadamune's father) and Tsunamune Date (Tadamune's son).

Extraction of genomic DNA

The specimens were digested by proteinase K in the presence of 2.0 % SDS and 40 mM DTT (Pääbo and Willson, 1988). DNA was extracted by organic solvent followed by purification using cetyl trimethyl ammonium bromide (CTAB) (Murray and Thompson, 1980).

PCR amplification of HLA-DQA1 genes

The second exon of the HLA-DQA1 gene was amplified by polymerase chain reaction (PCR) with the Eleventh International Histocompatibility Workshop primers (Kimura et al., 1992). To obtain high sensitivity, part of the PCR product was further amplified using a new semi-nested primer set inside of a 5' sided primer (Fig. 8). The precise data will be published elsewhere.

Determination of HLA-DQA1 genotypes

The HLA-DQA1 genotypes were determined as described previously (Uchihi et al., 1992) using a dot blot hybridization method with sequence specific oligonucleotide (SSO) probes recommended by the Eleventh Histocompatibility Workshop (Kimura et al., 1992).

The genotype of Tadamune was DQA1*0101/DQA1*0301, and those of his father and his son were both DQA1*0301/DQA1*0301. Since both his father and his son share the same allele, DQA1*0301, with Tadamune, there are no contradictions about their kinship (Fig. 9).

Discussion

Preservation of the soft tissue

The result of the gross observations indicated that the specimens in question were fragments of human thoracic viscera, possibly the lung. The presence of human albumin was confirmed by immunological techniques. Electron microscopy proved that the sample contained a dense connective tissue resembling the pleura. Cross-striated fibrils similar to collagen were also observed. Human DNA was extracted from the tissue pieces. All these results together tell us that in the present case human soft tissue, possibly part of the lung, had been preserved for more than 320 years.

In Japan, the environment for buried tissue is unfavourable for the preservation of organic materials, because of the highly humid and relatively hot weather and low pH level of the soil. Even skeletons are rather uncommonly discovered after being buried for more than 50 years (Suzutani, 1985). Exceptions are from shell mounds, lime stone regions and a few limited areas, where the soil pH level is sufficiently high.

As to the preservations of the soft tissues, several "mummified" bodies of Buddhist priests and those of the four governor-generals of Ohshu Fujiwara Clan (Hiraizumi in Iwate prefecture, the 11th and the 12th century) have been known, but in those cases there were possibilities of human intention to make bodies last eternally (Matsumoto, 1993; Suzuki, 1950). On the other hand, as far as the family history had documented, "mummification" of the corpse was not intended in the present case (Taira, 1974b).

Presumably the lime used in the burial had increased pH level in the sarcophagus and contributed to preserve the skeletal elements. Eventually, the local environment in the coffin was similar to those of the shell mounds. However, this does not explain why the soft tissue had also been preserved.

What was remarkable in the present case was the presence of an extremely large amount of mercury in the tissues and around the chest region. Mercury itself or a mercurial compound of some form must have functioned as an antibiotic agent.

SSO	Specificity DQA1 allele 0101	0102	0103	0201	0301	0401	0501	0601	MASAMUNE	TADAMUNE	TSUNAMUNE
DQA2501	+	+	-	-	-	+	+	-	-	+	-
DQA2502	-	-	+	+	-	-	-	+	-	-	-
DQA2503	-	-	-	-	+	-	-	-	+	+	+
DQA3401	+	+	-	-	-	-	-	-	-	+	-
DQA3402	-	+	-	-	-	+	+	-	-	-	-
DQA3403	-	-	-	-	+	-	+	-	-	-	-
DQA4102	-	-	+	-	-	-	-	-	-	-	-
DQA5501	+	+	+	-	-	-	-	-	-	+	-
DQA5502	-	-	-	+	-	-	-	-	-	-	-
DQA5503	-	-	-	-	+	-	-	-	+	+	+
DQA5504	-	-	-	-	+	+	+	-	-	-	-
DQA6901	+	+	+	-	-	-	-	-	-	+	-
DQA6902	-	-	-	+	-	-	-	-	+	+	+
DQA6903	-	-	-	-	-	+	-	-	-	-	-
DQA6904	-	-	-	-	+	-	+	-	-	-	-
DQA7502	-	-	-	+	+	-	+	-	-	-	-
DQA7504	-	-	-	-	-	+	-	-	-	-	-

Fig. 9. Specificities of SSO probes and results of dot blot hybridization

MASAMUNE* DQA1*0301/DQA1*0301
TADAMUNE* DQA1*0101/DQA1*0301
TSUNAMUNE* DQA1*0301/DQA1*0301

The family history of the Date Clan says, "his body was dressed in a formal costume and put in a wooden tub (coffin) with oyster ash (lime) and 'Kou', within twelve hours after his death" (Taira, 1974b). The actual condition of their use was not described in the document, however, supposedly similar use of "Kou" was mentioned in another series of documents on his mother's burial (Taira, 1974a). His mother, Mego-Hime (literally Lady Lovely), had been buried with this "Kou" of approximately 1,200 grams poured into the mouth. Assuming that Tadamune was buried in a similar manner as his mother had been, this exceptionally well preserved soft tissue was an indication of the effect of the "Kou". Because of the anatomical disposition of the bronchi, a substance poured into the human pharynx is most likely to fall into the right lower lobe of the lung through the larynx, if the oesophagus is occluded. This general description coincides with the fact that the soft tissue preserved was found from the right side of the thoracic cavity. No trace of the soft tissue other than the pieces mentioned above was detected from the trunk region. The amount of mercury collected, ca. 1,000 grams, approximates to that of his mother's case.

Considering these conditions, the present case could be interpreted as follows. More than 1,000 grams of mercury was poured into the mouth while the body was set sitting in the coffin. It fell through the pharynx, the larynx, the bronchi and finally into the lung. Because of the extremely large specific gravity, the mercury accumulated in the lower lobe of the lung, probably in the right lower lobe. Since the oesophagus was occluded as a result of the cadaveric stiffening, mercury did not drop into the alimentary tract lower than the pharynx. This mercury in the limited part of the lung worked as an antibiotic and several pieces of the lung were preserved. After a while, most of the soft tissues including the rest of the lung were decomposed and the mercury in the lung fell out into the thoracic cavity.

Yajima et al. (1986) pointed out the effect of mercury as an antiseptic. The mercurial ointment is effective only when mercury is ground into fine particles (Shimonaka, 1972). Generally, the toxicity of swallowed metallic mercury remains not so serious. However, once mashed into fine particles, they transform into positively charged ions, and are lethally poisonous. It is probable that what preserved Tadamune's soft tissue was the metallic mercury. The ideograph "Kou" used in the document literally means mercury, at least according to the recent dictionary, but the usage of it in those days is not certain. It is possible that this ideograph was also used for mercurial compounds in other forms.

The possibility of other mercurial compounds was considered. Toxicity, i.e., effectiveness as a preservative, of a mercurial compound depends on the solubility in water (Bozman, 1967). Yajima et al. (1986) employed X-ray diffraction analysis on the lime which was contained in the material and detected a very small amount of calomel (mercurous chloride), lead carbonate and tin oxide as major substances as well as the mercury.

It may be possible to conclude that calomel (Hg_2Cl_2) was used in the burial and the mercury had been deposited from it, but the amount confirmed was too small to be convincing. What is more, the solubility and toxicity of calomel is not so significant. Since calomel had been used as a cosmetic called "Ise Oshiroi" for the noble people in those days, it would be reasonable to conclude that the calomel detected had resulted from a cosmetic used for a final make up or daily use of it before the death.

Cinnabar (HgS) can be the source of the mercury, and also the cause of this phenomenon because it had often been used for burials of the nobles from the Jomon period to the Edo era (Ichige, 1975), e.g., in the cases of Tokugawa Shoguns' (Yamanobe & Kamiya, 1967). In these cases, droplets of mercury were sometimes found as a product of the reduction of cinnabar. In the present case, however, no trace of cinnabar itself or sulphur was found. It would be very unlikely that the entire amount of cinnabar had been reduced to mercury and sulphur, and then every particle of sulphur had somehow disappeared completely. Thus, cinnabar was excluded from the alternatives.

The corrosive sublimate ($HgCl_2$), with remarkable solubility and toxicity, was not liable to have been used, because a reasonable amount of chloride was not detected from the environment.

Our conclusion is, therefore, that what preserved the pieces of tissue of Lord Tadamune Date was metallic mercury.

DNA analyses

When we want to analyze kinship between males, we have to use genomic DNA that exists as a single copy in each cell. Mitochondrial DNA which exists as thousands of copies in each cell cannot be used because it comes only from its mother.

As the original specimens are limited, we had to use a highly sensitive semi-nested PCR for successful analysis. The present results suggest that the genomic DNA of tissue remains or hairs as old as some 300 years could be typed, and the method described here seems useful for DNA analysis of ancient materials.

The HLA class II molecules are highly polymorphic cell surface molecules that have an important function in the generation and regulation of the immune response.

The HLA-DQA1 genotype can be determined using a very small amount of tissue material DNA, if we apply the HLA-DQA1 typing method using the PCR procedure and the SSO probes. The method is also very useful to analyze archaeological samples that are often degraded, because it only requires a limited region of human genome DNA (242 base pairs). In this study we applied the method to examine a parent-child relationship of the Japanese feudal lords of the Family Date.

Using the ancient DNAs as template, we could amplified 212 base pairs of HLA-DQA1 gene by PCR (Fig. 8). Arrows indicate the sites of primers used in this study.

Fig. 9 shows SSO probes used in the present study. In HLA-DQA1 locus, Masamune and Tsunamune were determined as a homozygous type of DQA1*0301, and Tadamune a heterozygous DQA1*0101/DQA1*0301.

From our experiments, a parent-child relationship is confirmed between Masamune and Tadamune, and between Tadamune and Tsunamune.

Mummified human bodies in Japan

History

As is stated by Matsumoto (1993), the dead human body had been regarded as filthy and untouchable in Japan, and an idea of artificial mummification was most unlikely. Usually bodies were simply buried after death, although there were several secondary burials in certain areas. The climate in this country is not favourable for natural mummification. Because of the acidity of the soil, high humidity and fairly high temperature of this country made the preservation of organic matters uncommon. More than twenty mummies, however, were found and scientific examinations of these mummified bodies have been made.

In 1950, the first scientific investigation of mummies in Japan was done in Konjiki-do (a golden palace) of

Chusonji temple in Hiraizumi, Iwate prefecture. Four mummified bodies of the governor generals of the "Ohshu Fujiwara clan" were examined. Since the 1960's further investigations were made on about 20 more mummified bodies from Yamagata, Niigata and other prefectures.

Morimoto (1993) and Matsumoto (1990) classified the mummies in Japan into four categories. This classification is good enough for understanding Japanese mummies, but if we think about the chronological order and the "phylogeny" of the Japanese mummies these groups are aligned in order of a)–c)–b)–d).

a) Mummies that had much to do with the Amitabha sect of the Buddhism, which worships the Supreme Buddha residing in the West. b) Sokushinbutsu mummies of the Shingon sect priests in Yamagata and Niigata prefectures. c) Nyujo mummies of the priests of the Maitreya sect of the Buddhism. Maitreya is the Buddha of the Pure Land in the North. d) Other mummies. The mummified bodies found in Japan are listed and classified into these four categories, in Table 1. Several relevant remarks will be made in the followings:

Among mummies of the group a) are those of the four governor-generals of the Fujiwara clan in the 12th century, those are preserved in Chusonji (Hiraizumi, Iwate prefecture).

Although the mummies of group c) are few, Kochi was exceptionally important. He died in 1363. The mummified body of Kochi had become known to the people since the 17th century. His body had been kept in a small temple called Saisyoji for about 300 years without being known widely. In the late 17th century. the story of his mummified body became introduced, for example in a drama (anonymous author, 1685. The only copy of this was brought to London by Engelbert Känpber) or in a travel diary (Sora, 1689). Because of the mummy the small temple of Saisyoji became very prosperous and could gather donations. It was possible that the temples of Yudonosan area work out a program of making mummies for the financial prosperity. All mum-

Table 1. List of mummified bodies in Japan (including adipoceres)

		Age	Died in	Examined in	Temple		Locality		Remarks
Well preserved mummies									
a) Mummies that had much to do with the Amitabha sect									
Kiyohira	清衡	73	1128	1950	Chusonji	中尊寺	Hiraizumi, Iwate	平泉, 岩手	
Moyohira	基衡	54?	1157	1950	Chusonji	中尊寺	Hiraizumi, Iwate	平泉, 岩手	
Hidehira	秀衡	66	1187	1950	Chusonji	中尊寺	Hiraizumi, Iwate	平泉, 岩手	
Tadahira	忠衡	23	1189	1950	Chusonji	中尊寺	Hiraizumi, Iwate	平泉, 岩手	Yasuhira (泰衡) ?
Syungi	舜義	78	1686	1960	Myohoji	妙法寺	Iwase, Ibaragi	茨城, 岩瀬	
b) Sokushinbutsu mummies of the Shingon sect priests									
Hommyokai	本明海	61	1683	1960	Honmyoji	本明寺	Asahi, Yamagata	山形, 朝日	
Zenkai	全海	85	1687	1959	Kan'nonji	観音寺	Kanose, Niigata	新潟, 鹿瀬	
Chukai	忠海	58	1755	1960	Kaikoji	海向寺	Sakata, Yamagata	山形, 酒田	
Sinnyokai	眞如海	96	1783	1960	Dainichibo	大日坊	Asahi, Yamagata	山形, 朝日	
Enmyokai	円明海	55	1822	1960	Kaikoji	海向寺	Sakata, Yamagata	山形, 酒田	
Tetsumonkai	鉄門海	62	1829	1960	Churenji	注連寺	Asahi, Yamagata	山形, 朝日	Diaphragm and Lung preserved
Myokai	明海	44	1863		Myojuin	明寿院	Yonezawa, Yamagata	米沢, 酒田	
Tetsuryukai	鉄龍海	62	1878	1960	Nangakuji	南岳寺	Tsuruoka, Yamagata	山形, 鶴岡	Only case of extracting the viscera
c) Nyujo mummies of the priests of the Maitreya sect									
Kochi	弘智	82	1363	1959	Saishoji	西生寺	Teradomari, Niigata	寺泊, 新潟	
d) Other mummies									
Tansei	弾誓	63	1613	not examined	Amidaji	阿弥陀寺	Sakyoku, Kyoto	京都, 左京	
Myoshin	妙心	?	1817	not examined	Yokokuraji	横蔵寺	Tanigumi, Gifu	岐阜, 谷汲	
Present case of mummified tissue									
Tadamune Date	伊達忠宗	59	1658	1981			Sendai, Miyagi	宮城, 仙台	Usage of Mercury
Mummies preserved in the University of Tokyo									
Adipocere		?	1885		Cemetry	墓地	Yanaka, Tokyo	東京, 谷中	Being buried for 23 years (1885-1908)
Adipocere		male 71	1855		Cemetry	墓地	Ibaragi	茨城	Being buried for 43 years (1855-1898)
Adipocere			1752		Cemetry	墓地	Fykagawa, Tokyo	東京, 深川	Being buried for 187 years (1752-1909)
Adipocere		male 42	1837		Cemetry	墓地	Fykagawa, Tokyo	東京, 深川	Being buried for 72 years (1837-1909)
Mummified body		female 96	1841		Cemetry	墓地	Fykagawa, Tokyo	東京, 深川	
Poorly preserved mummies									
b) Sokushinbutsu mummies of the Shingon sect priests									
Junkai	淳海	78	1636		Gyokusenji	玉泉寺	Tsugawa, Niigata	新潟, 津川	
Komyokai	光明海	?	1854		Zokoin	増光院	Shirataka, Yamagata	山形, 白鷹	
Kankai	観海	?	1878		Daienji	大円寺	Niigata, Niigata	新潟, 新潟	
Bukkai	佛海	76	1903	1961	Kanzeonji	観世音寺	Murakami, Niigata	新潟, 村上	
c) Nyujo mummies of the priests of the Maitreya sect									
Shukai	秀快	62	1780	1990	Shinjuin	真珠院	Kashiwazaki, Niigata	新潟, 柏崎	Documented, 63 years after death, as being totally mummified
d) Other mummies									
Yutei	宥貞	92	1686	1974	Kanshuji	貫秀寺	Asakawa, Fukushima	福島, 浅川	Documented, 103 years after death, as being totally mummified
Gyojun	行順	45	1687	1971	Zuikoin	瑞光院	Anan, Nagano	長野, 阿南	Documened as being mummified several years after death
Chiei	智映	30's	1736		Shoryuji	紹隆寺	Suwa, Nagano	長野, 諏訪	

mies of the group b) are preserved in temples in Yudonosan (Mt. Yudono) and the surrounding area (Yamagata and Niigata prefectures). The Japanese word "Sokushinbutsu" means to make someone's body directly into Buddha (mummy) after his death according to the will of himself, mostly for being worshipped. A priest who wishes to be a mummy restricts what he eats until he dies, in the first step, abstinence from five major grains (rice, barley, soy beans, adzuki beans and sesame seeds), finally from five more cereals. Several years after the death the buried body is taken out from the grave and given a finishing touch.

Matsumoto (1993), in his article on the religious background of mummies in Japan, stated his hypothesis as follows: The mummified body of Kochi of the group c) mummies was one of the oldest and had been so well known throughout Japan in Edo era (17th C.) and his name "Kochisan" was used as a synonym of a mummified body. Kochi's story was dramatized and was introduced to Europe by Engelbert Känpber. It was worshipped widely and the temple where his body was kept could gather large amount of donations. This stimulated priests of the temples in Yudonosan area to encourage mummification, not only for the promotion of Buddhism itself but also for raising financial supports.

Those mummies of other fractions of Buddhism or of unknown religious context are grouped in d).

Acknowledgements

We are greatly indebted to Mr. Yasumune Date, the 18th heir of the clan for permitting us to study the material. We wish to express our sincere thanks to Prof. Tsuneo Fujita of Niigata University for the organization of the present work, to Prof. Sumiasaku Yajima of Waseda University for his valuable advice on metallurgic problems. Thanks are also due to Prof. Katsuji Kumaki of Niigata University and Prof. Kaoru Sagisaka of Tohoku University for their support.

Summary

Preserved soft tissue of a Japanese feudal Lord, Lord Tadamune Date (1599–1658) is described. His body had been buried in a lime-filled wooden sarcophagus buried in an underground stone chamber. It appeared that his body was completely skeletonized. A considerable amount of mercury used for the burial was also found. During the course of the osteological investigation, dried-out soft tissue pieces were found in the thoracic cavity. Gross-anatomically, the tissue was identified as pieces of human thoracic viscera. Immunological analyses detected human albumin in the tissue. Electron microscopy showed that they contained dense connective tissue composed of collagen fibrils. Genomic DNA was extracted from this tissue piece, and compared with those extracted from his father's and son's hairs.

Since the soil in Japan is quite unfavourable for the preservation of organic matters, even skeletons are rarely discovered after being buried for a long time. The present case is remarkable because part of the soft tissue was preserved for more than 320 years. The coexistence of mercury is concluded as the cause of this exceptionally long preservation of the soft tissue.

Zusammenfassung

Es wird das erhaltene Weichteilgewebe des japanischen Feudalherren Tadamune Date (1599–1658) beschrieben. Sein Leichnam war in einem kalkgefüllten Holzsarkophag in einer unterirdischen Steingruft beigesetzt worden. Von dem Körper schien nur mehr das Skelett übrig zu sein. Es fand sich auch ein beträchtlicher Gehalt an Quecksilber, das für das Begräbnis verwendet worden war. Bei der osteologischen Untersuchung fand man in der Brusthöhle vertrocknete Weichteilgewebe, die in der grobanatomischen Untersuchung als Teile menschlicher Brustorgane identifiziert wurden. Immunologische Analysen wiesen in dem Gewebe Humanalbumin nach. Die elektronenmikroskopische Untersuchung zeigte dichtes Bindegewebe aus kollagenen Fibrillen. Genomische DNA wurde aus dem Gewebestück entnommen und mit Proben aus den Haaren seines Vaters und seines Sohnes verglichen.

Da der Boden in Japan für die Konservierung von organischen Substanzen sehr ungünstig ist, werden gewisse Zeit nach der Beisetzung sogar Skelette nur mehr selten gefunden. Der vorliegende Fall ist bemerkenswert, weil sich auch ein Teil der Weichgewebe mehr als 320 Jahre lang erhalten hat. Es wird angenommen, daß der Quecksilbergehalt der Grund für die außerordentlich lange Konservierung der Weichteilgewebe ist.

Résumé

On décrit ici les tissus organiques du seigneur féodal japonais Tadamune Date (1599–1658). C'est dans un caveau souterrain que sa dépouille avait été inhumée, reposant dans un sarcophage en bois dans lequel se trouvait du calcaire. Il ne semblait rester que le squelette du corps. On a aussi trouvé une teneur considérable en mercure qui avait été utilisé pour l'enterrement. Lors des examens ostéologiques, on a trouvé dans la cage thoracique des tissus organiques desséchés qui furent identifiés comme des parties d'organes humains du thorax. Des analyses immunlogiques ont montré qu'il y avait dans le tissu de l'albumine humaine. L'examen effectué au moyen d'un microscope électronique a mis en évidence un tissu conjonctif dense fait de fibres de collagène. De l'ADN a été prélevé du tissu et a été comparé aux épreuves résultant de l'analyse des cheveux du père et du fils du défunt.

Etant donné qu'au Japon le sol est peu propice à la conservation de substances organiques, il est même rare de trouver des squelettes de personnes ayant été inhumées quelques temps auparavant. Le cas présent est exceptionnel parce que même une partie du tissu organique a été conservée pendant plus de 320 ans. On suppose que la présence de mercure est à l'origine d'une conservation aussi longue des tissus organiques.

Riassunto

Si descrive il tessuto delle parti molli del Signore Feudale Giapponese Tadamune Date (1599–1658). Il cui cadavere era stato sepoltovi in una tomba sotteranea di pietra in un sarcofago ligneo riempito di calcio. Del corpo si è conservato solo lo scheletro. Nella tomba si sono trovate anche considerevoli quantità di mercuria utilizzato nel nel rito del funderale. Gli esami osteologici hanno rivelato la presenza di tessuti essicati di parti molli nella cassa toracica – individuati come parti di organi umani del turace. Le analisi imunologiche e hanno dei tessuti videnziato la presenza di albumine umane. In base all'esame al microscopio elettronico si è potuta dimostrare la presenza di tessuto connetivo denso di fibrille al collagene. È stato prelevato dal tessuto il DNA genomico che stato comparato con quello di campioni di capelli del padre e del figlio del defunto.

Essendo il suolo del giappone molto sfavorevole alla conservazione di sostanze organiche di frequente non si ritrovano nemmeno piu le ossa di cadaveri inumati qualche tempo prima. Il presente caso è eccezionale perchè si sono conservati anche parti di tessuti molli do po un lasso di tempo di ben 320 anni. Si suppone che tale lunga conservazione sia dovuta alla utilizzo del mercurio.

References

Anonymous author (1685). Kochi hoin odenki (A bibliography of Rev. Kochi) (in Japanese, only copy is in the British Museum).

Bozman, E. E. (1967). Everyman's encyclopaedia, Vol. 8, 5th Ed. London: Dent.

Brothwell, D., Sandison, A. T. & Gray, P. H. K. (1969). Human biological observation on a Guanche Mummy with Anthracosis. American Journal of Physical Anthropology 30, 333–348.

Brothwell, D. (1986). The bog man and the archaeology of people. London: British Museum.

Bruns, R. R. (1976) Supramolecular structure of polymorphic collagen fibrils. Journal of Cell Biology 68, 521–538.

Ichige, I. (1975). Archaeology of the cinnabar. Tokyo: Yuzankaku (in Japanese).

Kimura, A., Dong, R P., Harada, H., and Sasazuki, T. (1992). DNA typing of HLA class II genes in B-lymphoblastoid cell lines homozygous for HLA. Tissue Antigens 40, 5–12.

Lewin, P. K. (1967). Palaeo-electron microscopy of mummified tissue. Nature 213, 416–417.

Matsumoto, A. (1991). Ideological background of Japanese mummies. J. Anthrop. Soc. Nippon 99, 185.

Matsumoto, A. (1993). Japanese Nyujo mummies. In (Group for Research of Japanese Mummies Ed.), Worship of mummies in Japan and China.), p. 17–97. Tokyo: Heibonsha (in Japanese).

Morimoto, I. (1991). Physical characters of Buddhist mummies in Japan. J. Anthrop. Soc. Nippon, 99: 185.

Morimoto, I. (1993). Buddhist mummies in Japan. Acta Anatomica Nipponica, 68, 381–398 (in Japanese with English summary).

Murray, M. G. & Thompson, W. F. (1980). Rapid isolation of high molecular weight plant DNA. Nucleic Acids Res., 8, 4321–4325.

Pääbo, S., Gifford, J. A. & Willson, A. C. (1988). Mitochondrial DNA sequences from a 7000 year old brain. Nucleic Acids Research, 16 (20), 9775–9786.

Sakurai, K. & Ogata, T. (1980). Japanese mummies. In (Cockburn, A. & Cockburn, E. Eds.) Mummies, disease and ancient cultures, pp. 211–223 Cambridge, Cambridge University Press.

Sandison, A. T. (1969). The study of mummified and dried human tissues. In (D. Brothwell & E. Higgs Eds.) Science in archaeology, pp. 490–502. London: Themes & Hudson.

Sato, T. (1968). A modified method for lead staining of thin sections. Journal of Electronmicroscopy 17, 158–159.

Shimonaka, K. (1972). World encyclopaedia, Volume 16. Tokyo: Heibonsha (in Japanese).

Smith, G. E. & Dawson, W. R. (1924). Egyptian mummies. London: George Allen & Unwin.

Sora (1689). Okuno hosomichi Zuiko Nikki (Diary of one of Basho's suite for the Haiku travel to the North) (in Japanese).

Suzuki, H. (1950). Anthropological observations of the bodies. In (H. Suzuki, Ed.) Chuson-Ji Temple and four generations of Ohsu Fujiwara Clan, the Governor Generals, pp. 23–44. Tokyo: Asahi Shimbun (in Japanese).

Suzuki, H. & Hayama, S. (1979). Skeletal remains. In (N. Itou, Ed.) Mausoleum and mementos of Masamune Date, Zuihoden, Sendai: Zuihoden (in Japanese).

Suzuki, H. & Yamada, T. K. (1985). Skeletal remains. In (N. Itou, Ed.) Mausolea and mementos of Tadamune Date, Kansenden and Tsunamune Date, Zen-noden, Sendai: Zuihoden (in Japanese).

Suzutani, T. (1985). Diagnostics in Forensic Medicine. Tokyo: Nankodo (in Japanese).

Taira, S. (1974a). Household Records of the Lord Tadamune Date, Vol. 8. In (S. Taira, comp.) Compiled documents of Sendai Han, Date Jikakiroku, 5, p. 46. Sendai: Houbundou (in Japanese).

Taira, S. (1974b). Household Records of the Lord Tadamune Date, Vol. 10. In (S. Taira, comp.) Compiled documents of Sendai Han, Date Jikakiroku, 5, p. 67. Sendai: Houbundou (in Japanese).

Uchihi, R., Tamaki, K., Kojima, T., Yamamoto, T., and Katsumata, Y. (1992). Deoxyribonucleic acid (DNA) typing of human leukocyte antigen (HLA)-DQA1 from single hairs in Japanese. Journal of Forensic Sciences 37(3), 853–859.

Yajima, S., Nakamura, T. & Tsutsumi, S. (1986). The grave of Tadamune Date, Second Lord of Sendai Han. Journal of the historical study group of mining and metallurgy, Japan 26, 1–12 (in Japanese).

Yamanobe, T. & Kamiya, E. (1967). Coffin contents. In (H. Suzuki et al. Eds.) Studies on the graves, coffin contents and skeletal remains of the Tokugawa Shoguns and their families at the Zojoji Temple, pp. 90–117. Tokyo: University of Tokyo (in Japanese with English summary).

Correspondence: Dr. Tadasu K. Yamada, Department of Zoology, National Science Museum Tokyo, Hyakunincho 3-23-1, Tokyo 169, Japan.

Arctic mummies

Mummies of the Arctic regions

M. R. Zimmerman

The Mount Sinai Medical Center, New York, NY, U.S.A.

Mummified bodies occasionally found in the frigid regions of the world include animals, such as mammoths in Alaska and seals in Antarctica, and Eskimos and Aleuts in Alaska, Canada and Greenland. The animal and Eskimo mummies are naturally frozen, while the Aleuts practiced artificial mummification. Tissue from such remains can be examined by dissection and rehydration in Ruffer's solution (water, alcohol and sodium carbonate) for microscopic study, providing information on the evolution and prevalence of disease in these remote areas.

The oldest preserved bodies from Alaska are mammals of the late Pleistocene (15,000–25,000 years B.P.). Specimens examined from the collection of the American Museum of Natural History in New York City were collected from the area of modern Fairbanks during gold mining operations (Zimmerman and Tedford, 1976). The remains included the face and right forefoot of an immature woolly mammoth (*Mammuthus primigenius*), the nearly complete body of a rabbit (*Lepus sp.*), a lynx (*Lynx sp.*), a lemming or vole and marrow from a horse cannon bone (*Equus sp.*). The mammoth was radiocarbon dated at 21,300±1,300 years and the other animals probably fall within the range of 15,000 to 25,000 years based on stratigraphic evidence.

The animals were dry and leather like, with skin and hair well preserved. Dissection of the mammoth head revealed preservation of the eyes as globoid structures filled with soft white cheesy material. The viscera of the rabbit were easily identifiable and appeared to be well preserved. The viscera of the lynx were totally autolyzed and the marrow of the horse bone was reduced to a small amount of greasy yellow material.

Representative specimens of the various structures were selected for rehydration, overnight immersion being sufficient. Of interest was the failure of the solution to turn dark brown, a change usually seen in the rehydration of human tissues. The lemming or vole was rehydrated *in toto* for a week, in an effort to facilitate identification.

After rehydration, the specimens were fixed in absolute alcohol and processed as would be fresh tissue. The sections were stained with hematoxylin & eosin, trichrome, PTAH and the Fontana stain for melanin. Histologic structure was found to be preserved in several of the specimens. The mammoth eye showed preservation of the extra-ocular skeletal muscles, which retained their affinity for the Masson trichrome. The PTAH stain revealed preservation of the cross striations characteristic of skeletal muscle (Fig. 1). Other structures and the melanin of the retina were not identified. The general architecture of the rabbit liver was preserved, the fibrous tissue of the portal areas being clearly visible. The hepatocytes had completely disintegrated, being replaced by masses of bacteria. The wall of the bowel remained as strands of tissue containing well preserved vegetable contents. No ova or parasites were seen.

No trace of histologic structure was seen in the other tissues examined, including the heart and spleen of the rabbit, skin and muscle of the mammoth and lynx and the horse marrow. The rehydration of the lemming (or vole) was only partially successful and did not aid in its identification.

This study demonstrated the preservative effect of freezing and subsequent mummification to last much longer than had previously been suspected. It has been suggested that most human infections originated as zoonoses and studies of ancient animal specimens might be able to identify pathogens dating back for many millennia.

On the other hand, a considerable degree of tissue destruction did indicate that a significant period of time had elapsed between the death of the animals and their entombment in the permafrost, countering a popular notion that Arctic mammals had been killed and preserved instantaneously by a catastrophic climate change. The rarity of complete mummies of the larger species also indicates that after death these mammal remains were usually dismembered and partially decomposed by the normal depositional processes of a periglacial environment.

Naturally frozen bodies of ancient humans have also been found in Alaska. The oldest one, dating to about 400 A.D., was that of a middle aged Eskimo woman found on St. Lawrence Island (Zimmerman and Smith, 1975). In October of 1972 a frozen body washed out of a low beach cliff at Kialegak Point on St. Lawrence Island in the Bering Sea. The Kialegak site is on the Southeast Cape of the island, which has been occupied for more than 2,000 years and is about 40 miles from Russia and 130 miles from mainland Alaska.

The body was found by 3 Eskimo hunters, the Gologergen brothers of the village of Savoonga. Believing that the body would be of interest to scientists, they reburied it in the tundra, below the permafrost layer, which in that area is 2 to 4 inches below the surface. In the summer of 1973, a visiting National Park Service (NPS) naturalist, Zorro Bradley, was notified of the find and taken to the burial site. With the permission of the Eskimos of the island, the body was exhumed and transported to Fairbanks, where it was stored in the freezer facilities of the federal Arctic Health Research Center.

Using the facilities of the research center, George Smith of the NPS and University of Alaska and I performed a complete autopsy. The body was thawed at room temperature in 24 hours. Tattooing noted on the arms indicated some degree of antiquity, as this practice had been discontinued on St. Lawrence Island by the 1930s. Muscle tissue was later radiocarbon dated at ca. 400 A.D., placing the body in the Old Bering Sea Phase on St. Lawrence Island (A.D. 200–500). The woman was determined to have been 53±5 years of age at the time of death, by studies of bone and teeth (Masters and Zimmerman, 1978).

Examination of the tattoos, confined to the arms, was undertaken in an effort to provide an archeological date (Smith and Zimmerman, 1975). The arms were photographed with infrared film, which has the effect of lightening the skin and darkening the tattoos. The tattooing on the right forearm, much clearer than the left, was visible on the dorsal aspect of the forearm, hand and fingers, starting 90 mm below the elbow. The tattoos consisted of rows of dots with alternating lines covering about 100 mm on the forearm.

Tattooing on the dorsal aspect of the right hand was much less clear, but the infrared photographs disclosed a "flanged heart" design attached to a horizontal line. There were also two rows of dots on the 2nd and 3rd fingers. The coloration of the tattooing was dark blue to black.

The tattooing on the left arm was more elaborate than that found on the right, covering 146 mm, but was more difficult to decipher due to desiccation. The flanged heart was seen. A design on the back of the left hand seemed to consist of a series of ovals with rows of dots on the fingers.

The process of tattooing on St. Lawrence Island is described by Otto Geist in a letter to Dr. Charles Bunnell dated 1928, a portion of which follows:

"Some of the St. Lawrence Island Eskimo women and girls have beautifully executed tattoo marks. These are made free hand although sometimes an outline is traced before the tattooing takes place. The pigment is made from the soot of seal oil lamps which is taken from the bottom of tea kettles or similar containers used to boil meat and other food over the open flame. The soot is mixed with urine, often that of an older woman, and is applied with steel needles. Two methods of tattooing are practiced. One method is to draw a string of sinew or other thread through the eye of the needle. The thread is then soaked thoroughly in the liquid pigment and drawn through the skin as the needle is inserted and pushed just under the skin for a distance of about a 32nd of an inch when the point is again pierced through the skin. A small space is left without tattooing before the process is again repeated. The other method is to prick the skin with the needle which is dipped in the pigment each time."

Illustrations of decorative motifs of Old Bering Sea Style 2 show a gorget-like ornament with a line and dot motif very similar to the tattoos, as well as designs along the same lines as the "flanged heart". The artistic motifs of the tattooing thus correlate with the radiocarbon dates in placing this individual within the Old Bering Sea phase of Alaskan prehistory.

As the body was well preserved, autopsy was performed in a standard fashion, with Y shaped and intermastoid incisions. The internal organs were somewhat desiccated but generally comparable in appearance to those of cadavers used for anatomical dissection. The body was that of a post menopausal woman, as atrophic internal genitalia were identified. Gross pathological changes were found in several organs. There was a moderate degree of coronary atherosclerosis but no evidence of myocardial infarction, acute or healed. The well preserved valves and chambers were normal. The lower lobes of both lungs showed fibrous adhesions to the chest wall and diaphragm and the lungs contained heavy deposits of anthracotic pigment. The smaller bronchi of both lungs were packed with moss (later identified as *Meesia triquetra*), forming casts of the bronchi. A calcified carinal lymph node was found. Moderate scoliosis and aortic atherosclerosis were present. The brain was a crumbling brown mass.

The tissues were somewhat desiccated, a process that continues even in the frozen state. Tissues were rehydrated with Ruffer's solution, embedded in paraffin and sectioned. Slides of the coronary arteries showed the atheromatous deposit that had been seen on gross examination (Fig. 2). The myocardium was less well preserved; striations and, as is usual in mummified tissue, nuclei were not seen. The lungs showed the patchy deposition of anthracotic pigment observed in modern patients with centrilobular emphysema. The alveolar architecture was generally preserved. Many of the alveoli appeared coalescent but this could have been postmortem change. Some moss fibers were seen in the bronchi associated with hemorrhage (Fig. 3). The liver showed distinction between the parenchymal cells and the portal triads, particularly with the trichrome stain. The cells contained a brown pigment that failed to stain for iron or bile, representing lipofuscin, an aging pigment. The thyroid contained well preserved follicles and the colloid took the specific iron stain.

The grossly calcified carinal lymph node contained numerous concentric areas of fibrosis with central calcification, interpreted as healed granulomas. Identical lesions were seen in the spleen and possibly the meninges (although the last were smaller and may represent phleboliths). Examination with polarized light revealed only minute and insignificant amounts of silica. Stains for acid fast bacilli were negative and stains for fungi showed only saprophytic *Candida sp.* Fluorescen labeled *Histoplasma capsulatum* antiglobulins that had been absorbed with cells of *Candida sp.* did not demonstrate *Histoplasma capsulatum*.

Examination of the feces revealed the ova of a fish trematode, *Cryptocotyle lingua*. The ova of this parasite have been reported in modern Eskimos but the adult helminth has not been identified in humans.

The finding of this well-preserved Eskimo body afforded us a unique opportunity to perform a complete autopsy. Our conclusion was that this middle aged woman had been trapped in her semi-subterranean house by a landslide or earthquake and had been buried alive and asphyxiated. This conclusion was based on several findings. The body was unclothed and Eskimos are unclothed only in their houses; when burial is deliberate, they are clothed. In view of the preservation of the body, one would have expected any clothing to have been preserved as well.

Aspiration of foreign material into the bronchi is known to occur in accidental inhumation and has been demonstrated in persons who fall into or are buried in coal heaps. The microscopic finding of hemorrhage associated with the moss fibers in the bronchi is consistent with asphyxiation. It is not unusual for red blood cells to be preserved for extended periods. Preserved erythrocytes have been reported in the tissues of Peruvian and North American Indian mummies and in 6,000 year old Egyptian mummies (Zimmerman, 1973). Microscopic fracture of the right temporal bone was also seen, with associated hemorrhage indicating a true antemortem fracture, confirming the role of trauma in this woman's death.

This woman, far removed from the stresses of modern technological society, suffered from coronary artery disease, a process that has been well documented as far back as dynastic Egypt by both historical and anatomical evidence. The present case not only confirms the antiquity of coronary atherosclerosis but also its occurrence in a preliterate society. The finding of severe anthracosis can be attributed to a lifetime spent around open cooking and heating fires. Similar findings have been reported in many mummies (Cockburn and Cockburn, 1980) and demonstrate that air pollution, at least on a local level, is not a recent phenomenon.

Several of the organs also showed a healed granulomatous process (Zimmerman, 1981). Tuberculosis is considered to have been nonexistent in Alaska prior to its introduction by the Russians in the early 18th century. Of the fungi pathogenic for man, only *H. capsulatum* is thought to occur in Alaska, although less than 1% of modern Eskimos show a positive cutaneous reaction to histoplasmin. *H. capsulatum* was not demonstrated in the tissues but the distribution of the granulomas was most consistent with the diagnosis of healed histoplasmosis. The *Candida sp.* found in the granulomas and elsewhere in the body is undoubtedly a postmortem invader. The weak staining of the fungi indicates contamination some considerable time in the past, perhaps shortly after death.

In summary, this middle aged woman is thought to have suffered a traumatic death some 1,600 years ago. There was gross and microscopic evidence of skull fractures and the finding of aspirated moss associated with hemorrhage suggests that accidental burial and suffocation played a significant role in her death. Other pathologic changes included coronary atherosclerosis, scoliosis, pulmonary adhesions, anthracosis and emphysema and probable healed histoplasmosis.

Several frozen bodies, radiocarbon dated to ca. 1500 A.D., were recovered from a site in Barrow, the northernmost point of Alaska. Dr. Wayne Myers, Dean of the University of Alaska Medical School informed me in 1980 that an archeological team for the State University of New York at Binghampton had uncovered the remains of five or more Eskimos in a crushed winter house on a bluff overlooking the Arctic Ocean. I invited Dr. Arthur Aufderheide of the University of Minnesota-Duluth to join me in the study of this find and my wife, Barbara Zimmerman, to assist in taking pictures and notes (Zimmerman and Aufderheide, 1984; Zimmerman, 1985).

The site was in Barrow, the northernmost community in Alaska and the bodies were thought be several hundred years old. The archeological team had been doing salvage archeology in the area and thus was available when one of the natives found a foot sticking out of the ground just west of the town, in the ancient village site of Utqiagvik. He notified the local police officer, who realized the antiquity of the find and asked the archeologists to excavate the site. They found a Pompeii-like situation, as an entire family had been trapped in their house on a bluff overlooking the Arctic Ocean. Spring storms can break up the ice and force it onto the shore with tremendous destructive force, a phenomenon referred to by the Eskimo word *ivu*. The crushing of houses in the area of Barrow and Utqiagvik is well known. As we shall see, this family was trapped while asleep, crushed and frozen in their house.

Dr. Aufderheide and I were confronted with a frozen mass of sleeping robes, bodies and bones. Thawing out of the remains, which took about 24 hours at room temperature, followed by X-rays and CT scans, allowed us to sort the remains into 5 bodies. These radiological studies also demonstrated the difficulties of interpretation in ancient bodies, with frozen pleural fluid being interpreted as aerated lung tissue (Fig. 4). Three of the bodies, a 20 year old female, a 13 year old male, and an 8 year old female, had been reduced to skeletons. The other two bodies were intact and extraordinarily well preserved. Based on where they were found in the house, the intact bodies were named the Northern Body and Southern Body, while the 3 skeletons, found on the floor, were simply labeled as skeletons 1, 2, and 3. A radiocarbon date on the femur of the Northern Body was 1,520 A.D.±70, well before white contact in the area (and in most of the New World).

The three skeletons were sorted for aging but showed no pathologic change other than a general disarticulation and mixing, so the study concentrated on the two intact bodies.

The Northern Body (NB) was an adult female, about 25 to 30 years old. She was found on the sleeping platform in the back of the house, wrapped in her sleeping robes but otherwise unclothed. When the body was thawed out, she was found to be very well preserved, with the skin still soft and pliable. The body was essentially complete, missing only the forearms and hands and weighing about 40 pounds. An eiderdown blanket was removed, revealing the right hand, which had been severed at the wrist. The entire right hemithorax was crushed, with multiple rib fractures extending in a line 3 cm lateral to the sternum.

With this degree of preservation, a relatively standard autopsy was possible, including the usual Y shaped incision. The organs were removed individually, the Virchow technique, as is generally done in examining mummies. The right pleural cavity contained approximately 250 cc of frozen clear yellow fluid and the right lung was completely collapsed, lying next to the spinal column, while the left lung was still partially inflated. The other remarkable finding was that the lungs and hilar lymph nodes were pitch black. Houses in Utqiagvik were semi-subterranean and entered through a tunnel below the floor of the house, entry to the house being through a hole in the floor. The houses were heated with small seal oil lamps, which made it warm enough that the Eskimos can be unclothed in their houses. The tunnel acts as a cold trap, since the hot air will not sink into the tunnel when the door is opened. However, this arrangement also traps smoke in the house and the effects of breathing this atmosphere are clearly seen in the black lungs and hilar lymph nodes. It was the duty of the women to trim the lamp at night and sleeping next to the lamp increased their exposure to smoke, resulting in severe anthracosis at an early age. Although anthracotic (carbon) pigment is relatively innocuous, with the introduction of cigarette smoking to Alaska during World War II a synergistic effect has developed and lung cancer is a major health concern for modern Eskimo women.

This young woman's heart showed a slight dilatation of the right side, probably related to obstructed pulmonary blood flow through the crushed and collapsed lungs. Her coronary arteries were free of disease.

The abdominal viscera were easily identified and appeared to be well preserved but microscopic examination of the intestines was somewhat disappointing as there was considerable loss of cellular detail and fungal contamination. The bones were markedly osteoporotic. The urinary bladder was markedly dilated and the stomach empty. Similar findings in the other body led to the conclusion that the catastrophe occurred early in the morning, trapping the sleeping family.

The other adult did, however, try to escape. Her body, the Southern Body, was found in the doorway, with her boots in one hand and a fish shaped dish or tray, apparently part of her sewing kit, in the other. She was found with a roof beam across her chest and there were multiple fractures of both right and left ribs. As mentioned above, the CT scan seemed to indicate that the lungs were inflated but autopsy revealed the lungs to be collapsed, with bilateral frozen pleural effusions with many bubbles in the ice. Analysis of the fluid revealed traces of hemoglobin.

The body was so well preserved that we were able to use the Rokitansky autopsy technique, removing the organs in one block. This approach allows for detailed examination of the organs and their interconnections and has not been used in any other mummy.

The SB was aged at 42–45 years, so she had more time to acquire evidence of disease. The lungs and lymph nodes were even more severely anthracotic. She had atherosclerosis, involving the aorta and coronary arteries. The mitral valves showed focal calcification (Fig. 5), for which the differential diagnosis includes rheumatic valvulitis, unlikely as there was no shortening or fusion of the chordae tendinae, and calcific mitral stenosis, again unlikely as the calcification was out on a leaflet rather than in the ring, and she was too young for this disease. The most likely choice is a healed bacterial endocarditis, even considering the odds against survival in the pre-antibiotic era. Two other points support this diagnosis. Pleural adhesions suggest a previous bout of pneumonia, as a source for bacteremia, and the kidney show changes of healed tubular necrosis, concretions that stain positively for iron and calcium, indicating that she had survived a serious illness earlier in life.

Her breast were prominent on gross examination and histologic examination showed the hyperplastic lobules of lactation (Fig. 6). Although the vessels in the uterus showed the calcifications we see in modern peri-menopausal women, it would appear the SB was still reproductively active. A corpus luteum was also found in one of the ovaries (Fig. 7), indicating delivery within 6 months prior to death. A baby was not found and may have been completely destroyed or was not in the house at the time of the catastrophe (either being elsewhere or having died previously), or perhaps it survived.

Examination of the diaphragm revealed several ovoid structures highly suggestive of trichinosis. This disease is fairly common among Eskimos, who eat polar bear meat and rarely cook it thoroughly, as fire wood is scarce in the Arctic.

Like the St. Lawrence Island Eskimo, a healed granuloma was seen in the lung, probably due to histoplasmosis. However, stains for microorganisms and antigen studies by the Centers for Disease Control in Atlanta were negative, so the diagnosis is based on the microscopic size and the absence of tuberculosis before white contact. This sample of two is obviously is too small to state that the disease was common in ancient Alaska. The fungus *Histoplasma capsulatum* today has a worldwide distribution and is seen occasionally in native Alaskans.

Both the NB and SB showed severe osteoporosis, the bone spicules being remarkably thinned and decalcified (Fig. 8). Osteoporosis is a major health problem for modern Eskimos and the gaining of a historical perspective on this disorder was one of the reasons the village elders in Barrow allowed us to do this study. Knowing now that this disorder of some antiquity, we are at least relieved of the responsibility of a modernized western diet as a causative factor. The long, dark Arctic winter has been suggested as a factor, but the resultant Vitamin D deficiency would result in osteomalacia, with broad bands of uncalcified osteoid, rather than the thin spicules seen in these osteoporotic bones. The most likely cause is the traditional high protein diet, resulting in a metabolic acidosis and consequent calcium loss from the bones.

A number of special studies, done by other investigators, can only briefly be summarized. No parasites were found in the fecal material and there was absolutely no lead in the bodies. Aging was done by examination of bones and teeth with good agreement by several different methods.

The bodies were reburied in accordance with the wishes of the elders and the artefacts and records of the studies have formed a modest museum display in the town hall in Barrow, as a memorial to this family and to the human ability to survive in hostile environments.

Another group of frozen Eskimo bodies was found in Greenland in 1972, near the abandoned settlement of Qilakitsoq. Dr. Hart Hansen's report on this remarkable 15th century find describes the value of the interdisciplinary approach to such rare finds (Hart Hansen and Gullov, 1989).

Another Greenland Eskimo mummy is in the collection of the Peabody Museum at Harvard University. This mummy, collected in 1929, is a clothed adult of undetermined sex and has not been examined, remaining in storage in the museum.

Much more recent, although not strictly speaking of the Arctic region, are the artificially prepared 18th century mummies of the Aleutian Islands. The Aleuts of that time were anatomically sophisticated, performing autopsies on humans and comparative studies on sea otters. The dead were mummified by drying the cadaver in the air or over a fire, wiping it frequently. The mummy was then dressed, wrapped in sea lion skins and placed in a burial cave.

A number of mummies were removed from Kagamil Island, in the mid-Aleutian chain, in the late 19th and early 20th centuries. 50 of these were donated to the Smithsonian Institution and one to the Peabody Museum at Harvard University. Unfortunately, almost all of the Smithsonian mummies were skeletonized by the museum in the 1920s. The mummy of a middle aged male from the Smithsonian (Zimmerman et al., 1971) and the mummy from the Peabody, a middle aged female (Zimmerman et al., 1981), have been examined.

The Smithsonian mummy was collected in 1938 by Ales Hrdlicka. This Aleut was probably from the immediate pre-Russian era, prior to 1740. There were insufficient archeological data for more precise dating, and the material was felt to be too recent for radiocarbon dating.

The cold damp climate of the Aleutian Islands would appear to be ill suited to the practice of mummification, which is generally based on desiccation. Hrdlicka attributed the development of mummification by the Aleuts to a reluctance to depart with the deceased, while Laughlin points to the anatomic interests of the Aleuts in conjunction with their desire to preserve and use the spiritual power residing in the human body. The Aleuts studied comparative anatomy, using the sea otter as the animal most like man. They conducted autopsies on their dead and had an extensive anatomic vocabulary. Mummification as an Aleut funerary practice was an extension of their pragmatically oriented society.

The technique of mummification varied with the social status of the deceased. The bodies of tribal leaders and hunters were eviscerated through an incision in the pelvis or over the stomach. No chemicals were used but fatty tissues were removed from the abdominal cavity, which was stuffed with dry grass. The body would then be put in running water, which completed the removal of fat, leaving only skin and muscle. The body was then bound with the hips, elbows and knees flexed. This posi-

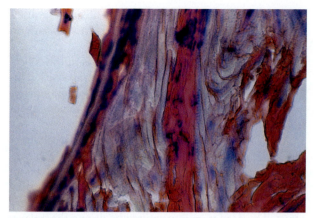

Fig. 1. Periorbital muscle from a 21,300 year old mummified mammoth shows preservation of characteristic cross striations. PTAH, ×700

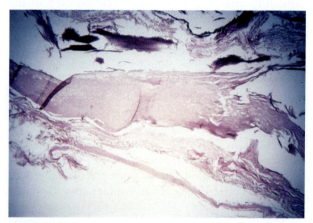

Fig. 2. Coronary atherosclerosis was seen in the St. Lawrence Island Eskimo mummy. H & E, ×95

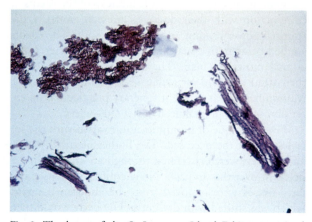

Fig. 3. The lungs of the St. Lawrence Island Eskimo contained moss fibers associated with hemorrhage, as is often seen in cases of asphyxiation. Masson's trichrome, ×95

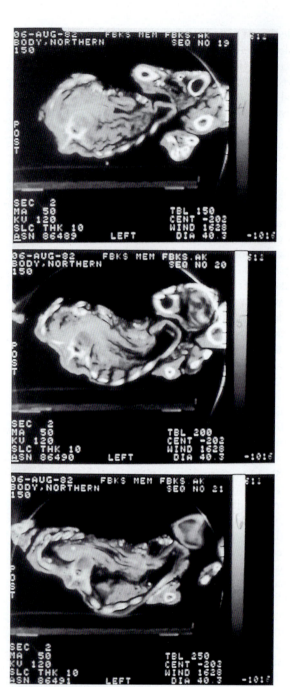

Fig. 4. A CT scan of the Northern body showed a hemopneumothorax, which was erroneously interpreted as aerated lung tissue

Fig. 5. Southern body. The mitral valve shows central calcification indicative of healed bacterial endocarditis. H & E, ×8

Fig. 6. Southern body. The breast showing marked lobular proliferation, indicating lactation. H & E, ×32

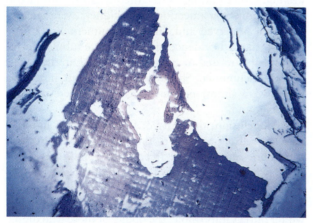

Fig. 7. Southern body. A large corpus luteum of the ovary, indicating a recent pregnancy. H & E, ×8

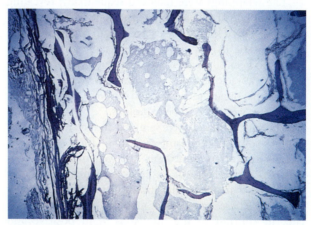

Fig. 8. Both mummies from Barrow, Alaska showed osteoporosis, as seen in the thinning of the cortex and cancellous bone. This disorder remains a problem in modern Arctic populations. H & E, ×8

tion has variously been explained as an imitation of the fetal position, an attempt to economize on space, or an effort to prevent the dead from returning and harming the living. A more likely possibility is that the flexed position is the habitual leisure position of the Aleuts. The binding of the mummy bundle is considered to be an effort to maintain the deceased in a comfortable position.

The flexed body was then air dried by carefully and repeatedly wiping off extruded moisture. When drying was complete, the cords were removed and the mummy wrapped in the deceased's best clothes, usually a coat of aquatic bird skins. The mummy was then encased in a waterproof coat of seal lion intestines and then various layers of seal, sea lion or otter skins and perhaps some matting. The entire bundle would then be tied with a braided sinew cord and removed to a burial cave, where the mummies were placed on platforms or suspended from the ceiling to avoid contact with the damp ground. The cave in which this mummy was found was heated by a volcanic vent, creating a preservative warm dry atmosphere.

These caves were probably used only for a few hundred years before the Russian contact of the early 18th century. An Aleut tale, explaining the use of the warm cave, tells of a rich headman, Little Wren, who lived near the cave on Kagamil Island. His young son was accidentally killed by his brother-in-law in a boating accident. In the subsequent funeral procession, the boy's pregnant sister slipped on a rock and suffered a fatal miscarriage. As the season was snowy and cold, the chief decided the place the bodies in the nearby cave, which had previously been used for storage. The chief declared that the cave would become a mausoleum for his entire family and when he died of grief shortly afterward, he was interred there with all his possessions.

It is thought that the body of Little Wren was removed from the warm cave in 1874 by Captain Hennig of the Alaska Commercial Company, and Dr. Hrdlicka removed some 50 mummies from the cave in 1938. Other than blood group determinations, the mummies remained uninvestigated at the Smithsonian until a group I directed examined one in 1969. It was fortunate that the mummy selected was apparently that of a common man, as it had not been eviscerated.

Radiological examination of the 112 cm long coffin shaped fur wrapped bundle revealed the outlines of the heart and lungs. The brain presented as an occipital opacity. Pathology was limited to minor arthritic changes in the vertebrae and evidence of dental attrition and periodontal disease. A number of radio opaque masses were seen in the left side of the abdomen.

The wrappings were removed sequentially, revealing an adult male of indeterminate age. The weight was approximately 10 kg and the overall length of the body was 165 cm. The hair appeared singed, suggesting that the body had been suspended over a fire for desiccation. The anterior thoracic and abdominal walls were removed and the internal organs identified and removed. The abdominal masses were seen to be coprolites.

The various tissues were sampled and rehydrated with Ruffer's solution for histologic examination. While the architecture of the heart was not well preserved, seen scattered throughout the tissue were small aggregates of crystalline material ranging from 50 to 200 microns in diameter. These contained numerous Gram negative bacilli, in contrast to the numerous gram positive and negative cocci scattered throughout the rest of the heart and were considered to be abscesses.

The pulmonary architecture was generally well preserved, with a moderate amount of anthracosis and some fibrosis, coalescence of alveoli and increase in elastic tissue. A number of abscesses, again containing Gram negative bacilli, were seen, particularly in the pleural areas. A section of the right lower lobe showed complete loss of the normal architecture, the tissue consisting of amorphous material that stained diffusely red with a bacterial stain. Many free Gram negative bacilli were present and small abscesses were scattered through the section.

The trachea was well preserved and showed sinus tracts containing Gram negative bacilli extending from the surface. The abdominal organs were generally not preserved but a few abscesses were seen in the area of the kidneys. Examination of the aorta and other arteries showed atherosclerotic plaques.

A number of special studies were done, including blood typing (type O), bacterial cultures, which were negative, and protein and enzyme analysis, showing total preservation of protein but loss of enzymatic activity. Neutron activation analysis of various organs revealed no evidence of poisoning and examination of coprolites found in the abdomen was negative for parasites and ova.

The conclusion from this autopsy was that this middle aged male had died of a right lower lobe pneumonia, probably caused by *Klebsiella pneumoniae*, with sepsis and spread to the heart, other lobes of the lungs and kidneys. Other findings included anthracosis, due to open cooking and heating fires in the home. Tobacco may be ruled out as a cause, as this was introduced later by the Russians. There was significant periodontal disease, atherosclerosis and no evidence of any neoplastic process.

The Aleut mummy from the collection of Harvard University's Peabody Museum was a middle aged woman who also showed evidence of pneumonia, in this case pleural adhesions and pulmonary fibrosis indicative of healing. Other findings were atherosclerosis, anthracosis and a chronic ear infection. The kidneys were remarkably well preserved and, like the Southern Body from Barrow, showed evidence of healed acute tubular necrosis, probably related to an episode of shock in the course of her pneumonia. She also suffered from head lice, which were almost perfectly preserved. There were no tumors found.

In conclusion, the rarity of frozen bodies is unfortunate, as freezing appears to preserve bodies in excellent condition. Although the freezer-like conditions of the Arctic would seem to provide for excellent preservation of soft tissues, bodies are in fact preserved only under extraordinary circumstances. The frozen ground makes winter burials impossible; the bodies of those who die are put out for disposal by animals. The permafrost layer is only a few centimeters below the surface, discouraging deep burials even in summer. Cycles of freezing and thawing tend to bring summer burials to the surface, exposing remains to the ravages of climate and animals.

These studies of Arctic mummies point out a major focus of paleopathology, the reconstruction of ancient disease patterns. Rare finds, such as those described above, give us a glimpse into the prehistoric Arctic and show health hazards shared by past and present inhabitants of a once remote area. Diseases such as pneumonia, anthracosis, osteoporosis, histoplasmosis and trichinosis have well known natural histories and are relatively easily explained in the context of the Arctic and Aleutian ecosystems, while the lack of evidence of cancer is consistent with paleopathologic studies in other geographic areas and suggests that the factors important in the pathogenesis of cancer are confined to the modern world (Zimmerman, 1977).

Specific recommendations for the study of the Iceman, based on these experiences, are that representative specimens be retrieved from the organs by endoscopy and submitted to a panel of pathologists for histologic examination. A coordinator should be responsible for the production of a single final report, incorporating the findings of the pathologist team. When we are presented with a unique find such as the Iceman, a careful multi-disciplinary international study will allow us to use all facets of modern science in extending our knowledge of the archeological aspects of the individual case and the larger scope of the study of human disease.

Summary

Frozen bodies of mammals of the late Pleistocene from Alaska and ancient Eskimo bodies found in Alaska and Greenland reveal remarkable histologic preservation. The Alaskan Eskimos died of trauma, with pneumonia, anthracosis and atherosclerosis common diseases. Severe osteoporosis, a major health problem for modern Eskimos, was also noted and related to dietary reliance on meat. The Greenland Eskimos showed anthracosis, Down's syndrome, osteoporosis and nasopharyngeal carcinoma, another relatively common disease among the modern Eskimos. Artificially prepared 18th century mummies from the Aleutian Islands show pneumonai, atherosclerosis and anthracosis.

Bodies in the Arctic, preserved only under extraordinary circumstance, give us insight into the health hazards of phrehistoric Arctic ecosystems. The application of techniques for the examination of frozen mummies to the Neolithic mummy found in the Alps is discussed.

Zusammenfassung

Gefrorene Säugetierleichen aus dem späten Pleistozän und frühzeitliche Eskimomumien aus Alaska und Grönland weisen einen – auch histologisch – erstaunlich guten Erhaltungszustand auf. Die in Alaska gefundenen Eskimos starben an Verletzungen, aber auch Lungenentzündung, Anthrakose und Atherosklerose waren weit verbreitet. Schwere Osteoporose, ein größeres Gesundheitsproblem bei den heutigen Eskimos, wurde ebenfalls festgestellt und der hauptsächlich auf Fleisch basierenden Ernährung zugeschrieben. Bei den grönländischen Eskimos wurden Anthrakose, Down-Syndrom, Osteoporose und Nasenrachenkarzinom, eine weitere bei den heutigen Eskimos relativ häufige Krankheit diagnostiziert. Künstlich mumifizierte Leichen aus dem 18. Jahrhundert von den Aleuteninseln weisen Lungenentzündung, Atherosklerose und Anthrakose auf.

Die unter außergewöhnlichen Bedingungen erhaltenen Körper aus der Arktis gewähren uns einen Einblick in die Gesundheitsrisiken der arktischen Ökosysteme der Vorgeschichte. Es wird die Möglichkeit diskutiert, die Methoden für die Untersuchung der gefrorenen Mumien auf den neolithischen Eismann der Alpen anzuwenden.

Résumé

Les cadavres congelés de mammifères du pléistocène touchant à sa fin trouvés an Alaska, ainsi que les momies préhistoriques d'Esquimaux, détectées en Alaska et au Groenland, sont assez bien conservés en ce qui concerne leurs structures histologiques. Les Esquimaux trouvés en Alaska succombèrent à des blessures, mais aussi la pneumonie, l'anthracose et l'artériosclérose étaient largement répandues. Des cas d'ostéoporose au stade avancé, un problème actuel majeur pour la santé des Esquimaux, ont été également détectés et attribués à l'alimentation principalement carnée des Esquimaux. Chez les Esquimaux du Groenland des cas d'anthracose, de mongolisme, d'ostéoporose et d'épithéliome du rhinopharynx, une autre maladie assez répandue parmis les Esquimaux, ont été diagnostiqués. Les cadavres momifiés des îles Aléoutiennes, datant du XVIIIe siècle, présentent également les symptômes de la pneumonie, de l'artériosclérose et de l'anthracose. Les cadavres trouvés dans la région arctique et conservés à cause des conditions extraordinaires nous permettent de conclure aux risques pour la santé liés à l'écosystème de la région arctique à l'époque préhistorique. Il est fort question d'appliquer les méthodes employées dans l'examination des momies congelées à l'homme néolitique de l'Hauslabjoch.

Riassunto

I cadaveri congelati di mammiferi del tardo Pleistocene ritrovati in Alasca e le mummie preistoriche esquimesi dell'Alasca e della Groenlandia presentano, dal punto di vista istologico, un eccellente stato di conservazione. Tra le cause di morte degli Esquimesi ritrovati in Alasca rientrano, oltre a ferite, polmoniti, antracosi e aterosclerosi nonché l'osteoporosi che rappresenta tuttora una delle più gravi malattie della popolazione esquimese e che è ascrivibile soprattutto ad un'alimentazione a base di carne.

Negli esquimesi di provenienza groenlandese sono state diagnosticati antracosi, sindrome di Down, osteoporosi e il carcinoma rinofaringeo, un altro disturbo tuttora particolarmente diffuso tra gli esquimesi. I cadaveri mummificati artificialmente, risalenti al XVIII secolo e ritrovati nelle isole Aleutine, presentano polmoniti, aterosclerosi e antracosi.

I cadaveri di provenienza artica conservatisi in condizioni climatiche estreme, testimoniano quanto fosse minacciata la salute umana negli ecosistemi artici preistorici. Si discute la possibilità di applicare i metodi di analisi delle mummie congelate all'uomo dei ghiacciai del Neolitico alpino.

References

Cockburn, A. and E. Cockburn, eds. 1980 *Mummies, Disease and Ancient Cultures.* Cambridge Univ. Press, Cambridge.

Hart Hansen, J. P. and H. C. Gullov, eds. *The Mummies from Qilakitsoq – Eskimos in the 15th Century. Meddelelser om Gronland,* Man & Society 12.

Masters, P. M. and M. R. Zimmerman 1978 Age Determination of an Alaskan Mummy: Morphologic and Biochemical Correlation. Science 201: 811–812.

Smith, G. S. and M. R. Zimmerman 1975 Tattooing Found on a 1,600 Year Old Frozen Mummified Body from St. Lawrence Island, Alaska. Amer Antiq 40: 434–437. Reprinted in: *Soot noshehenie drevnikh Kultur Siberia – Kulturami Coprodelnikh Territorii.* Akad Nauk SSSR, Novosibirsk.

Zimmerman, M. R. 1973 Blood Cells Preserved in a 2,000 Year Old Mummy. Science 180: 303–304.

Zimmerman, M. R. 1977 An Experimental Study of Mummification Pertinent to the Antiquity of Cancer. Cancer 40: 1358–1362.

Zimmerman, M. R. 1981 The Diagnosis of Granulomatous Disease in Mummies. In: *Prehistoric Tuberculosis in the Americas.* J. E. Buikstra, ed. Northwestern Univ. Archeol Progr., Evanston, IL, pp. 63–68.

Zimmerman, M. R. 1985 Paleopathology in Alaskan Mummies. Amer Sci 73: 20–25.

Zimmerman, M. R. and A. C. Aufderheide 1984 The Frozen Family of Utqiagvik: The Autopsy Findings. Arctic Anthrop 21: 53–63.

Zimmerman, M. R. and G. S. Smith 1975 A Probable Case of Accidental Inhumation of 1,600 Years Ago. Bull NY Acad Med 51: 828–837.

Zimmerman. M. R. and R. H. Tedford 1976 Histologic Structures Preserved for 21,300 Years. Science 194: 183–184.

Zimmerman, M. R., E. Trinkaus, M. LeMay, et al. 1981 The Paleopathology of an Aleutian Mummy. Arch Pathol Lab Med 105: 638–641.

Zimmerman, M. R., G. Yeatman, H. Sprinz, et al. 1971 Examination of an Aleutian Mummy. Bull NY Acad Med 47: 80–103.

Correspondence: Dr. Michael R. Zimmerman, 732 E. Weadley Road, Wayne, PA 19087-2851, U.S.A.

The palaeoimaging and forensic anthropology of frozen sailors from the Franklin Arctic expedition mass disaster (1845–1848): a detailed presentation of two radiological surveys

D. Notman[1] and O. Beattie[2]

[1] Palaeoimaging and Forensic Radiology, Director, Musculoskeletal Imaging, Minneapolis, MN, U.S.A.
[2] Department of Anthropology, University of Alberta, Edmonton, Alberta, Canada

Introduction

Natural and human caused disasters are not limited only to our time. Over the last few decades a branch of the forensic sciences has been developed which involves the collaboration of forensic researchers and specialists with archaeologists and physical anthropologists in the investigation and interpretation of historic and prehistoric disasters causing human death, and in the investigation of the deaths of historical figures (Beattie, 1993). This paper outlines the current findings of a multidisciplinary team of specialists investigating a 19th century mass disaster, the Sir John Franklin Arctic expedition of 1845–1848, with a focus on radiological analysis.

Part of our research into the causes of the Franklin expedition mass disaster involved the temporary exhumation and autopsy of the bodies of three sailors from the expedition that had been continually frozen in Arctic permafrost for nearly 140 years. The exhumations were conducted in 1984 and 1986, and both preliminary and final results of various aspects of these autopsies and related research have been previously published (Amy et al., 1986; Beattie, 1993; Beattie and Amy, 1991; Beattie and Geiger, 1993; Kowal et al., 1990; Kowal et al., 1991).

In 1984, the exhumation and autopsy of one expedition crew member was completed at Beechey island, but on-site radiological examination was not conducted. In 1986, and in addition to full autopsy, a portable radiographic examination was performed on the other two sailors from the expedition buried at Beechey Island. Preliminary radiological findings of the 1986 field season were published the following year (Notman et al., 1987). A detailed analysis of the complete radiographic surveys is now presented here, along with their pathological and historical correlations.

Historical background

Until its eventual discovery nearly 150 years ago, Europeans had spent centuries searching for a Northwest Passage across the top of the North American continent. The motives for these ship-based and land-based searches were both economic and nationalistic: the goal was to locate a short-cut trade route to the exotic goods and lucrative markets of the western Pacific, as well as to extend colonial influences to parts of the world less accessible to Europe by traditional routes. Many expeditions were sent across the north Atlantic into the waters west of Greenland in search of a passage way that was thought to exist even further to the west. The rigours of the Arctic environment, the short sailing season, and the ice-choked condition of the existing waterways during most of the year posed insurmountable challenges to these early explorers. The cost in ships and men proved too high a price to pay to continue pursuing the Northwest Passage as a viable trade route.

However, in the early part of the 19th century, the British continued their Arctic exploring tradition with a series of expeditions intent on completing a Passage for the scientific and military benefits that could be gained, as well as for national pride. None of these expeditions succeeded in traversing the Passage. By 1845, ship-based and land-based explorers had visited and mapped extensively through the Arctic islands and northern parts of continental North America, though the combined geographical discoveries still did not provide a single continuous route. In one location in the south-central region of the Arctic islands only a stretch of coastline of about 90 km required charting to complete a Passage (Figure 1).

In early 1845, a prominent 59 year old British explorer, Sir John Franklin, was given command of two Royal Navy ships with crews and supplies for three years,

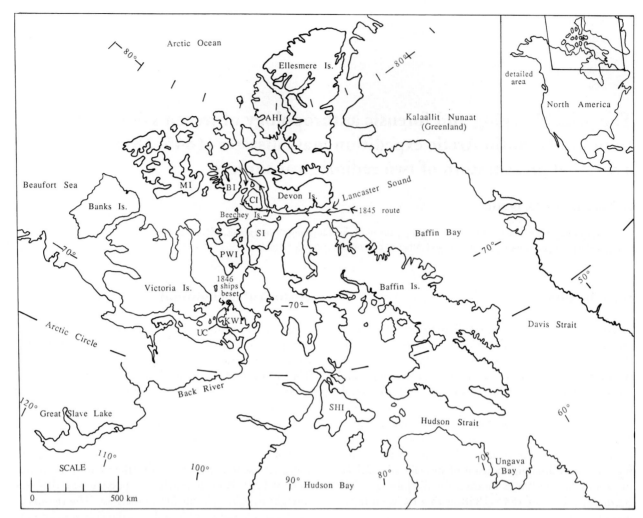

Fig. 1. Map showing the Canadian Arctic islands. Part of the route taken by Franklin's expedition is marked, including Beechey Island and the location off King William Island where the ships were beset in 1846. Abbreviations: BI: Bathurst Island; CI: Cornwallis Island; KWI: King William Island; MI: Melville Island; PWI: Prince of Wales Island; SI: Somerset Island; UC: indicates the 90 km stretch of the Northwest Passage that was still uncharted in 1845. Map adapted from Beattie, 1993

with the goal of completing the Northwest Passage (Cyriax, 1939, Neatby, 1970, Beattie and Geiger, 1993). By the summer of 1848, Franklin and his men were dead and the ships lost deep in the Arctic. Claiming 129 lives, it was the greatest disaster in the history of polar exploration.

The Franklin Arctic expedition is an example of an historic mass disaster with a total loss of all participants. Attempts to understand and explain the disaster began in 1848 and continue to this day. So much time, resources, energy, and deep personal involvement have been invested in the search for the fate of Franklin that he and his expedition have taken on a true mythological dimension totally disproportionate to the real importance and accomplishments of the expedition. The traditional explanations of the disaster range from the interpretation that vitamin C deficiency and starvation were the ultimate causes of death for the 129 crewmembers (Cyriax, 1939), to the suggestion that homesickness and a "mental attitude change" were ultimately responsible (CBC, 1994). Further details on the expedition can be found in a multitude of books and articles (e.g. M'Clintock, 1908; Cyriax, 1939; Neatby, 1970; Woodman, 1991; Beattie and Geiger, 1993).

Ironically, one of the greatest exploratory accomplishments of the Franklin expedition was the creation of a need for rescue and search missions. From 1848 to 1854, dozens of expeditions were dispatched to discover the fate of the lost ships and men. Other significant search expeditions continued in the late 1850's, and again in the late 1860's and 1870's. While much of the information we know today regarding the Franklin expedition was pieced together by these succeeding expeditions, their most significant accomplishments were the exploration and charting of vast areas of the Arctic archipelago previously unknown to Europeans.

The most significant of the discoveries made by these searchers relating to the Franklin expedition are as follows (Beattie, 1993, n.d.) (see Figure 1):

– in 1850 the Franklin expedition winter camp of 1845–6, and the graves of three crew members, were found at Beechey Island in the centre of the Arctic archipelago (Osborn, 1865)

– in 1854 the location of most of the disaster was discovered over 500 km to the south of Beechey Island, at King William Island (Rae, 1855)

– in 1859 British explorer Leopold M'Clintock (1908), in searching for evidence of the disaster in the King William Island area, discovered a note from the expedition describing the following events: in September of 1846 the expedition's two ships had become beset in the ice approximately 25 km off the northwest coast of the Island; on June 11 of 1847 Sir John Franklin died; the ships were not released from the ice in the summer of 1847; in late April of 1848, and after an additional 20 deaths, the surviving crews deserted the ice-bound ships; they proceeded to man-haul heavily laden sledges and lifeboats towards the mouth of the Back River to the southeast of King William Island (this river, and its associated tributaries and neighbouring water ways, would have led them to an Hudson Bay fort – and rescue – over 1000 km inland from their starting location; none reached the river); human skeletal remains from some of the expedition members were discovered on King William Island by the searchers

– in 1869 an American explorer (C. F. Hall; see Nourse, 1879) discovered more skeletal remains on King William Island; and in 1879 another American explorer (F. Schwatka; see Gilder, 1881; Stackpole, 1965) discovered additional sites and skeletons from the expedition.

Though a number of skeletons presumed to be from Franklin expedition crewmembers had been found on King William Island, none were examined to provide information on the health or physical state of the crews. Also, until recently, no scientifically based investigation had been conducted on the discoveries and compared to the interpretations and opinions of the searchers. Our own research on the disaster addresses this last observation.

Project overview and principle results

As the intent of our research was to examine the human remains from the expedition, two different approaches were required. First, human skeletal remains on King William Island would need to be located through archaeological foot survey; and second, the known graves of three expedition members at Beechey Island would require excavation. During 1981 and 1982, foot surveys on King William Island located the incomplete and fragmentary skeletal remains of from 7 to 15 sailors (Beattie and Savelle, 1983; Beattie, 1983). These remains were found on the surface and probably represent crewmembers who died during the process of man-hauling the sledges taken from the ships. Analysis of these remains did not allow personal identifications to be made. Evidence for vitamin C deficiency was observed (Beattie, 1983). Cut marks made by a metal implement were found on one femur, providing convincing evidence that intentional dismemberment and cannibalism occurred during the last stages of the disaster. ICP-AES (inductively coupled plasma atomic emission spectrometry) analysis of the bones revealed moderately to highly elevated bone lead levels (range 87–223 micrograms/gram, mean 138.1 micrograms/gram), indicating that some degree of lead intoxication was possible in a number of these individuals (Beattie, 1985; Kowal et al., 1989). When the various negative physiological, neurological, and psychological effects of lead were considered, it was not unreasonable at this stage of the research to hypothesize that lead may have had serious detrimental health effects on the crews, possibly contributing in some significant manner to the disaster.

This hypothesis was further tested in 1984 and 1986 during the exhumations of the three sailors buried at Beechey Island. The important feature about these particular individuals was that they were buried and preserved in permafrost, and therefore data on the occurrence and amounts of lead in soft tissues could be gathered. The bone lead data (because of the very slow turnover rate of lead in bone) could not be interpreted regarding the actual period of exposure (Ratcliffe, 1981). The bone lead could have originated some number of years prior to the deaths of the sailors (Barry, 1978; also see Aufderheide, 1991), and therefore would not necessarily have had a negative impact on the health of the sailors. Analysis of soft tissues, with turnover rates of weeks (Ratcliffe, 1981), could establish whether there were lead contaminating sources on the expedition.

In August of 1984, the body of Petty Officer John Torrington (d. January 1, 1846, aged 20 years) was temporarily exhumed and an autopsy performed (Figure 2); and in June of 1986 the bodies of able-bodied seaman John Hartnell (d. January 4, 1846, aged 25 years)(Figure 3), and Royal Marine private William Braine (d. April 3, 1846, aged 33 years)(Figure 4), were temporarily exhumed and medical autopsies (including x-rays) performed (Figures 5 and 6). The excavations were conducted with the goal of restricting the degree of disturbance to the site while guaranteeing the maximum gain of information. This approach also invested considerable effort into reburying the sailors, and reconstructing the grave sites to pre-excavation conditions.

Fig. 2. Twenty year old John Torrington; the material surrounding his head is wood shavings, and his kerchief has been tied around his head and lower jaw. Tissue dessication has resulted in the opening of the eye lids and mouth

Fig. 3. Twenty-five year old John Hartnell; the damage to his right eye occurred in September of 1852 during the partial exhumation conducted by Inglefield and Sutherland

Fig. 4. Thirty-three year old William Braine; the material on top of his head is his kerchief, which had been draped over his face when he was buried. Progressive changes due to decomposition can be observed in his orbits, nasal aperture, and mouth

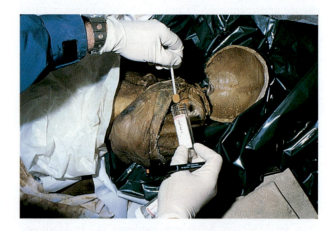

Fig. 5. Autopsy of John Hartnell, showing the collection of autolysed brain tissue for bacteriological analysis. Photograph by Brian Spenceley

Fig. 6 (top). Autopsy of William Braine, showing an example of the excellent preservation of tissues and anatomical detail (lung). Photograph by Brian Spenceley. Fig. 7 (right). John Hartnell, photographed during the initial stages of our autopsy after x-ray. Note the inverted "Y" autopsy incision; the suturing has been removed prior to the taking of this photograph. Also note the postmortem loss of facial and scalp hair

The prevailing permafrost conditions at Beechey island guarantees that any object buried more than ca. 20 cm below the surface will remain continually frozen. Therefore, though the depths of the three bodies varied from 0.9 m below the surface for Hartnell to 1.9 m below the surface for Braine, we can conclude that the permafrost conditions responsible for the preservation of these bodies were consistent and unviolated since their burials in 1846. However, the body of John Hartnell had been partially exposed and thawed in September of 1852, when a British search expedition financed by Franklin's widow made a short stop at Beechey Island (Inglefield, 1852, 1853). During a period of time that may have been as short as four hours, they conducted an hurried and incomplete superficial examination of the sailor's body. We confirmed that Inglefield's group thawed only a part of Hartnell's face and right arm, damaging his right eye in the process (Figure 3). When the doctor observing the exhumation examined the emaciated arm, and looked at the face of the sailor, he concluded that tuberculosis played a role in his death (Sutherland, 1852; Murchison, 1853). The results of our autopsies provide clear evidence for the presence of tuberculosis in all three men (Amy, 1994; Amy et al., 1986; Beattie and Amy 1991; Notman et al., 1987).

ICP-AES and atomic absorption spectrometry analyses of the hard and soft tissues from the three sailors revealed very high levels of lead (Beattie, 1985; Kowal et al., 1989). The levels seen in John Torrington were particularly high, with nape hair lead averaging 565 micrograms/gram. Nape hair lead levels in John Hartnell and William Braine averaged 326 micrograms/gram and 225 micrograms/gram, respectively. The magnitude of these levels strongly supports the conclusion that these three men, and the men from King William Island referred to previously, would have been adversely affected by exposure to lead (Kowal et al., 1991). With all individuals tested in our project affected by lead, the probability that most or all of Franklin's men were similarly burdened is high. Therefore, the hypothesis that lead played some significant role in the disaster is strengthened.

After establishing that the men were being exposed to lead on the expedition, questions arise regarding the source or sources of the lead. Previous reviews of foods and materials supplied to the expedition indicated that the approximately 8,000 tins of preserved food were the most likely source (Beattie, 1985). Mass spectrometry analysis was used to test if there was an association between the tinned foods and the lead in the sailors. Lead isotope data was gathered for the solder lead from Franklin expedition food tins collected at Beechey Island, and for the lead found in the hard and soft tissues of the human materials from King William and Beechey Islands. The results of the analysis indicate that the solder lead and body lead originate from a common single European geological source location (Kowal et al., 1990, 1991). This is very strong evidence that the lead in the bodies of the sailors came from tinned foods contaminated by solder. The high degree of similarity in the lead isotope data for all of the human remains provides indirect evidence that the men were not exposed to any significant amounts of lead before leaving on the expedition. The men were from all parts of the United Kingdom and exposed environmentally to a number of different societal and geological sources of lead. If these exposures were anything but trivial, then the lead isotope data would have been more diffuse for each sailor, indicating exposures to more than a single dominating source. Analyses of ice and artifacts from within the graves and coffins, as well as the materials making up the tools used during our excavations, confirms that there was no detectable contamination of tissue or bone samples from either the burial environment or the excavation process.

Of course, exploratory expeditions preceding and succeeding the Franklin expedition also relied heavily on tinned foods, as did the military. In fact, the Franklin expedition may provide a model for understanding the extent of the effects felt by the use of this new food technology (e.g. the 1848 Franklin search expedition of James Clark Ross; see Beattie and Geiger, 1993: 167).

It is important to add that alloyed materials containing lead were commonly used in food preparation utensils and containers, as well as in the tin-plate manufacturing for food tins (for example, see Carlson, 1993). Two further historical examples of the many available from the contemporary literature are as follows (Magruder, 1883):

1. the Arctic expedition of the *Jeannette*, in an attempt to sail and steam to the North Pole, was affected in 1881 by the effects of lead poisoning from the tinned tomatoes supplied to that expedition, and

2. a military regiment of 150 men in the South Tyrol in 1883 suffered various degrees of lead poisoning (including the death of one person) from the use of a tin-lined copper vessel in which they cooked their food (the tin coating was an alloy of tin and lead).

During the second half of the 19th century, as the technology became more affordable, tinned foods increased in popularity with the general populace in industrialized countries. Not surprisingly, cases of lead intoxication were reported in the medical literature with increasing frequency, and doctors were beginning to link these cases in their patients to the consumption of food from internally soldered tins (Beattie, 1993). By the end of the century, manufacturing methods (specifically, the internal soldering of food tins) in Great Britain and the United States were changed to reduce the potential for food contamination. This occurred slightly later in Canada. Today, soldered tins have been almost completely phased out of use in North America, and replaced by tins with welded seams (U.S. Department of Health and Social Services, 1988).

Radiological analysis

Materials and methods

A Port-a-Ray 9100A (Deer Park, New York) portable x-ray unit was erected inside a large, floorless canvas laboratory tent on the gravel beach of Beechey Island directly adjacent to the graves of the three Franklin expedition crew members. A darkroom tent, constructed from a metal frame covered by heavy gauge opaque black plastic sheets, was assembled next to the x-ray examination site and within the canvas tent. The darkroom permitted basic but essential film developing, allowing us to make necessary adjustments in imaging technique. Power for the x-ray and darkroom equipment was provided by a Honda 4.5 kilowatt generator. This simple facility worked well, and radiographic quality was satisfactory in spite of the extreme conditions (air temperatures ranged from ca. –15 °C to 0 °C).

Results

Detailed summaries of the imaging parameters for the studies of John Hartnell and William Braine are listed in table form in the Appendix. Discussion of these data is as follows:

John Hartnell
– skull

The first AP radiograph demonstrates a completely opaque skull. Imaging parameters had to be adjusted several times before optimum technique was achieved. Ice artifact increased the complexity of the problem slightly until it was recognized as a contributing factor.

Some collapse of the brain away from the inner table of the vault is present, although the convolutions of gyri and sulci are still identifiable. Dural venous sinuses are accentuated in the pa-

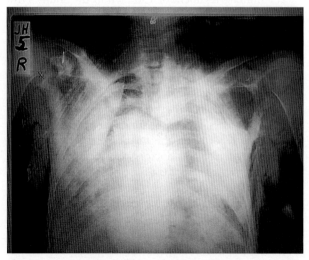

Fig. 8. AP chest x-ray of John Hartnell showing the disturbed anatomy due to the 1846 autopsy

rieto-occipital regions, as a post mortem phenomenon caused by autolysis. No intracranial calcifications are detected. Bony anatomy is normal. Paranasal sinuses and mastoid air cells are adequately pneumatised. There are no signs of trauma. Dentition is unremarkable except for impaction of both mandibular first molars.

– cervical spine

The upper cervical spine is unremarkable. C7 is completely visualized, due to the overlying shoulders. Distraction of the arms fails to demonstrate the cervicothoracic junction adequately in the lateral projection, but this level appears normal on oblique and AP projections. However, there does appear to be some compression of the superior end-plate of the body of C6, consistent with mild fracture, probably representing subacute injury. Cervical alignment is normal. There are no arthritic changes.

– upper extremities

The shoulder articulations appear intact. A vacuum phenomenon beneath the left acromial process is noted; in the living patient this intra-articular air (containing nitrogen) is a common transient finding of no clinical significance. After 140 years, its gaseous composition is uncertain. On some projections, the left acromion appears slightly tapered, possibly due to old trauma. A small spur off the posterior aspect of the left olecranon is also likely to be post-traumatic.

A benign 9 mm sclerotic density is incidentally noted in the distal shaft of the right humerus and is consistent with a small bone island. There are no fractures. In 1852, Hartnell's body had been briefly and superficially exposed by Inglefield and Sutherland, and during our examination there was some damage noted to the right arm. The shirtsleeves of the right forearm were torn, and a small portion of the skin and musculature overlying the posterior surface of the radius had been pierced and damaged, with exposure of the bone. This damage was likely due to an unexpected overpenetration of the coffin with a pickax during the 1852 excavation.

The wrists and hands are unremarkable, except for diffuse muscle wasting. There is mild bilateral negative ulnar variance (the distal ulna lies more proximal than the distal radius at the distal radioulnar articulation; if severe, this malalignment may predispose to avascular necrosis of the lunate). No evidence for fractures or arthritis was found.

– chest

One of the most interesting discoveries was the fact that John Hartnell had been autopsied in 1846 (Figure 7). Initial chest x-rays were taken while the body was still clothed, presenting a confusing pattern of ice density and disrupted anatomy (Figure 8). Only when the three shirts were removed did the extensive autopsy incision reveal itself and explain the bizarre appearance on the films. The decision to perform an autopsy must have been influenced in part by circumstances surrounding Hartnell's death, but there are no written records which contain this information.

A metallic ring overlying the right upper thorax turned out to be a button, delicately woven and stitched. Small initials reading "T.H. 1844" were found stitched in red thread on the outer shirt tail, suggesting that the shirt belonged to his brother, Thomas, who was also a member of the expedition.

The ribs are intact. Hartnell, like Braine, had evidence of pulmonary tuberculosis, but there is no obvious bony involvement.

– abdomen/pelvis

Except for the irregular density and unusual cleavage planes created by freezing of the internal organs, there are no abnormalities noted in the abdomen and pelvis. Although poorly nourished, Hartnell's condition in general appears better than that of Braine.

– thoracolumbar spine

There are no signs of trauma in the thoracic and lumbar spine. Vertebral bodies and disk spaces appear intact.

– lower extremities

The femurs appear normal. Hip and knee articulations are preserved. However, there is a slight widening of the medial aspect of the left ankle joint, raising the possibility of ligamentous injury (Figure 9). No fractures are noted. The talar dome and tibial plafond show no evidence for osteochondritis dissecans, an occasional sequel to subchondral injury.

A few growth recovery lines (Harris lines) are noted in the lower tibial shafts. There is also some spurring off the anterior articular lip of the distal right tibia, and probably post-traumatic in origin.

A more curious finding is a possible focus of cortical demineralization in the plantar aspect of the 4th metatarsal shaft of the left foot, best demonstrated on oblique projections. Actual disruption of the cortex may be present and could represent osteomyelitis. However, there were no overlying superficial ulcers or open wounds.

Aside from minor degenerative changes at the articulation between the first cuneiform and first metatarsal base, the remainder of the feet is unremarkable. There are no stress fractures.

William Braine (Figures 10 and 11)

– skull

There are no fractures or evidence of disease. The brain and dura are partially collapsed away from the inner table of the vault. Normal venous sinuses and vascular channels or grooves are present. No intracranial calcifications are identified. Calcification of the pineal gland and habenular commissure are often encountered and rarely carry any clinical significance, but calcification of the basal ganglia would have been consistent with chronic lead intoxication. The sella appears normal. Incidentally noted is minimal bathrocephaly, or developmental protrusion of bone from the external occiput. Para-nasal sinuses are adequately aerated. There is some frontal bossing. Maxillary and mandibular dentition shows slight uniform wear but no abscess cavities. Temporomandibular joints are intact.

– cervical spine

Normal.

– upper extremities

No fractures. There is narrowing of the acromiohumeral spaces bilaterally, which may be seen with degeneration or tear of the rotator cuff. However, this is probably due to postmortem changes, because both coracoclavicular spaces are also narrowed. The glenohumeral articulations are intact. The humeri appear normal. There is minimal spurring of the posterior aspect of the right olecranon process consistent with old trauma.

There are no acute changes in the hands. The left 4th metacarpal is short (with positive metacarpal sign), which is almost certainly a simple anomaly and is unlikely to be of any clinical significance. There is no deformity of the 4th metacarpal itself to suggest foreshortening from old fracture – the right 4th metacarpal is normal – though a tiny ossicle at the radial aspect of the left 5th metacarpophalangeal joint probably does represent old trauma. Some spurring is noted off the radial aspects of the right 3rd and 4th distal phalangeal tufts and could be related to occupational stresses.

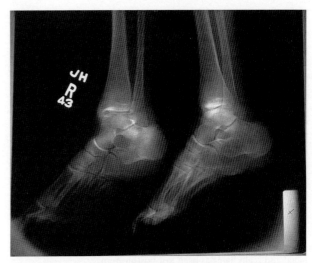

Fig. 9. Bilateral ankles (oblique) x-ray of John Hartnell, showing a slight widening of the medial compartment of the left ankle joint

– chest/abdomen/pelvis

There are no definite soft tissue abnormalities, although ice artifact creates multiple cleavage planes and ridges within the thoracic and abdominal cavities. It should be noted that William Braine was not autopsied in 1846. A marked paucity of adipose tissue confirms the state of emaciation. Thoracic and lumbar spinal findings are discussed below.

– thoracolumbar spine

The upper thoracic spine is unremarkable. There is an angular kyphoscoliosis centered at the T11 vertebral body, which is considerably wedged anteriorly (Figures 12, 13, and 14). The posterior height of the body is preserved. Moderate narrowing of the T10–T11 and T11–T12 inter-spaces is also present, with irregularity and sclerosis of the articular end-plates. These abnormalities are consistent with tuberculous spondylitis (Pott's disease), which is usually caused by hematogenous spread from a primary infection in the lung or other site.

– lower extremities

The femurs appear normal. Small bilateral symmetrical exostoses project from the medial aspects of the proximal tibial metaphyses. These are consistent with benign osteochondromas and are not found elsewhere in the axial and appendicular skeleton. Some irregular cortical thickening involves both anterior tibial tuberosities in a pattern suggesting old Osgood-Schlatter's disease, a common self-limiting osteochondrosis in children and adolescents.

The left knee joint was aspirated under "sterile" conditions, using a 19-gauge needle and a 25-cc plastic syringe. Approximately 25 ml of cloudy, straw-colored, foul-smelling fluid was withdrawn for routine cytology and bacteriology. No bacterial strains could be cultured from this sample. A simple arthrogram was then performed by injecting an equal quantity of water-soluble contrast material (Urovist) into the knee joint, via the same puncture (Figure 15). AP and lateral radiographs demonstrate a normal appearing joint, with opacification of both suprapatellar and popliteal bursas (Figure 16). After 140 years, the synovial lining remained intact.

The ankle joints are normal. In the feet, bilateral hallux rigidis deformities are present, with mild degenerative changes involving the first metatarsophalangeal joints. The great toes are disproportionately long. There are no fractures or stress injuries. No other significant arthritic changes are noted in the feet.

Fig. 10. William Braine, visible to pelvis. The x-raying and autopsies were conducted on plastic sheets directly on the ground. All facial and scalp hair has been lost postmortem

Fig. 11. William Braine, positioned on his right side (lateral view), illustrating his emaciated appearance and the greenish colouration over abdomen characteristic of early stages of decomposition

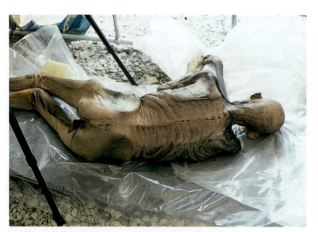

Fig. 12. William Braine, positioned on his right side (view from back), clearly illustrating the angular kyphoscoliosis

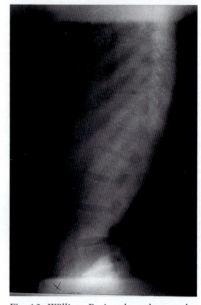

Fig. 13. William Braine, lateral x-ray thoracolumbar spine, showing angular kyphoscoliosis

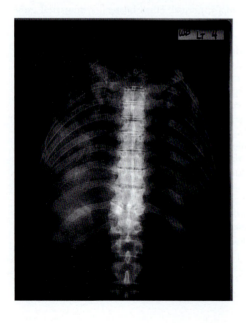

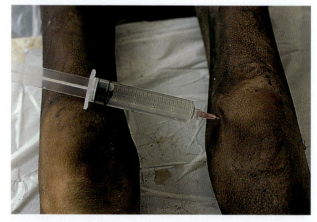

Fig. 14 (left). William Braine, AP x-ray thoracolumbar spine, showing angular kyphoscoliosis. Fig. 15 (top). William Braine, arthrogram, with the needle entering the medial compartment of the left knee

Discussion

John Hartnell died only three days after John Torrington, on 4 January, 1846. Events surrounding his death evidently aroused sufficient curiosity among the medical staff to warrant an autopsy. Neither Torrington nor William Braine was autopsied. There are no written records to explain these circumstances. However, we will attempt to integrate the pertinent radiographic findings with the superficial inspections, post mortem examinations, and laboratory data. From these it is possible to construct a plausible scenario which would fit with chronological events.

The suspected mild compression fracture of the C6 vertebral body probably represents subacute injury and would not have been fatal. Hartnell may have fallen while on board, straining his left ankle joint at the same time. Both injuries would have confined him to sick bay, where he would have been given tinned foods as a medical comfort. Superimposed lead intoxication would have contributed to further debilitation. A final acute illness, probably a pneumonia (reactivation of pulmonary TB?), would have proved fatal. It is possible that there was seeding of a blood-borne pathogen to the left fourth metatarsal, resulting in a tiny focus of osteomyelitis not associated with overlying soft-tissue ulceration.

Even if this postulated clinical history had occurred as described, it would probably not have been enough to persuade the surgeon on the ship (Dr. S. Stanley) to conduct an autopsy without some additional compelling factor(s). Plumbism may have generated a behavioral change (lead encephalopathy) that influenced his decision. In his final hours, his brother Thomas would certainly have kept vigil at his side.

It must be remembered that these were tough men, accustomed to prolonged hardship and hand-picked to join the expedition. Over the span of a normal two to three year voyage, the spread of pulmonary TB below decks would have eventually affected most of the crew. If those sailors who suffered acute exacerbations of the chronic disease were treated in sick bay, then the insidious onset of lead poisoning might be masked until later in the voyage. Failing appetite and strength, impaired judgment, and harsh conditions would have weakened resolve as well as body.

William Braine, 33 year old Royal Marine private, died 3 April, 1846. Judging from his emaciated appearance, he must have suffered considerable hardship in his final days. The presence of ulcerations over the anterior aspects of both shoulders strongly suggests abrasion from an activity like sledge-hauling, though these lesions may be the result of blistering related to the progress of putrefaction. It is possible that Braine could have participated in a sledging journey in search of fresh game, open leads of water, and scientific observations. He may have suc-

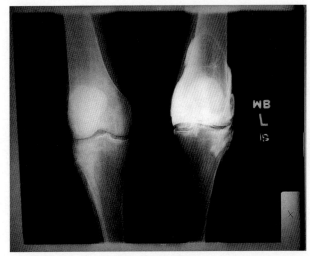

Fig. 16. William Braine, AP knee arthrogram, with contrast material in left knee joint

cumbed to illness on the trek and barely managed to return to the ships, or died before reaching them.

Evidently, his body was not buried immediately after death, as is suggested by the characteristic greenish discoloration and decomposition of superficial soft tissues. Under conditions expected for their situation, it is estimated that three to five days may have elapsed from the time of his death to the time of his burial. Also visible on the body were groupings of rat bites. These could only have occurred on board ship. It is quite possible that a late winter storm was responsible for the delayed burial.

The x-ray study of William Braine reveals an angular kyphosis of the lower thoracic spine which is quite consistent with tuberculous spondylitis (Pott's disease). These changes do not suggest acute bony infection. There is no evidence for trauma. It is likely that Braine, too, died from an acute viral or bacterial illness, perhaps a fatal tuberculous crisis. One can only speculate why there was delay between death and burial, and why there was no autopsy.

Although a plain radiographic survey of John Hartnell and William Braine did not reveal any signs of chronic lead intoxication or dietary deficiency, it must be emphasized that *the absence of such findings does not exclude the presence of disease.* This well-accepted and absolutely fundamental concept in clinical medicine appears to have been misinterpreted in a recent publication (Farrer, 1993), in which the author incorrectly concludes that the lack of x-ray evidence proves that no significant lead poisoning could have occurred in Franklin's men.

Other research results

The permafrost conditions at the Beechey Island site also resulted in excellent preservation of objects associated with the bodies of the three sailors. The mahogany coffins provide information on 19th century carpentry methods, even allowing comparisons of the materials and workmanship of the carpenter and carpenter's mate on the *Terror* (Torrington's ship) with their counterparts on the *Erebus* (Hartnell's and Braine's ship). The textile and clothing inventory for the three sailors is as follows:

Torrington: kerchief, shirt, pants, binding material

Hartnell: shroud, three shirts, cap, pillow, blanket, binding material

Braine: shroud, kerchief, three shirts, one pair of stockings

The sailors' clothing is important historically because few examples of expedition wear from men of these ranks exist in museum collections. The research on the clothing is progressing, and some results have been published (Kerr and Schweger, 1989; Schweger and Kerr, 1987).

During the autopsies on Hartnell and Braine, samples of tissues were collected for microbiological studies. This procedure involved the dissection of tissues using flamed scalpels and forceps, with the samples then being placed in sterile containers and kept frozen until appropriate processing in labs at the University of Alberta. Six strains of the genus *Clostridium* have been cultured from samples of gut tissues from Hartnell and Braine. Tissue samples from other locations throughout their bodies did not produce these bacteria, indicating that modern contamination from our research team or the burial environment was unlikely. The potential significance of these findings relates to the discovery that three strains appear resistant to modern antibiotics. The results of this study have been reported at conference (Kowalewska-Grochowska et al., 1988) and a paper is presently in preparation for publication. Future field research is planned to investigate these observations further.

Conclusions and comments

There can be little doubt that the toxic metal lead was a major factor affecting the health and performance of many, if not all of the men on the Franklin expedition. The magnitude of these effects cannot be interpreted with certainty, and likely varied considerably from one individual to the next. Our data incontrovertibly indicates that the men were chronically, and occasionally acutely exposed to one or more major sources of lead during the period of the expedition. Lead isotope analysis points to the tinned foods as the primary, if not totally dominant source. Our rebuttal to a questioning of our data and interpretations is presently in progress, and will not be discussed here (see Farrer, 1993). Indirect evidence provided by the lead isotope analysis also indicates that the sailors tested in our research were not exposed to significant amounts of lead prior to the expedition.

One final comment must be made regarding the role of lead intoxication and the Franklin expedition: the disaster was likely not the result of a single cause, but occurred when the compounding effects of scurvy, lead intoxication, entrapment in the ice, eventual food deficit, and other unknown stresses overwhelmed the sailors. Considering our findings relating to the isotopic linking of food tin lead with body lead, questions can be raised regarding the effects that this form of food preserving technology may have had on the health of millions of people in the latter half of the 19th century.

The x-ray results represent a study which was as detailed as possible, given the considerable technical limitations of the portable equipment and the challenging field conditions. In general, we were pleased with the basic performance of the x-ray unit, generator, and dark room facility. The x-ray analysis does not add significantly to the conclusions regarding the loss of the expedition, but it does provide important insight into the lives of two individual sailors. Healed and unhealed trauma and pathological lesions indicate the physical risks and quality of life experienced by sailors involved in exploration. It is clear, however, that computed tomographic (CT) capability would have been very helpful in dealing with superimposed densities.

As for future Arctic palaeoimaging research, digital radiography, digital fluoroscopy, and CT would considerably enhance field work, especially with remote image acquisition and satellite transmission to advanced diagnostic centers. In particular, CT will be helpful in detecting the subtle intracranial calcifications which can occur with chronic lead intoxication. Dual energy CT could be used for bone mineral analysis and lead content. CT-guided biopsy and three-dimensional CT reconstruction are additional capabilities which conventional radiographic techniques do not possess.

Siemens Medical Systems has agreed to lend us a new Somatom ARC CT scanner for our continued forensic research in the Arctic. Siemens equipment, including CT, has already been extensively employed in the excellent study of the "Iceman" by Prof. zur Nedden and his colleagues at the University of Innsbruck (see Höpfel et al., 1992). We look forward with great interest to further studies of frozen human remains and to the development of complex tele-imaging methods which will distribute this information worldwide.

The Franklin Expedition was a mass disaster. As such, it must be classified and investigated as an unresolved forensic case, with important historical ramifications. The fact that the events are temporally remote, and the involved individuals long dead, does not alter the status of an unsolved forensic investigation. If Franklin and his men disappeared today, the tragedy would be fully investigated as a matter of course. Autopsies would be a routine and absolutely essential part of such an inquiry: Franklin's men deserve no less than a meticulous scientific examination by qualified experts, just as their modern counterparts could expect today.

Finally, our research has provided a new perspective on an historical mass disaster which has been pursued by countless searchers over the decades to the point of obsession. This perspective does not perpetuate the preferred and comfortable idea held by many that the Franklin expedition explorers could have succumbed only to overwhelming circumstances of "heroic" proportions:

"The romance of the Franklin era of exploration, and the emotional response it evokes, is enduring. What is not is the assumption that great men die only of great causes. For all the hostile forces of climate and geography the region might represent, it was something else which had a catastrophic effect on the Franklin expedition – something human." (Beattie and Geiger, 1993: 168–169)

Appendix

John Hartnell, Radiology notes

EXAM	KVP	MA	SECS	+/- GRID	REMARKS
AP SKULL	80	15	0.7	+	8:1 wafer-type grid; dense ice artifact
AP skull	100	13	1.5	+	slight improvement
AP skull	100	13	3.0	+	1st attempt: circuit overload (due to 200-watt fishtank heater)
AP skull	100	13	3.0	+	2nd attempt: unplug heater; now overpenetrated
AP skull	80	15	2.5	+	good technique; off center
AP skull	80	15	2.5	+	good
Towne	80	15	2.5	+	good
Rt. lateral skull	80	15	1.2	+	good
CERVICAL SPINE-AP	80	15	1.0	+	good; note absent sternum; polar bear
C-spine Rt post. obl.	80	15	0.2	+	good
C-spine Lt post. obl.	80	15	0.2	+	good
C-spine Lt lateral	80	15	0.2	+	C7 obscured by shoulder
C-spine Lt lateral	80	15	0.2	+	C7 still not seen with shoulder distracted
UPPER EXTREMITIES: Rt. humerus (external rotation)	80	15	0.1	-	satisfactory
Rt humerus (internal rotation)	80	15	0.1	-	good
Right forearm: lateral elbow	80	15	0.1	-	good
Rt forearm: external rotation	80	15	0.1	-	good
Rt hand PA	80	15	0.1	-	overpenetrated
Rt hand PA	60	16	0.2	-	good
Left humerus (external rotation)	80	15	0.1	-	good
Lt humerus (internal rotation)	80	15	0.1	-	good
Lt forearm: external rotation	80	15	0.1	-	good
Lt forearm: internal rotation	80	15	0.1	-	good
Lt hand PA	80	15	0.1	-	overpenetrated
Lt hand PA	60	16	0.2	-	good
CHEST AP	80	15	1.0	+	good; multiple artifacts
ABDOMEN AP	80	15	1.0	+	good bone detail; multiple artifacts
PELVIS AP	80	15	1.0	+	good
THORACIC SPINE AP	80	15	1.0	+	good
T-spine - Lt lateral	80	15	3.5	-	overpenetrated
T-spine - Lt lateral	80	15	1.5	-	overpenetrated
T-spine - Lt lateral	80	15	2.7	+	satisfactory
LUMBAR SPINE AP	80	15	1.0	+	good
Lumbar spine - Lt lateral	80	15	3.0	-	satisfactory
Lumbar spine - Lt lateral	80	15	3.0	+	slight overpenetration
LOWER EXTREMITIES: AP Bilateral femurs	80	15	0.6	-	good
Lateral Lt femur	80	15	0.5	+	grid cut-off
Lateral Rt femur	80	15	0.5	+	good
Lateral Lt femur	80	15	0.3	-	good
Lateral Rt femur	80	15	0.3	-	good
Bilateral knees AP	80	15	0.5	+	good
Bilateral tib/fib AP	80	15	0.2	-	good
Lateral Lt tib/fib	80	15	0.2	-	slight overpenetration
Lateral Rt tib/fib	80	15	0.2	-	slight overpenetration
Bilateral feet AP	80	15	0.1	-	good; feet held for correct position
Bilateral feet lat.	80	15	0.1	-	good

Began x-ray examination of John Hartnell at 2000 hours on 17 June, 1986, and completed study at 1125 hours on 18 June (15 hours 25 mins of continuous work). Total of 46 radiographs taken and developed by hand.

William Braine, Radiology notes

EXAM	KVP	MA	SECS	+/-GRID	REMARKS
AP SKULL	100	13	2.0	+	quite overpenetrated
AP skull	80	15	2.5	+	slightly overpenetrated
AP skull	80	15	1.2	+	good bone detail
Towne 25	80	15	1.6	+	good
Rt lat skull	80	15	1.2	+	slightly overpenetrated
Rt lat skull	80	15	0.8	+	good; normal sella
CERVICAL SPINE – AP	80	15	1.0	+	slightly overpenetrated
Rt lat c-spine	80	15	0.4	-	OK; big sella?
UPPER EXTREMITIES: Rt humerus int. rotate	80	15	0.1	-	good
Rt humerus ext. rotate	80	15	0.1	-	arm held during exposure
AP Rt forearm	80	15	0.1	-	arm held; good
Lt humerus- int. rotate	80	15	0.1	-	good
Lt humerus- ext. rotate	80	15	0.1	-	good
AP Lt forearm	80	15	0.1	-	good
Lat Rt forearm	80	15	0.1	-	good; arm held
Lat Lt forearm	80	15	0.1	-	good; arm held
Rt hand	60	16	0.2	-	overpenetrated
Lt hand	60	16	0.2	-	overpenetrated
Rt hand	60	16	0.1	-	good
Lt hand	60	16	0.1	-	good
CHEST AP	80	15	1.0	+	overpenetrated
AP Chest	80	15	1.0	-	good; good bone detail
ABDOMEN AP	80	15	1.0	+	overpenetrated
AP abd.	80	15	1.0	-	good
PELVIS AP	80	15	1.0	-	good; less ice artifact (see AP chest)
THORACIC SPINE AP					
Rt lat. T-spine	80	15	3.9	+	underpenetrated
Rt lat. T-spine	100	13	3.9	+	good
Rt post oblique T-spine	80	15	1.0	-	good
Lt post. oblique T-spine	80	15	1.0	-	good
Rt lat. T-L spine	100	13	9.0	+	3 x 3.0 sec with 45-sec cool-down intervals
Rt lat. T-L spine	80	15	1.6	-	OK
Rt lat. T-L spine	80	15	2.5	-	OK
LUMBAR SPINE AP					(see AP abdomen)
Rt lat. L-spine	100	13	3.0	-	underpenetrated
Rt lat. L-spine	100	13	7.0	+	2 x 3.5 sec with 45-sec cool-down intervals

William Braine, Radiology notes

EXAM	KVP	MA	SECS	+/-GRID	REMARKS
LOWER EXTREMITIES:					
AP Femurs	80	15	0.6	-	good
lat. Rt femur	80	15	0.2	-	good
lat. Lt femur	80	15	0.2	-	good
AP knees	80	15	0.5	-	good
AP knees	80	15	0.5	-	needle aspiration and arthrogram Lt
lat Rt tib-fib	80	15	0.2	-	good
lat Lt tib-fib	80	15	0.2	-	with 20cc Urovist
AP tibias	80	15	0.2	-	good
AP fee	80	15	0.1	-	slightly overpenetrated
AP feet	60	16	0.2	-	good
Lateral feet	60	16	0.2	-	held feet by hand

Examination of William Braine, age 34, died 3 April, 1846. Began at 1100 hours on 19 June, 1986, and completed exam at 2125 hours on 19 June. 46 x-rays were taken and developed by hand, working continuously for 10 hours 25 minutes.

Acknowledgements

The research reported in this paper has been supported in part by the Social Sciences and Humanities Research Council of Canada, the Polar Continental Shelf Project, the Boreal Institute for Northern Studies, and the University of Alberta.

Summary

Recent exhumations of the frozen and preserved bodies of three members from the Franklin Arctic expedition of 1845 have provided new insight into the causes of the disaster. Trace element analysis of hard and soft tissues has indicated that lead poisoning occurred on the expedition, and played some role in the cause of the disaster. Through stable lead isotope analysis and deduction, the origin of the contaminating lead has been traced to the solder used in the manufacture of the large quantities of tinned foods supplied to the expedition. X-ray analysis of two of the sailors revealed the presence of Pott's disease as well as indications of moderate localized skeletal trauma. An unexpected discovery was the existence of viable bacteria recovered from the frozen bodies of two sailors from the expedition. Research into the apparent resistance of some strains of these bacteria to modern antibiotics continues.

Zusammenfassung

Die kürzlich erfolgte Exhumierung der gefrorenen, konservierten Leichen von drei Teilnehmern der Franklin-Expedition von 1845 hat neue Erkenntnisse über die Ursachen des katastrophalen Endes der Expedition gebracht. Die Untersuchung von Knochen und Weichteilen auf Spurenelemente weist auf eine Bleivergiftung der Expeditionsteilnehmer hin, die das Unglück zumindest mitverursachte. Durch die Analyse stabiler Blei-Isotope und entsprechende Schlußfolgerungen konnte die Herkunft des kontaminierenden Bleis auf das Lötmittel zurückgeführt werden, das in der Herstellung der großen Menge Konserven, mit der die Expedition ausgerüstet war, verwendet wurde. Eine weitere unerwartete Entdeckung war das Vorhandensein lebensfähiger Bakterien, die von den gefrorenen Leichen zweier Matrosen entnommen wurden. Die offensichtliche Resistenz einiger dieser Bakterienstämme gegen moderne Antibiotika sind Gegenstand weiterer Untersuchungen. Vergleiche zwischen den Gräbern der Franklin Expeditionsteilnehmer und der Fundstätte des „Eismanns" weisen zahlreiche Ähnlichkeiten auf, bis hin zu den Bedingungen, die zur Konservierung der Leichen führten.

Résumé

La récente exhumation des cadavres, conservés par congélation, de trois membres de l'expédition de Franklin en 1948 a révélé de nouvelles découvertes quant aux circonstances étant à l'origine de la fin tragique de la susdite exposition. Les éléments de trace détectés dans des os et des parties molles au cours des examens nous signalent une intoxication saturnine des participants de l'expédition qui aurait au moins contribué à provoquer le malheur. L'analyse d'isotopes stables du plomb a permis de déduire que le plomb contaminant serait provenu de la soudure utilisée pour la production de la grande quantité de conserves dont le groupe de l'expédition avait été muni. Une deuxième découverte inattendue était la présence de bactéries viables prélevées des cadavres congelés de deux marins. L'évidente résistance d'une partie de ces souches bactériennes aux antibiotiques modernes font l'objet d'ultérieures études. En confrontant les sépultures des membres de l'expédition avec le lieu ou la momie du Hauslabjoch a été trouvée, on remarque de nombreuses ressemblances, jusqu'aux conditions ayant permis la conservation des cadavres.

Riassunto

La recente esumazione delle salme di tre partecipanti dell'espedizione Franklin del 1845, ha fornito nuove conoscenze sulle cause della fine catastrofica di tale spedizione. La ricerca di oligoelementi in ossa e organi ha fornito la prova che si trattasse di una intossicazione da piombo, che avrebbe provocato come concausa la fine della spedizione. Sulla base dell'analisi degli isotopi stabili di piombo e le rispettive conclusioni trattene, si e potuta constatare l'origine del piombo contaminato. Si tratta di un legante utilizzato nella produzione di grandi quantita di conserve, utilizzate dall' espedizione come approvvigionamento. Un' ulteriore scoperta inaspettata e stata la presenza di batteri capaci di sopravvivere, prelevati dalle salme gelate di due marinai. L' evidente resistenza di alcuni di questi batteri ad antibiotici moderni formano l'oggetto di ulteriori analisi. Paragoni tra le tombe del partecipanti all' espedizione Franklin e il sito di ritrovamento dell'uomo del Similaun, presentano numerosi parallelismi tra cui anche le condizioni che portarono alla conservazione di cadaveri.

References

Amy, R (1994). Unpublished report on the autopsies of John Hartnell and William Braine. Edmonton: University of Alberta.

Amy, R., Bahatnagar, R., Damkjar, E. and Beattie, O. B. (1986). Report of a post-mortem exam on a member of the last Franklin Expedition. Canadian Medical Association Journal 135: 115–117.

Aufderheide, A. (1991). Lead analysis. In S. Saunders and R. Lazenby (eds.): The Links That Bind: The Harvie Family Nineteenth Century Burying Ground. Dundas: Occasional Papers in Northeastern Archaeology No. 5, pp. 71–74.

Barry, P. S. I. (1978). Distribution and storage of lead in human tissues. In J. D. Nriagu (ed.): The Biogeochemistry of Lead in the Environment. New York: Elsevier/North-Holland Biomedical Press, pp. 97–150.

Beattie, O. B. (n.d.). The Results of Multidisciplinary Research into Preserved Human Tissues from the Franklin Arctic Expedition of 1845. Proceedings of the First World Congress On Study of Human Mummies, Museo Arqueologico Y. Etnografico, Cabildo de Tenerife.

Beattie, O. B. (1983). A report on newly discovered human skeletal remains from the last Sir John Franklin Expedition. The Muskox 33: 68–77.

Beattie, O. B (1985). Elevated bone lead levels in a crewman from the last Arctic Expedition of Sir John Franklin. In P. Sutherland (ed.): The Franklin Era in Canadian Arctic History: 1845–1859. Ottawa: National Museum of Man, Mercury Series Archaeological Survey of Canada Paper No. 131, pp. 141–148.

Beattie, O. B. (1993). Applying modern forensic anthropology to historical problems. In: Proceedings of the International Symposium on the Forensic Aspects of Mass Disasters and Crime Scene Reconstruction. Washington: Federal Bureau of Investigation, Department of Justice, United States Government Printing Office. Pages 79–91.

Beattie, O. B. and Amy, R. (1991). A report on present investigations into the loss of the third Franklin expedition, emphasizing the 1984 research on Beechey Island. Inter-Nord 19: 77–86.

Beattie, O. B. and Geiger, J. (1993). Frozen In Time. London: Bloomsbury.

Beattie, O. B. and Savelle, J. M. (1982). Discovery of human remains from Sir John Franklin's last Expedition. Historical Archaeology 17: 100–105.

Carlson, A. (1993). Lead Analysis of Human Skeletons from the 19th Century Seafort Burial Site, Alberta. Unpublished M.A. thesis, Department of Anthropology, University of Alberta.

CBC (1994) Primetime documentary, 12 September.

Cyriax, R. J. (1939). Sir John Franklin's Last Arctic Expedition. London: Methuen.

Farrer, K. T. H. (1993). Lead and the last Franklin expedition. Journal of Archaeological Science 20: 399–409.

Gilder, W. H. (1881). Schwatka's Search. New York: Scribner's Sons.

Höpfel, F., Platzer, W. and Spindler, K. (Eds.), (1993). Der Mann im Eis, Band 1. Innsbruck: Veröffentlichungen der Universität Innsbruck 187.

Inglefield, E. A. (1852). Unpublished letter to Sir Francis Beaufort, dated September 14, 1852. UK: Hydrographic Department, Ministry of Defense.

Inglefield, E. A. (1853). A Summer Search for Sir John Franklin with a Peep into the Polar Basin. London: Thomas Harrison and son.

Kerr, N. and Schweger, B. F. (1989). Survival of archaeological textiles in an arctic environment. Biodeterioration Research 2: 19–37.

Kowal, W., Beattie, O. B., Baadsgaard, H. and Krahn, P. M. (1990). Did solder kill Franklin's men? Nature 343: 319–20.

Kowal, W., Beattie, O. B., Baadsgaard, H. and Krahn, P. M. (1991). Source identification of lead found in tissues of sailors from the Franklin Arctic Expedition of 1845–48. The Journal of Archaeological Science.

Kowal, W., Krahn, P. and Beattie, O. B. (1989). Lead levels in human tissues from the Franklin forensic project. International Journal of Environmental Analytical Chemistry 35: 119–126.

Kowalewska-Grochowska, K., Amy, R., Lui, B., McWhirter, R. and Merrill, H. (1988). Isolation and sensitivities of century-old bacteria from the Franklin expedition. Interscience Conference of Antimicrobial Agents and Chemotherapy, Los Angeles, California, 24 October.

M'Clintock, F. L. (1908). The Voyage of the "Fox" in Arctic Seas. London: John Murray.

Magruder, W. E. (1883). Lead-poisoning from canned food. Medical News xliii: 261–263.

Murchison, R. (1853). Commander E. A. Inglefield – Royal Awards. Journal of the Royal Geographical Society 23: pp. 1x–1xi.

Neatby, L. H. (1970). The Search for Franklin. London: Barker.

Notman, D., Anderson, L., Beattie, O. B. and Amy, R. (1987). Arctic paleoradiology: portable radiographic examination of two frozen sailors from the Franklin expedition (1845–48). American Journal of Roentgenology 149: 347–350.

Nourse, J. E. (ed.) (1879). Narrative of the Second Arctic Expedition of C. F. Hall. Washington: United States Naval Observatory.

Osborn, S. (1865). Stray Leaves from an Arctic Journal; or, Eighteen Months in the Polar Regions, in Search of Sir John Franklin's Expedition, in the Years 1850–51. London: Longman, Brown, Green, and Longmans.

Rae, J. (1855). Arctic exploration, with information respecting Sir John Franklin's missing party. Journal of the Royal Geographical Society 25: 246–256.

Ratcliffe, J. M. (1981). Lead in Man and the Environment. Toronto: John Wiley.

Schweger, B. F. and Kerr, N. (1987). Textiles collected during the temporary exhumation of a crew member from the Third Franklin Expedition. Journal of the International Institute for Conservation – Canadian Group 12: 9–19.

Stackpole, E. A. (ed.) (1965). The Long Arctic Search: The Narrative of Lieutenant Frederick Schwatka, U.S.A., 1878–1880. Mystic: Marine Historical Association.

Sutherland, P. C. (1852). Journal of a Voyage in Baffin's Bay and Barrow Strait in the Years 1850–51. London.

U.S. Department of Health and Social Services (1988). The Nature and Extent of Lead Poisoning in Children in the United States: A Report to Congress. Georgia: Public Health Service.

Woodman, D. C. (1991). Unraveling the Franklin Mystery: Inuit Testimony. Montreal: McGill-Queen's University Press.

Correspondence: Dr. O. Beattie, Department of Anthropology, University of Alberta, Edmonton, Alberta, Canada T6G 2H4.

The mummy find from Qilakitsoq in northwest Greenland

J. P. Hart Hansen[1] and **J. Nordqvist**[2]

[1] Department of Pathology, Gentofte Hospital, University of Copenhagen, Hellerup, Denmark
[2] Department of Conservation, National Museum of Denmark, Brede, Lyngby, Denmark

Introduction

The mummified bodies of some deceased Eskimos were incidently discovered in two graves in a rock cleft near the abandoned settlement Qilakitsoq in the Uummannaq district of Northwestern Greenland in 1972. At first, nothing happened besides registration of the find in the files of the Greenland Museum. Neither the extent of the find was known, nor the number of buried persons. In early 1978, however, the graves were opened and samples were taken for radio carbon dating. The mummified bodies of two children, who were encountered just below the cap stones, were sent to Copenhagen with the samples.

According to the radio carbon dating the find dated from about 1,475 AD (± 50 years). Thus, the find consisted of the oldest, well preserved bodies in the whole arctic region, all fully dressed in equally well preserved skin clothing. Besides, a number of garments and skin pieces were found. The find offered a unique opportunity to obtain close insight in the clothing of the Thule culture giving information about the style and the complicated cut of men and women's garments and expanding our knowledge about the development of the Eskimo culture.

The mummies had lived during the same period as the last Norse descendants of Erik the Red and his followers from Iceland and Norway who then still inhabited the Eastern Settlement (Østerbygden) in Southwestern Greenland. The mummies could be regarded as true representatives in 'flesh and blood' for the 'Skrellings' whom the Norse people had encountered on journeys in Greenland according to the Icelandic and Norwegian Sagas.

The result of the dating resulted in a thorough investigation of the site the same year and the graves were emptied. It turned out that the find consisted of eight fully clothed, mummified bodies, two children and six

Fig. 1. Qilakitsoq with the island of Uummannaq in the background.
Photo: Greenland National Museum

adult females, many loose garments (78 pieces all together) and a number of unprepared skins. The director of the Greenland Museum at that time, Jens Rosing, has given his description of Qilakitsoq and the find in an artistic book illustrated with his own drawings (Rosing 1986).

The reason for the excellent condition of the find, the well preserved human bodies as well as the totally intact garments and hides was favourable local conditions. Low temperatures, low air humidity, and protective sheltering against water (rain and snow) enhanced an extensive drying of the material (natural mummification) and stopped the decaying processes, which need water and a temperature some degrees above the freezing point.

The Greenland Museum decided to bring the whole find to Copenhagen for restoration and conservation in the laboratories of the National Museum of Denmark. This procedure was adopted first and foremost because of the garments. Simultaneously, however, many questions were put forward as to the background and origin of the deceased persons. How old were they? Were they males or females? What did they die from? Did they die simultaneously and were they buried at the same time? Did they suffer from any chronic diseases? Could we learn anything about their life-ways and culture through investigating the bodies? What did they eat and how were their life influenced by their environment when compared to present days?

This prompted the initiation of an extensive, interdisciplinary program of scientific investigations utilizing the different methods of archaeology, ethnography, natural science, and medicine in a stimulating collaboration. An overview of the many different investigations was published in 1985 by the Greenland Museum in a popular book (Hansen et al. 1985) and the greater part of the scientific results were published in a special volume of Meddelelser om Grønland, Man & Society series (Hansen & Gulløv 1989). In this volume the background of the find and the archaeological facts are given in detail (Andreasen 1989), and a comprehensive catalogue of the garments is presented (Møller 1989). The garments seem to have reached the ultimate in technical terms already in the 15th century. They were the most excellent dresses to use for living, hunting and travelling in the harsh climate of arctic Greenland. The women knew how to utilize the different types of animal skins available to give the highest degree of isolation while at the same time giving the body the opportunity to let surplus heat escape so that dangerous sweating did not occur. The Eskimo dresses express better than many words the experience and adaptability of the Eskimos acquired through hundreds of years in the Arctic.

Many questions were answered by the many investigations but many new turned up. The results can not be called sensational in any context. The studies gave, however, rise to important interdisciplinary contacts and collaboration. During the work a few new methods were developed of importance for future investigations on similar ancient material, for example an indirect method of tissue typing mummified tissue (Hansen & Gürtler 1983), biochemical analyses on small animal skin pieces of diagnostic importance for the determination of the proper species (Ammitzbøll et al. 1989) and a photographic method for disclosing tattooings in mummified skin (Kromann et al. 1989). It must also be mentioned that through work on material from the mummies and supplementary present day material from Greenland and Denmark it was possible to disprove the often stated importance of diatoms for the medico-legal diagnosis of drowning (Foged 1989).

The engagement of other scientists like botanists, zoologists and geologists in the investigations was indispensable. It was shown for example how stones and mineral fragments can give information about the mobility or social contacts of a local group of people within an area when the necessary background knowledge of the geologic conditions is available (Ghisler 1989).

One of the most important results of the whole Qilakitsoq investigation is without doubt that a 500-year-old reference material has been obtained for the evaluation of present and future environmental conditions in Greenland (Hansen et al. 1989, Grandjean 1989). The investigations support the supposition that toxic substances are taken in by the Greenlanders today at a much higher level than in the days of the Qilakitsoq mummies and that important micronutrients are not as abundant in the daily diet as previously. It seems that modern technology on a global plane and changes of the Greenlandic cultural pattern toward the living habits of the so-called modern westernized world may influence the state of health in Greenland in the long run. The Qilakitsoq reference material will be of the greatest value in future evaluations of the environmental impact on the population of the eastern Arctic.

Background

Mummified human remains from the Arctic are rare. The dry and cold arctic climate enhances natural mummification, but relatively few ancient bodies have been discovered so far. In former days the special geographic conditions of the Arctic often made a safe burial impossible and the body disappeared due to exposure to the weather and animals.

Naturally mummified Eskimo bodies have been discovered in Alaska. Thus, a frozen female body dating from about 400 AD was found on a beach on St. Lawrence Island in 1970. The body had most probably been washed out of the bank side because of erosion. The body was extremely well preserved and examination revealed that she must have succumbed to suffocation being trapped in a landslide or earthquake (Zimmerman & Smith 1975, Masters & Zimmerman 1978).

Another extraordinary find was made in Barrow, Alaska, in 1982 when two Eskimo bodies and three skeletons dating from AD 1,510 (± 70 years) were discovered. The persons had been crushed and killed in a winter house by overriding sea ice, and the bodies were frozen. An excellent interdisciplinary investigation was carried out (Newell 1984, Zimmerman & Aufderheide 1984).

In Western Greenland graves with mummified bodies have been encountered from time to time. The graves date most probably from the period from the 16th century up to the introduction of the Christian burial custom some time in the 18th century.

Two important finds of mummified bodies have been made in this century in rock caves in South Greenland. Inhabitants on the island of Uunartoq north of Nanortalik knew of caves containing many dead bodies (Bak 1971). Local legends tell that the bodies were of Eskimos who were not able to fly when Norsemen planned to attack some time in the 14th or 15th century. They hid in the caves close to the settlement and starved to death. An American anthropologist, Martin Luther, visited the area in 1930 on a study tour financed by The Association of American Meat-Packers. Luther obtained permission from Copenhagen to open ancient graves and bring back a few skeletons to Harvard University in Boston to study. He discovered by chance the caves at Uunartoq and removed about 15 mummified bodies (Hooton 1930). A few of these mummies are still preserved in the Peabody Museum at Harvard University.

In 1934 the Danish archaeologist Therkel Mathiassen visited the place and found the disturbed remains of stone graves in the caves. Only one grave was found intact with three infants, all of whom a few months old (Mathiassen 1936). These bodies are preserved in Copenhagen.

Mummified bodies have also been found beneath the Pisissarfik mountains in the Nuuk/Godthaab district. In this area Norse and ancient Eskimo ruins are found. Many graves lie on the mountainside among fallen boulders, and they have been investigated in 1945 and 1952 (Meldgaard 1953). The well preserved bodies of three adults and seven infants and children were found with well preserved clothing and garments. The find has not so far been radio carbon dated but is most probably from the 16th or 17th century. These Greenlandic finds are now being investigated.

Artificial mummies are not known from Greenland or other parts of the Eastern Arctic but the habit has been practised by the inhabitants along the West coast of Alaska and on the Aleutian Islands. A few artificially mummified bodies from the Aleutian Islands have been subjected to investigations (Zimmerman et al. 1971, Zimmerman et al. 1981).

The find

The bodies from Qilakitsoq were preserved due to desiccation in the arid Arctic environment with low temperatures, a sheltered position of the graves with no admittance for snow, rain, direct sun and animals, dry air and some ventilation around the bodies.

In the graves were six adult women aged from about 18 to 50 and two children of six months and about four years, respectively. They wore outer jackets, trousers and boots of sealskin, an inner shirt of bird skins and stockings of caribou skin. The ability of the Arctic people to survive in the harsh climate is due to their skilful use of animal skins for their protection.

In order to obtain as much information as possible about the people, their culture, environment and diseases an extensive multidisciplinary program of investigations was carried out. This type of investigations can be said to be particularly important when no written history is available like in Greenland during this period of time.

The completed investigations up to now have comprised extensive radiological (Eiken 1989); biological-anthropological (Jørgensen 1989); and odontological investigations (Pedersen & Jakobsen 1989); pathological and forensic examinations with light and electron microscopy (Hansen 1989, Kobayasi et al. 1989, Myhre et al. 1989); skin, nail and hair investigations for diseases, tatooings, and traces of work and strains (Kromann et al. 1989, Bresciani et al. 1989, Kromann et al. 1989); blood grouping and tissue typing (Hansen & Gürtler 1983); DNA-analysis (Thuesen & Engberg 1990, Nielsen et al. 1994); examination of the papillary lines of the skin; investigations of eye remains (Andersen & Prause 1989) and the middle and inner ears; toxicological investigations; morphological investigations of faecal material (Lorentzen & Rørdam 1989); analyses of bone mineralisation (Gotfredsen et al. 1989, Thompson et al. 1989); biochemical and electron microscopical investigations of human and animal skin with regard to the degree of preservation and species determination on isolated pieces of skin (Ammitzbøll et al. 1989); microbiological, mycological and parasitological investigations (Hansen 1989, Svejgaard et al. 1989, Bresciani et al. 1989); analyses of heavy metals and trace elements in humans and animals (Grandjean 1989, Hansen et al. 1989); examinations for diatoms (kisel algae) (Foged 1989); investigations of irradiation conservation technics (Johansson 1989); geological and mineralogical investigations of stone and minerals found on the bodies and in the garments (Ghisler 1989); zoological analyses of animal bones, insects, bird skins and feathers etc.; botanical investigations of plant material and pollen found in connection with the bodies (Fredskild 1989); chemical analyses of fat tissue and abdominal content; and paleo-climatological, archaeological and ethnographical investigations.

Mummification

The preservation of the bodies, the garments and the skins and hides in the two graves at Qilakitsoq is a result of the particular local conditions at the place. The annual mean temperature at the site is well below the freezing point. The climate is arctic. The fluctuations are probably considerable with temperatures below –40 °C in the winter and up to +15 °C in the summer. The temperature in the graves can, however, hardly ever have reached above +5 °C. The graves faced North and an overhanging rock protected the capstones against direct sunlight.

The projecting rock also sheltered the graves from direct snow and rain so that only minimal amounts of water came into direct contact with the bodies (Fig. 2). Any water and moisture was drained off through the bottom of the graves. In addition air could pass between the stones around the bodies thus enhancing evaporation and the desiccating process. The air in the area is very dry with low humidity. Animals could not obtain admittance to the bodies.

It is a process of natural mummification which has taken place stopping the normal postmortem decomposition. Immediately after death, when the cessation of breathing and circulation stops the nourishment of the organs and tissues, the processes of decomposition start. Factors such as low temperature and desiccation can stop or delay the processes, which otherwise gradually lead to the disappearance of the soft tissues of the body. A deceased adult person buried in well-drained soil in, for example, Denmark normally becomes a skeleton in about ten years; for a child, it takes about half the time. Some parts of the body are particularly resistant to decomposition, i.e. the bones, teeth, cartilage, hair, nails, and crystalline lenses.

The processes of decomposition are complicated. Decomposition is primarily real putrefaction with decomposition of organic material due to bacterial activity. Bacteria from the intestines and also the air passages spread through the dead body, first along the circulatory system. Blood is an excellent nutritive medium for bacteria. Later, bacteria from the environment and soil may join in. Decomposition is also due to autolysis, caused by the enzymes of the dead body. These are bound in the cells during life but escape after death.

The decomposition of a dead body can take a varied period of time. Temperature is a very significant factor. Thus, a dead body decomposes rapidly when the temperature of the environment is high as in the Tropics, or when the deceased has had fever. Low temperatures delay or completely stop decomposition, which is minimal at temperatures below + 4 °C. It is the complicated chemical processes of decomposition which are slowed down or stopped by low temperatures.

Water is necessary for the processes. The growth of bacteria and to a lesser degree of fungal organisms depends upon the presence of water. As water freezes into ice crystals it is not available for the processes. If water disappears by freezing or desiccation bacterial growth and thus putrefaction is hindered. Alternating periods of freezing and thawing have a particularly desiccating effect. The cell membranes are destroyed so that the water of the cell is released and can evaporate.

Mummification by desiccation is not characteristic of cold regions alone. The drying process is rather favoured by high temperatures, dry air, and draughts which enhance evaporation. Mummification is normally rare in temperate climates. It may, however, be encountered in certain parts of the body such as the hands, which have a relatively large surface and therefore dry easily. Finds of mummified infants are also known. These infants can have been born clandestinely, done away with, and hidden in a warm and dry attic. In warm and dry desert regions it is not unusual to find mummified bodies in the sand, sometimes more than one thousand years old (i.a. Nubia). These desert people have died in the open and rapidly dried up due to the high temperature and low humidity of the air. These bodies are preserved for an unlimited length of time.

Fig. 2. The graves under the protruding rock. Photo: Greenland National Museum

The bodies

A detailed description of the eight mummified bodies can be pieced together from particularly the papers on the anthropological (Jørgensen 1988), radiological (Eiken 1988), dermatological (Kromann et al. 1988) and odontological (Pedersen & Jakobsen 1988) investigations (Figs. 3 & 4).

The range of the investigations were restricted because the four best preserved bodies were restored with the garments *in situ* in order to be on permanent exhibition in the Greenland National Museum in the capital Nuuk/Godthåb. This meant that only non-invasive methods could be employed in the investigation of these bodies. Hairs from the heads were sampled and small specimens of mummified tissue and bone were, however, removed for different analyses through a small opening made in the garments in the back of all of them, except for the smallest child (Fig. 5).

Due to the optimal preservation and the small size this mummy was left untouched. It was decided that more extensive investigations than X-rays might damage the body without any chance of giving new information. Hence, even the sex of the child was not determined.

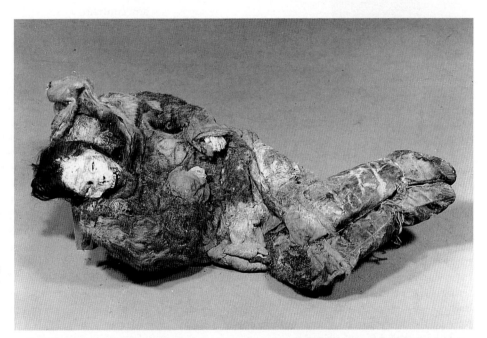

Fig. 3. Woman of about 50 years who died of a malignant tumour in the nasopharynx.
Photo: John Lee, National Museum, Copenhagen

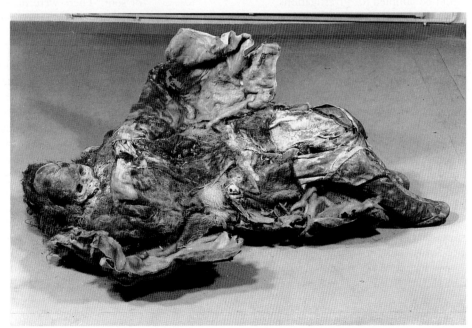

Fig. 4. Woman of 20–25 years.
Photo: John Lee, National Museum, Copenhagen

Fig. 5. The well-preserved body of a 6-month-old child. Photo: John Lee, National Museum, Copenhagen

Small details in the making of the hood of its skin jacket indicate, however, that the child was a boy (Rosing 1986).

The four remaining, lesser well preserved mummies were accessible for more thorough investigations after their garments were removed for conservation. None of these bodies were suitable for exhibition because of differing decay and destruction. The soft tissues of the bodies had disappeared in some places and some bodies were partly skeletonized. From all of the bodies material could be sampled for investigations.

Only in one mummy could internal organs be identified. In the other three bodies the mummified internal structures appeared as uncharacteristic brittle material. In the best preserved mummy the thoracic wall and part of the abdominal wall could be opened. In the thoracic cavity the desiccated heart and two shrunken lungs could immediately be identified. In the abdominal cavity part of the transverse and descending colon could be seen on the front side of the spine. On the left side a small structure was registered at the normal position of the spleen and on the right side a small shrunken liver below the diaphragm with a small structure looking like the gallbladder. White mould was encountered on all internal as well as external surfaces (Svejgaard et al. 1988). Eight mummified flies were also found, some of them embedded in mould. They were determined as representatives for *trichocera sp.* and *neoleria prominens* (Lyneborg 1984).

From the thoracic cavity the two lungs and the heart could easily be removed. It was, however, not possible to remove the abdominal content without using a saw due to the hard consistency of the mummified tissues. Thus, the abdominal cavity was rehydrated using the original solution of Ruffer according to Zimmerman et al. (1971). This procedure made the organs soft and easily removable.

During the removal of the descending colon and the sigmoideum several faecal lumps were discovered with discernible hairs (Lorentzen & Rørdam 1988, Fredskild 1988). After the removal of the colon the small intestine could be identified. The wall was thin as paper. No content was encountered. Also the ventricle could be identified with a paper thin wall without mucosal relief or any content.

The lungs and the heart were placed in the rehydrating solution of Ruffer. After 24 hours the consistency had changed. The tissues were now soft and permitted use of knife and scissors without doing any damage to the structures. The rehydrating fluid was strongly coloured dark brown. After rehydrating the heart measured $10 \times 7.5 \times 2$ cm. The pericardium was normal. It was possible to inspect the cavities and valves of the heart and no abnormalities could be registered. The openings of the coronary arteries were easy to locate. Other internal organs could not be identified, neither kidneys, nor other urinary or sexual organs.

Tissue samples were taken for histological examination (Myhre et al. 1988). Beside heavy anthracosis of the lungs no sign of disease or other disorder were found. The explanation of the relatively heavy anthracosis in a young woman aged 18–22 years (Fig. 6) having lived in the unpolluted arctic air is the fact that it was the duty of the Eskimo woman to tend the lamps and cooking fires (Kleivan 1984). By doing this she often inhaled particles of soot. There were no signs of tuberculosis, a disease which in recent generations has ravaged the Greenlandic population.

Even if tissue typing of the mummies turned out to be successful (Hansen 1988) it was not possible to demonstrate antibodies in eluent from mummy skin and muscle tissue using Western blot technique (Shand & Høiby 1987). This was in agreement with the finding that immunohistochemical investigations were negative applying a broad range of stained antibodies to sections of paraffin embedded tissue.

Bacteriological investigations were carried out by Sebbesen & Thomsen (1984). Material was secured from most of the bodies (skin, heart, lungs, lever, ileum, and colon). The results showed that up to ten different bacteria were cultured in some of the samples, all of them gram-positive and not different from what can be found in common soil. Most of the bacteria were found on the surface as well as in the tissues indicating contamination from the surroundings. In tissue taken from the pelvic cavity growth of *Clostridium perfringens* was found. The

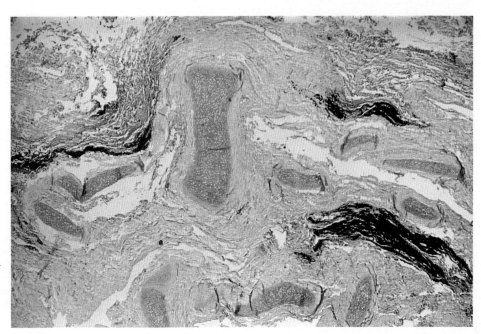

Fig. 6. Section from the root of a lung showing cartilage in the air passages, collapsed alveoli and heavy deposits of inhaled soot (anthracosis).
Photo: Gentofte Hospital

importance of this finding is difficult to evaluate. Most probable the find was made by chance. Gas gangrene seems not to have been recorded in Greenland in historic times.

Cause of death

It has not been possible to establish neither the cause nor the manner of death for several of the mummies. This is partly caused by the fact that thorough internal investigation was not permissible in the cases of the best preserved mummies. These bodies would have provided the best possibilities for identification of structures and possible pathological changes.

In some of the cases there are, however, firm indications of the cause of death. In other cases it is only possible to conjecture. Many likely causes could not be put to the test because access for internal examination was limited. In some cases the organs were poorly preserved. When considering the possible causes of death it must be kept in mind that the individuals may have died at the same occasion from the same cause, or at different times for different reasons, perhaps with intervals of several years.

The smallest child did not show any sign of disease or trauma by external examination and X-ray. The child may have been killed by being buried alive together with its deceased mother (Meldgaard 1953). This procedure was not unusual in ancient Greenland. In the small and secluded arctic communities it was not always possible to save a child whose mother had died. Instead of letting the child die gradually from hunger because no other woman could be found to nurse it, the father would put the child to death. Often it was suffocated and buried with the mother so that they could travel together to the Land of the Dead. In other cases the child was buried alive with the dead mother.

In the present case the child was found on the top of the other four bodies in one of the graves. The lower four bodies were completely covered by hides and the child was placed on the top without any covering itself. This indicates that the child was placed into the grave as the last act before the grave was covered with stones.

In the boy of about four years of age, the radiological examination offers certain indications concerning the cause of death (Eiken 1988). Most probably this child suffered from Legg-Calvés-Perthes' disease of the hip (aseptic necrosis of the femoral head). This must have caused pain and difficulties in walking normally. Moreover, the shape of the pelvis indicates that the child suffered from Down's syndrome. This syndrome is well known also in Eskimos. Most probably the boy have had great need for support being only able to limp or perhaps just crawl around. His mental development was probably poor in relation to his age. Rarefaction of his bones indicates that he was disabled and probably immobilized for some time. Surprisingly, however, the boots of the boy showed that they were normally worn. The soles were even repaired below the heals. This seems to somehow contradict the above mentioned. The explanation can, however, be that the child was given the boots of another child after death. This supposition is supported by the fact that the right and left boot were on the wrong feet. No signs of rickets were found.

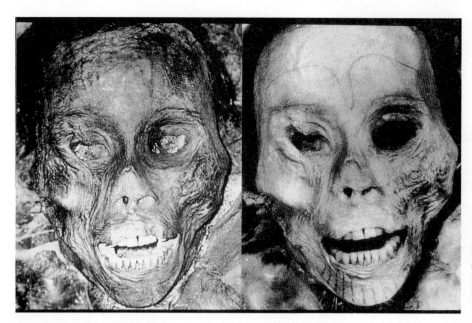

Fig. 7. An ordinary and an infra-red photograph of the face of a woman around 30 years of age. The infra-red photograph reveals tattoos on the forehead and the chin. Photo: E. R. Løytved

A disabled child like this boy must have had great difficulties surviving in ancient Greenland. Resistance to contagious diseases and hunger has been poor. Children suffering from Down's syndrome also have a higher death rate compared to normal children as a result of congenital heart diseases and blood cancer. No internal malformations or diseases could be discovered in the boy. The poor state of preservation of the body has, however, limited the extent of the investigations.

Fig. 8. Detail from a contempory painting by an unidentified artist of four Eskimos from Western Greenland captured by a Danish expedition in 1654. The tattoos have a close resemblance to those of the Qilakitsoq mummies. Photo: National Museum, Copenhagen

It is also well known that in ancient Greenland children and also adult disabled were sometimes done away with, either violently by strangulation or drowning or by exposure, due to the restricted resources in the small communities. It was, however, not possible to record any sign of strangulation or other violence in the boy, and death from exposure, cold or thirst can not be proven. Thus, the boy may have died a natural death as a result of congenital or acquired disease and general low resistance. It can not, however, be ruled out that he was done away with in one or another way.

The habit to do away with disabled and sick persons when provisions and resources were low was not regarded as an evil act but rather as an act of compassion. This is analogous to the behaviour of many old and disabled persons who left their homes and settlements on their own initiative and sought death in order not to be a burden to their families, particularly during periods of hunger (Egede 1939).

The finds made during the examination of *mummy I/3* included a kidney stone and a bone fragment from the temporal bone of either a seal or a polar bear (Møhl 1983), probably located in the gastrointestinal canal (duodenum). The kidney stone may have caused a malfunction of the kidney. The bone fragment must have been taken in with the food and was following the natural route through the body when death occurred. However, it is not possible to exclude entirely that the relatively big bone fragment may have torn a hole somewhere in the gastrointestinal canal or caused ileus by being lodged in the intestinal passage, both a possible cause of death.

Mummy I/4 presented herself with a distended abdomen. This gave immediately rise to speculations that she was pregnant at the time of death. Radiological examination outruled however this supposition. The distension of the abdomen was most likely caused by postmortem intestinal gas production. An ovarial cyst is, however, a possibility. Such a cyst only rarely causes death.

In *mummy I/5* the hair of the scalp was sparse appearing like the physiological type of baldness in men (Kromann et al. 1988). Baldness is infrequently seen in females, and physiological baldness is rare in Eskimos, even in men. The baldness may have been caused by a virilizing tumour of the ovary, and it can not altogether be excluded that such a tumour may have been of some importance for the occurrence of death. Smaller areas of baldness can be seen in Eskimos women in the temples caused by the wide spread habit of combing the hair with a very tight knotted top on the head. This type of baldness has been given the name *alopecia arctica s. Groenlandica*. The same type of baldness can be seen to-day in girls wearing their hair with a too tightly combed horse tail.

In *mummies II/6* and *II/7* no changes were observed of relevance to the cause of death. In *mummy II/8*, however, a woman of about 50 years, extensive destruction of the base of the skull was observed by radiological examination (Eiken 1989). This destruction was most probably caused by a malignant tumour spreading in the bone. The changes are identical to those which can be seen in patients with nasopharyngeal carcinoma. This cancer is particularly frequent among people of Eskimo origin in Greenland, Alaska and Canada (Nielsen et al. 1977). It is also frequent in certain regions of China and North Africa, although it is rare in Europe. In fact, the current incidence is nearly 25 times greater in Greenlanders than in Danes. Presumably, a nasopharyngeal carcinoma was the cause of death in this woman. This malignant tumour must have caused distressing symptoms during the latest period of the woman's life with blindness and pain. Cut marks in her left thumb nail indicate, however, that the woman was able to work to the very end (Kromann et al. 1988).

With regard to the establishment of the cause of death it must be taken into consideration that 500 years have elapsed since the death of the people from Qilakitsoq. The disease pattern in Greenland has changed profoundly during this period, particularly during this century. Many serious infectious diseases have been nearly irradiated. Nonetheless, the disease pattern in Greenland to-day still differs considerably from industrialized countries like Denmark (Bjerregaard 1991). For example, the cancer pattern in Greenland differs markedly with relatively high rates of certain types of cancer as for example nasopharyngeal and oesophageal cancer, salivary gland cancer, cervical cancer and lung cancer in women and relatively low rates of other types as breast cancer, uterine cancer and prostatic cancer (Nielsen 1986). No doubt, both hereditary and environmental conditions play a role for the occurrence of cancer and a number of other diseases.

The Eskimos have lived in relative isolation for thousands of years. This isolation was first really broken by World War II due to the increasing importance of the Arctic for strategic purposes and due to a growing exploitation of natural resources. The admixture of non-Eskimo genes to the gene pool is ever increasing and the living conditions and environment undergo radical changes. The proportion of European genes in West Greenland is to-day at least 25–30 per cent (Kissmeyer-Nielsen et al. 1971). These changes in heredity and environment are also reflected in the disease pattern.

One characteristic feature of the unique Eskimo disease pattern is that coronary thrombosis and both the juvenile and adult form of diabetes are rare (Bjerregaard 1991). In recent years coronary disease is four times more often diagnosed in Danes than in Greenlanders. This is probably due to the fact that the thrombocytes of Eskimo blood have a low ability to aggregate. For centuries Greenlanders have been known to bleed easily. Today there is a relatively high rate of cerebral haemorrhages, prolonged bleeding in connection with childbirth and a pronounced tendency for nosebleeding. The occurrence of atherosclerosis in Eskimos has not been conclusively investigated but seems to be lower than in non-Eskimos (Mulvad et al. 1993). This has been attributed to the traditional hunter's diet with a high content of mono- and polyunsaturated fatty acids. Atherosclerosis has, however, been found also in ancient Eskimo bodies (Zimmerman & Smith 1975, Zimmerman & Aufderheide 1984).

The bones of Eskimos are relatively deficient in calcium. This fact is well known from investigations of up to 1,500 years old Eskimo skeletons from the entire Arctic, e.g. Greenland, Canada, Alaska and the USSR, as well as of living persons (Thompson et al. 1982, Mazess 1974, Mazess & Mather 1975). The blood calcium content in Greenlanders is lower than in Danes (Jeppesen & Harvald 1983) but the content of vitamin D, which is essential for the formation and development of bone tissue, seems to be sufficient in Eskimos living on a traditional diet (von Westarp et al. 1982). As a result of the sufficient vitamin D content rickets is not found in Greenlanders, whereas the low calcium content of the blood tend to rarefy the bones with an increased risk of fractures. In several of the mummies the skeleton was rarefied (Eiken 1988), and small compression fractures of the lumbar vertebral bodies were encountered. Greenlanders who have moved to Denmark show a higher calcium level in the blood, probably due to calcium rich diary products. In this context an increased occurrence of lactase insufficiency in Greenlanders has to be kept in mind (Gudmand-Høyer et al. 1973).

Among the possible causes of death infectious diseases are important. Through centuries Greenland has been virgin country for many infectious organisms. Due to the secluded location of Greenland and the sparse, small settlements with little contact with the outside world infectious diseases often had a devastating effect. An infectious organism responsible for f. ex. influenza or measles, which were brought into the country from the outside by whalers or explorers to whom it may have been innocent, could spread in the community killing the greater part of the population in a short time. Many epidemics are on the record from historical time, also from diseases like smallpox and typhoid fever. Serious epidemics of measles and hepatitis are known from the latest decades and tuberculosis is still remembered as the great killer in the first half of this century. It has not proved feasible to apply any method on material from the mummies which with reasonable certainty could prove or disprove of infection as the cause of death.

Life in ancient Greenland was harsh, at least in periods. The dark winter period of the year with often very low temperatures was a threat if provisions had not been gathered during the period of the year with abundant hunting. Families and whole communities could be wiped out from starvation, and during such hunger periods the individuals were more exposed to the dangers of cold and exposure than normally.

Death by starvation, cold and exposure is difficult to trace in anthropological material as the present. It is, however, not likeable that the mummies died from starvation, at least not all of them. The intestinal content, which could be examined in mummy II/7 (Lorentzen & Rørdam 1988), showed that this person had a varied and abundant diet up to her death. There seems to have been enough food at the time of her death, which also seems to have occurred in the summer period when food normally is available. None of the bodies seemed lean; in the contrary mummy II/7 was rather stout. The garments and the loose skins and hides were all without traces of chewing or eating and this also contradicts the theory of death by starvation. When humans or their dogs starved in ancient Greenland pieces of skins were often cooked and chewed by the hungry humans and eaten by the dogs.

Another possible and rather frequent cause of death in Greenland is drowning. Thus, the theory of drowning was put forward immediately after the opening of the graves (Rosing 1986). It was supposed that all eight dead persons had sailed together in an *umiak*, the traditional Greenlandic skin boat for transporting many people and goods. This type of boats was always manned by women while the men followed in their *kayaks*. In the present case the *umiak* could have been sailing near the settlement of Qilakitsoq when an iceberg in the fjord capsized and raised a huge wave of water. This may have turned the boat upside down and thrown the women and the children into the sea where they drowned. The men managed the big wave due to their more maneuverable

Figs. 9–10. Parka of seal. Photo: John Lee, National Museum, Copenhagen

kayaks. The drowned may either have been taken from the sea or washed ashore and they might then have been buried together in the rock cleft. It has not been possible to find firm evidence to support this theory. Even today it can be difficult to find absolute proof of drowning in recently drowned persons.

It must be mentioned that a comparative geological investigation could not demonstrate in the garments of the mummies one single grain of some very characteristic mineral grains, which were abundant on the local beach (Ghisler 1989). Thus, the bodies could most probably not have been landed upon the beach.

Another theoretical cause of death is poisoning. Deadly epidemics and single fatal cases from food poisoning (botulism) are well known from the whole arctic area even to-day (Hansen & Bennike 1982). In Greenland deadly cases of poisoning from eating mussels are on the record but no cases from eating poisonous plants are known.

Did they die simultaneously?

It has not been possible to decide with any certainty if the mummies died and were buried simultaneously or if they were buried with may be intervals of decades. In the graves some indications were found pointing in the direction that there had been one burial initially in each of the graves, which later were filled up at one or several later occasions. Thus, in one of the graves a big flat stone was found standing vertically, most probably placed in its position after the burial of the first woman as the stone was standing on a heap of hides entombed with the woman. The stone may have had the purpose to increase the height of the grave for later burials and to secure that the opening of the grave was not too big when the covering stones had to be placed. In the other grave the bottom woman was difficult to discover as she was covered by hides and plants. There were many small stones and the grave seemed nearly filled up by loose skins, hides and stones when the two upper most bodies were removed.

At the first radiological examinations of the four year old boy seven deciduous teeth were discovered laying together on the back over the lumbar spine, seemingly under the garments. It was initially believed that the teeth was placed in a small, partly decayed skin purse resting on the surface of the skin, may be a kind of an amulet. Four other teeth were found firmly trapped in the mummified soft tissues on four different location of the trunk and neck. It was acknowledged that the teeth were all from the boy himself and that they had fallen out after death as they all had roots. It was also registered that the garments and the soft tissues on the lower part of the back were partly decayed and that the teeth in fact

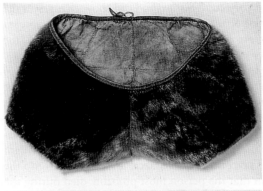

Figs. 11–12. A pair of short seal skin trousers. Each half is composed of ten pieces of skin. Photo: John Lee, National Museum, Copenhagen

were located inside the soft tissues below the skin surface and not in a purse.

It is most probable that the different positions of the child's teeth, which must have loosened and fallen out years after death and the burial, were caused by moving the body around. All the teeth were firmly trapped in the mummified tissues. This indicates that the teeth reached their final position before mummification was complete. After mummification it would have been impossible for the teeth to relocate. Tissues and the garments must still have been soft and decaying when the teeth were moved around on the body. Most probably, the child was moved from/in his grave, may be in order to have other bodies buried in the same grave, or with the purpose of burial in another grave together with his mother or other relatives. This may have happened long time after death, after the loosening of the teeth and before completion of the mummification. The total process of mummification may have lasted for decades taking the particular conditions of the grave into consideration.

It had been envisaged that modern scientific methods could help in deciding if the dead persons had died and were buried simultaneously or with intervals of may be decades. Carbon 14 dating is too inaccurate in this context. Longitudinal X-ray-fluorescence-spectrometry on hairs from the heads of the mummies was hoped to have given some clues with regard to the contemporary of the dead persons. This method can show changes in the concentration of different elements and these concentrations are known to fluctuate depending upon the character of the diet. By identical diet identical changes should be demonstrable in hairs from the individual mummies. Characteristic patterns could, however, not be demonstrated and the results could neither support contemporarity nor the opposite (Hansen et al. 1988).

It was only possible to recover food items from the intestines of one of the mummies (Lorentzen & Rørdam 1989). The content of the faecal lumps pointed into the direction that the person most probably died during the summer (July/August). Had it been possible to collect such material from several of the mummies it would have been possible to compare the food content in order to find out if the individuals had shared the same diet and if they might have died at the same time of the year or not.

Summary

Near the abandoned settlement Qilakitsoq in the Uummannaq district of Northwestern Greenland eight mummified Eskimo (Inuit) bodies were found in two graves. The people wore garments and were interred with many loose jackets, trousers and skins. They had been buried around 1475 AD and were preserved due to desiccation in the arid Arctic environment with low temperatures, a sheltered position of the graves, dry air and some ventilation around the bodies. These were the oldest, well preserved people and garments from the Thule culture, the immediate ancestors of the present Eskimo population of the Eastern Arctic.

In order to obtain as much information as possible about the peoples, their culture, environment, diseases etc. an extensive multidisciplinary program of investigations has been carried out applying methods of archaeology, modern medicine and natural science. The up to now completed investigations have comprised radiological; biological-anthropological; and odontological investigations; pathological and forensic examinations with light and electron microscopy; skin, nail and hair investigations for diseases, tatooings, and traces of work and strains; blood grouping and tissue typing; DNA-analysis; examination of the papillary lines of the skin; investigations of eye remains and the middle and inner ears; toxicological investigations; morphological investigations of faecal material; analyses of bone mineralisation; biochemical and electronmicroscopical investigations of human and animal skin with regard to the degree of preservation and species determination on isolated pieces of skin; microbiological, mycological and parasitological investigations; analyses of heavy metals and trace elements in humans and animals; examinations for diatoms (kisel algae); investigations of irradiation conservation technics; geological and mineralogical investigations of stone and minerals found on the bodies and in the garments; zoological analyses of animal bones, insects, bird skins and feathers etc.; botanical investigations of plant material and pollen found in connection with the bodies; chemical analyses of fat tissue and the abdominal content; paleoclimatological investigations; and archaeological and ethnographic investigations.

Most of the people were healthy. One of the older women had an extensive cancer of the nasopharynx, further a kidney stone, a few fractures, and parasites were found. The oldest child suffered from a disorder of the hip, probably Legg-Calvé-Perthes' disease, and probably also Down's syndrome.

It was impossible to uncover the cause of death in every case, and it was also impossible to determine whether the bodies had been interred at the same time or not. Tissue typing indicated that the persons may have been closely related representing three generations. DNA-analysis has so far been carried out on one person. Investigations showed that the food was primarily of marine origin. Bone material, teeth and hairs from the humans and hairs from the skins of the animals were analyzed for heavy metals which today are related to human pollution, e.g. mercury, cadmium and lead, and also for micronutrients like copper and selenium. Mercury and cadmium have increased about three times in humans when compared with present day Greenlanders while lead showed an increase of about 8 times. The content of these metals in the animals had also increased, although not to the same degree. Selenium had decreased in humans with no change in the animals. These results indicate that global pollution due to industrial activities and changes in the traditional lifestyle may influence health in the Arctic. The people from Qilakitsoq and their garments and skins are a valuable, 500-year-old material of reference for assessing the present and future living conditions in the Arctic.

Zusammenfassung

Nahe der aufgelassenen Siedlung Qilakitsoq im Bezirk Uummannaq in Nordwestgrönland wurden in zwei Gräbern acht mumifizierte Eskimo-(Inuit-)Leichen gefunden. Sie waren bekleidet und hatten eine große Zahl von Jacken, Hosen und Fellen als Beigaben bei sich. Sie waren um etwa 1475 v. Chr. begraben worden und waren durch Austrocknung in dem trockenen und kalten arktischen Klima, durch eine geschützte Lage der Gräber und die trockene Luft mit leichter Strömung um die Leichen sehr gut konserviert. Es sind dies die ältesten gut erhaltenen Leichen und Kleidungsstücke aus der Thule-Kultur, den direkten Vorfahren der heutigen Eskimobevölkerung in der östlichen Arktis.

Um so viele Informationen wie möglich über die Menschen, ihre Kultur, Umwelt, ihre Krankheiten etc. zu erhalten, wurde ein umfangreiches multidisziplinäres Untersuchungsprogramm durchgeführt, bei dem verschiedene Techniken der Archäologie, der modernen Medizin und der Naturwissenschaft zur Anwendung kamen. Bisher abgeschlossen wurden die folgenden Untersuchungen: radiologische, biologisch-anthropologische und odontologische Untersuchungen, pathologische und forensische Analysen mit Licht- und Elektronenmikroskop, Untersuchungen an Haut, Nägeln und Haaren auf Krankheiten, Tätowierungen und Spuren von Arbeit und Belastung, Blutgruppen- und Gewebebestimmungen, DNA-Analyse, Untersuchung der Papillarlinien der Haut, Untersuchungen der Überreste der Augen und der Mittel- und Innenohren, toxikologische Untersuchungen, morphologische Untersuchungen an Kotresten, Analyse der Knochenmineralisierung, biochemische und elektronenmikroskopische Untersuchungen der menschlichen Haut und der Tierhäute auf den Grad der Konservierung und zur Artenbestimmung mittels isolierter Hautstücke, mikrobiologische, mykologische und parasitologische Untersuchungen, Analysen von Schwermetallen und Spurenelementen bei Menschen und Tieren, Untersuchung auf Diatomeen (Kieselalgen), Untersuchung von Konservierungsmethoden durch Bestrahlung, geologische und mineralogische Untersuchungen der Steine und Mineralien, die an den Körpern und in den Kleidern gefunden wurden, zoologische Untersuchungen der Tierknochen, Insekten, Vogelhäute und Federn etc., botanische Untersuchungen von Pflanzenmaterial und Pollen, die an und bei den Körpern gefunden wurden, chemische Analysen von Fettgewebe und Bauchinhalt, paläoklimatologische sowie archäologische und ethnographische Untersuchungen.

Die meisten Individuen waren gesund. Eine ältere Frau wies ein großes Nasenrachenkarzinom auf; weiters wurden ein Nierenstein, einige Knochenbrüche und Parasiten gefunden. Das älteste Kind litt an einer Hüftläsion, wahrscheinlich Perthes-Calvé-Legg-Krankheit und möglicherweise auch an Down-Syndrom.

Es war unmöglich, für jeden einzelnen Fall die Todesursache festzustellen, und es war auch nicht möglich zu eruieren, ob die Leichen zur gleichen Zeit bestattet worden waren oder nicht. Gewebebestimmungen wiesen darauf hin, daß die Personen eng verwandt waren und wahrscheinlich drei Generationen einer Familie umfaßten. Eine DNA-Analyse wurde bisher an einer Leiche durchgeführt. Untersuchungen zeigten, daß die Nahrung vor allem aus dem Meer stammte. Knochenmaterial, Zähne und Haare der Menschen sowie Haare von den Tierhäuten wurden auf die Schwermetalle untersucht, die heute mit der Umweltverschmutzung durch den Menschen in Zusammenhang gebracht werden, z. B. Quecksilber, Cadmium und Blei, ebenfalls auf Mikronährstoffe wie Kupfer und Selen. Ein Vergleich mit den heute lebenden Grönländern zeigt eine Zunahme von Quecksilber und Cadmium um das Dreifache, während der Bleigehalt heute etwa achtmal so hoch ist. Der Gehalt dieser Schwermetalle in den Tieren hat ebenfalls zugenommen, allerdings nicht im selben Ausmaß. Selen ist bei den Menschen zurückgegangen, während es bei den Tieren keine Änderung gegeben hat. Diese Ergebnisse weisen darauf hin, daß die weltweite Verschmutzung durch die Industrie und durch Veränderungen im traditionellen Lebensstil sich wahrscheinlich auch auf den Gesundheitszustand der Menschen in der Arktis auswirken. Die Eskimos von Quilakitsoq, ihre Kleider und Felle sind wertvolles, 500 Jahre altes Vergleichsmaterial zur Beurteilung der jetzigen und künftigen Lebensbedingungen in der Arktis.

Résumé

Près de Qilakitsoq, une habitation abondonnée dans le district Uummannaq au nord-ouest de la Groënlande, huit Èsquimos mummifiés ont été trouvés dans deux tombesaux. Ces personnes portaient des vêtements et ont été enterrés avec plusieurs vestes, pantalons et fourrures séparés. Ils ont été enterrés vers 1475 AD et ont éte préservés par dessèchement dans les environnements arctiques dessertés avec des temperatures basses, un placement protegant des tombes, l'air sec et une certaine ventilation autour des corps. Ce sont les plus anciennes personnes et vêtements, bien conservés de la culture Thule, les antécédants directs de la population actuelle d'esquimos de l'arctique orientale.

Pour découvrir le plus d'informations possible sur le peuple, leur culture, leur environnement, leurs maladies etc. un programme multidisciplinaire élaboré a été poursuivi utilisant les méthodes d'archéologies, de médicine moderne et de sciences naturelles. Les investigations actuellement conclues comprennent des examens radiologiques, anthropolobiologiques, odentologiques, pathologiques et médicaux legaux par microscopie luminaire et électronique, examens de la peau, des ongles et des poils pour individuer des maladies, des tatouages et des consequences de travail et de charge corporelle; groupes sanguins et tissue typing; analyses de ADN, examens des lignes mammaeres; investigations des séquelles des yeux, des oreilles moyennes et internes; investigations toxicologiques; examinations morphologiques des selles; analyses des mineralisations osseuses; investigations biochimiques et microscopiques électroniques des peaux humaines et animales pour déterminer le degré de préservation et determiner les éspeces sur des morceaux de peau isolés; investigations microbiologiques, mycologiques et parasitologiques; analyses des métaux lourdes et des éléments marqueurs sur des humeins et des animaux; examens phycologique; investigations des techniques de conservations par radiation; investigations géologiques et minéralogiques des pierres et mineraux trouvés sur les corps et sur les vêtements; analyses zoologiques des os d'animaux, des insectes, des peaux d'animaux, plumes etc.; investigations botaniques des plantes et pollens trouvés proche des corps; analyses chimiques des tissues gras et contenus abdominaux; investigations paleo-climatologiques; et investigations archéologiques et etnografiques.

La plus part des personnes étaient en bonne santé. Une des plus vieilles femmes avait un cancer extensive de nasopharynx, en plus elle avait un calcul rénal, des fractues, et des parasites. L'enfant le plus âgé avait une maladie de la hanche, probablement une maladie de Legg-Calvé-Perthe, et probablement aussi un syndrome de Down.

Dans aucun des cas il n'était possible de déterminer la cause de la mort, ni de determiner si les corps avaient été enterrés simultanément. Tissue typing a indiqué que les personnes pouvaient être familièrement proches représentant trois génerations. Jusqu'à présent seulement une personne a été examinée par analyses de ADN. Les investigations ont montrés que les aliments étaient en priorité d'origine maritime. Echantillons des os, des dents et des cheveux des humains et poils des animaux ont été analysé pour leur contenu des métaux lourds qui aujourd'hui sont aliés à la polution humaine, e.g. mercure, cadmium, plomb, et leur contenu des micronutrients commes cuivre et selenium. Les taux de mercure et de cadmium etaient 1/3, les taux de plomb 1/6 des taux de concentration trouvés sur les Groënlandais de nos jours. La même tendance, avec moindre différence, était trouvée sur les animaux. Les taux de concentration de selenium étaient plus élevés sur les humains et identiques sur les animaux, en comparaison avec les mêmes analyses de nos jours. Ces résultats indiquent que la pollution industrielle globale et la changement de style de vie traditionnelle pouvaient avoir une influence sur le santé dans les arctiques. Les hommes de Qilakitsoq, leurs vêtements et leurs peaux, sont une précieuse référence pour l'appréciation des conditions de vie actuelle et à venir dans les arctiques.

Riassunto

Nei paraggi dell' insediamento abbandonato di Qilakitsoq nel distretto di Uummannaq nella Groenlandia Nordoccidentale otto mummie di esquimesi (Inuit) sono state trovate in due tombe. Le salme erano vestite e sotterrate con diverse giacche, pantaloni e pelli. Furono sepolte attorno al 1475 AD. La preservazione dei corpi era dovuta a dessiccazione a causa dell'arido ambiente Artico con basse temperature, la riparata posizione delle tombe, l'aria secca con ventilazione attorno ai corpi. Queste sono le più antiche, ben conservate salme ed abbigliamenti della cultura di Thule, gli immediati predecessori della popolazione esquimese d'oggi nell'Artico Orientale.

Per ottenere più informazioni possibili sugli abitanti, la loro cultura, l'ambiente, le malattie etc. etc., un esteso programma multidisciplinare di ricerche è stato messo in atto applicando metodi archeologici, di medicina moderna e scienza naturale. La serie di ricerche completate sino ad oggi hanno compreso: esami radiologici, biologici-antropologici e odontologici; esami patologici e di medicina legale con microscopia ottica ed elettronica; esami della cute, delle unghie e dei capelli alla ricerca di malattie, tattuaggi e tracce di lavoro ed usura; determinazione del gruppo sanguigno e tipo del tessuto; DNA-analisi; esame delle linee papillari della pelle: esami sui resti oculari e sull'orecchio medio e interno; esami tossicologici; esami morfologici del materiale fecale; analisi della mineralizzazione delle ossa; esami biochemici e microscopici-elettronici della pelle sia umana che animale con special riguardo al grado di preservazione e determinazione delle specie su singoli frammenti di pelle; esami microbiologici, micologici e parassitologici; analisi di metalli pesanti e tracce sia nei corpi umani che animali; esami alla ricerca di diatomee (Kisel algae); ricerche su techniche di conservazione per mezzo irradiazione; analisi geologiche e mineralogiche delle pietre e dei minerali trovati sui corpi e sugli abbigliamenti; esami zoologici delle ossa animali, insetti, pelle e piume di uccelli etc.; analisi botaniche del materiale vegetale, piante e polline trovato in connessione dei corpi; analisi chimiche del tessuto adiposo e del contenuto intestinale; analisi paleoclimatologiche ed etnografiche.

Gran parte delle persone erano sane. Una delle donne di media età aveva un cancro estensivo della rinofaringe, inoltre venne riscontrato un calcolo renale, alcune fratture e infezione parassitaria. Il più anziano dei bambini soffriva di un difetto articolare dell'anca, probabilmente la malattia Legg-Calvé-Perthes, e probabilmente anche il sindroma di Down.

Era impossibile accertare in ogni caso la causa di morte ed era impossibile determinare se i corpi fossero stati sotterrati allo stesso tempo o no. L'esame del tipo di tessuto indicò che le persone erano in stretta relazione famigliare rappresentando tre generazioni. DNA-analisi è stata sino ad ora effettuata in una sola persona. Le ricerche dimostrarono che il cibo alimentare era stato principalmente di origine marina. Ossa, denti e capelli degli esseri umani e peli della pelle degli animali furono esaminati per metalli pesanti, i quali oggi sono dovuti all'inquinamento dell'uomo, come p. e. mercurio, cadmio e piombo e inoltre oligo-elementi quali rame e selenio. La concentrazione di mercurio e cadmio nei confronti dei groenlandesi di oggi è nel corpo umano quasi triplicata, mentre la concentrazione del piombo è aumentata circa 8 volte. La concentrazione di detti metalli sono aumentati anche negli animali, anche se non nello stesso grado. La concentrazione del selenio è d'altra parte oggi minore nell'uomo mentre negli animali è invariata. Questi risultati indicano che l'inquinamento globale dovuto alle attività industriali e il cambiamento del modo di vivere può influenzare lo stato di salute nell'Artico. La popolazione di Qilaktisoq e i loro abbigliamenti e pelli sono un materiale con 500 anni di età valido e valutabile come referenza per adeguare le condizioni di vita presente e futura nell'Artico.

References

Ammitzbøll, T., Møller, R., Møller, G., Kobayasi, T., Hino, H., Asboe- Hansen G., Hansen J. P. H. (1989) Collagen and glycosaminoglycans in mummified skin. In Hansen J. P. H., Gulløv H. C. (Eds.) (1989) Meddr Grønland; Man & Soc. 12: 93–99.

Andersen S. R., Prause, J. U. (1989) Histopathological examinations of eyes. In Hansen J. P. H., Gulløv H. C. (Eds.) opera cit.: 109–111.

Andreasen, C. (1989) The archaeology at Qilakitsoq. In Hansen J. P. H., Gulløv H. C. (Eds.) opera cit.: 11–22.

Bak, O. (1971) Gravhulerne på øen ûnartoq (in Danish). Tidskr Grønland 18: 77–95.

Bjerregaard, P. (1991) Disease pattern in Greenland. Arc Med Res 50: suppl. 4 (62 pp).

Bresciani, J., Haarløv, N., Nansen, P., Møller, G. (1983) Head louse (Pediculus humanus subsp. capitis de Geer) from mummified corpses of Greenlanders AD 1475. Acta Entomol Fenn 42: 24–27.

Egede, N. (1939) Beskrivelse over Grønland (1769) (Description of Greenland). – In: Ostermann, H. (Ed.): Meddr. Grønland 120: 232–69.

Eiken, M. (1989) X-ray examination of the Eskimo mummies of Qilakitsoq. In Hansen J. P. H., Gulløv H. C. (Eds.) opera cit.: 58–68.

Foged, N. (1989) Diatoms in mummies from Qilakitsoq. In Hansen J. P. H., Gulløv H. C. (Eds.) opera cit.: 184–195.

Fredskild, B. (1989) Botanical investigations of the mummies. In Hansen J. P. H., Gulløv H. C. (Eds.) opera cit.: 179–183.

Ghisler, M. (1989) Significance of mineral grains in clothing and skins from the mummy graves. In Hansen J. P. H., Gulløv H. C. (Eds.) opera cit.: 172–178.

Gotfredsen, A., Borg, J., Christiansen, C. (1989) Bone mineral content in ancient Greenlandic Eskimos. In Hansen, J. P. H., Gulløv H. C. (Eds.) opera cit.: 147–150.

Grandjean, P. (1989) Bone analysis: Silent testimony of lead exposures in the past. In Hansen J. P. H., Gulløv H. C. (Eds.) opera cit.: 156–160.

Gudmand-Høyer, E., McNair, A., Jarnum, S. (1973) Laktosemalabsorption i Vestgrønland (Summary in English). Ugeskr læger 135: 169- 172.

Hansen, H. E., Gürtler, H. (1983) HLA types of mummified eskimo bodies from the 15th century. Amer j phys anthropol 61: 447–453.

Hansen, J. C., Toribara, T. Y., Muhs, A. G. (1989) Trace metals in human and animal hair from the 15th century graves in Qilakitsoq compared with recent samples. In Hansen J. P. H., Gulløv H. C. (Eds.) opera cit.: 161–167.

Hansen, J. P. H., Meldgaard, J., Nordqvist, J. (Eds.) (1985) Qilakitsoq – De grønlandske mumier fra 1400-tallet. Christian Ejlers forlag & Grønlands landsmuseum, Copenhagen & Nuuk 1985.

Hansen, J. P. H., Meldgaard, J., Nordqvist, J. (Eds.) (1991) The Greenland Mummies. British Museum Press, Smithsonian Institution Press & McGill-Queens University Press.

Hansen, J. P. H., Gulløv, H. C. (Eds) (1989) The mummies from Qilakitsoq – Eskimos in the 15th century. Meddr Grønland; Man & Soc 12: 1–199.

Hansen, J. P. H. (1989) The mummies from Qilakitsoq – Paleopathological aspects. In Hansen J. P. H., Gulløv H. C. (Eds.) opera cit.: 69–82.

Hansen, P. K., Bennike, T. (1989). Botulismus in Greenland Eskimos. In Harvald, B., Hansen, J. P. H. (Eds.). Circumpolar Health '81, Oulu: 438–441.

Harvald, B. (1982) Eskimo disease pattern – genes or environment? – Acta med scand 212: 97–8.

Hooton, E. A. (1930) Finns, Lapps, Eskimos, and Martin Luther. Harvard Alumni Bulletin 1930: 545–553.

Jeppesen, B. J., Harvald, B. (1983) Serum calcium in Greenland Eskimos. Acta med scand 214: 99–101.

Johansson, A. (1989) Final preservation of mummies by gamma irradiation. In Hansen J. P. H., Gulløv H. C. (Eds.) opera cit.: 134–136.

Jørgensen, J. B. (1989) Anthropology of the Qilakitsoq Eskimos. In Hansen J. P. H., Gulløv H. C. (Eds.) opera cit.: 56–57.

Kissmeyer-Nielsen, F., Andersen, H., Hauge, M. et al. (1971) HLA types in Danish Eskimos from Greenland. Tissue antigens 1: 74–80.

Kleivan, I. (1984) West Greenland before 1950. In Sturtevant, W. C. (Ed.). Handbook of North American Indians. Vol. V. Arctic, Damas, D.(Ed.), Smithsonian Institution, Washington: 612.

Kobayashi, T., Ammitzbøll, T., Asboe-Hansen, G. (1989) Electron microscopy of the skin of a Greelandic mummy. In Hansen J. P. H., Gulløv, H. C. (Eds.) opera cit.: 100–105.

Kromann, N. P., Kapel, H., Løytved, E. R., Hansen, J. P. H. (1989) The tatooings of the Qilakitsoq mummies. In Hansen J. P. H., Gulløv H. C. (Eds.) opera cit.: 168–171.

Kromann, N. P., Mikkelsen, F., Løytved, E. R., Hansen, J. P. H. (1989) Dermatological examination of the Qilakitsoq Eskimo mummies. In Hansen J. P. H., Gulløv H. C. (Eds.) opera cit.: 83–88.

Lorentzen, B., Rørdam, A. M. (1989) Investigation of faeces from a mummified Eskimo woman. In Hansen J. P. H., Gulløv H. C. (Eds.) opera cit.: 139–143.

Lyneborg, L. (1984) Personal communication.

Masters, P. M. & Zimmerman, M. R. (1978) Age determination of an Alaskan mummy: Morphological and biochemical correlation. Science 201: 811–812.

Mathiassen, T. (1936) The Eskimo archaeology of Julianehaab district. V. The mummy caves at Qerrortut. Meddr Grønland 118: 103–113.

Mazess, R. B. (1974) Bone mineral content of North Alaskan Eskimos. Am j clin nutr 27: 916–925.

Mazess, R. B., Mather, W. (1975) Bone mineral content in Canadian Eskimos. Hum biol 47: 45–63.

Meldgaard, J. (1953) Fra en grønlandsk mumiehule (From a Greenlandic mummy cave – in Danish). Fra Nationalmuseets arbejdsmark 1953 14–20.

Myhre, J., Svendstrup, L., Hansen, J. P. H. (1989) Histologic examination of the Qilakitsoq mummies. In Hansen J. P. H., Gulløv H. C. (Eds.) opera cit.: 106–108

Mulvad, G., Pedersen H. S., Jul E., Newmann W, Middaugh, J., Misfeldt, J. (1993) Atherosclerosis in Greenland Natives. An autopsy study. IX International Congress on Circumpolar Health, Reykjavik, Iceland, abstract D1–9.

Møhl, J. (1983) Personal communication.

Møller, G. (1989) Eskimo clothing from Qilakitsoq. In Hansen J. P. H., Gulløv H. C. (Eds.) opera cit.: 23–46.

Newell, R. R. (1984) The archaeological, human biological, and comparative contexts of a catastrophically-terminated Kataliguaq house at Utqiagvik, Alaska (BAR-2). Arctic anthropology 21: 5–51.

Nielsen, N. H. (1986) Cancer incidence in Greenland. Arct med res 43: 1–168.

Nielsen, H., Engberg, J., Thuesen, I. (1994) DNA in skin and bone samples from Arctic human burials. In B. Hermann (Ed.): Ancient DNA. Springer-Verlag.

Nielsen, N. H., Mikkelsen, F., Hansen, J. P. H. (1977) Nasopharyngeal cancer in Greenland. The incidence in an Arctic Eskimo population. Acta path microbiol scand sect A 85: 850–858.

Pallesen, G. (1987) Personal communication.

Pedersen, P. O., Jakobsen, J. (1989) Teeth and jaws of the Qilakitsoq mummies. In Hansen J. P. H., Gulløv H. C. (Eds.) opera cit.: 112–130.

Rosing, J. (1986) The sky is low. Penumbra Press, Ontario.

Sebbesen, O., Thomsen, V. F. (1984) Personal communication.

Shand, G., Høiby, N. (1987) Personal communication.

Svejgård, E., Stenderup, A., Møller, G. (1989) Isolation and eradication of fungi contaminating the mummified corpses from Qilakitsoq. In Hansen J. P. H., Gulløv H. C. (Eds.) opera cit.: 131–133.

Thuesen, I., Engberg, J. (1990) Recovery and analysis of human genetic material from mummified tissue and bone. J Arch Sci 17: 679–689.

Thompson, D. D., Cowen K. S., Laughlin, S. B. (1989) Estimation of age at death and histomorphometric analysis of cortical and trabecular bone from four Greenlandic mummies. In Hansen J. P. H., Gulløv H. C. (Eds.) opera cit.: 151–155.

Thompson, D. D., Harper, A. B., Laughlin, W. S., Jørgensen, J. B. (1982) Bone loss in Eskimos. In Harvald, B., Hansen, J. P. H. (Eds.). Circumpolar Health '81, Oulu: 327–330.

von Westarp, C., Outhet, D., Eaton, R. D. P. (1982) Prevalence of vitamin D deficiency in two arctic communities. In Harvald, B., Hansen, J. P. H. (Eds.), Circumpolar Health '81, Oulu: 331–333.

Zimmerman, M. R., Aufderheide, A. C. (1984) The frozen family of Utqiagvik: The autopsy findings. Arctic anthropology 21: 53–64

Zimmerman, M. R., Smith, G. S. (1975) A probable case of accidental inhumation of 1600 years ago. Bull NY Acad Med 51: 828–837.

Zimmerman, M. R., Trinkaus, E., LeMay M. et al. (1981) The paleopathology of an Aleutian mummy. Arch path lab med 105: 638–641.

Zimmerman, M. R., Yearman, G. W., Sprinz, H. et al. (1971) Examination of an Aleutian mummy. Bull NY Acad Med 47: 80–103.

Correspondence: Dr. J. P. Hart Hansen, Department of Pathology, Gentofte Hospital, University of Copenhagen. Niels Andersen Vej 65, DK-2900 Hellerup, Denmark.

South American mummies

Early mummies from coastal Peru and Chile

M. J. Allison

Department of Pathology, Medical College of Virginia, Richmond, Virginia, U.S.A.

Introduction

The countries of Peru, Bolivia, Ecuador and Chile in South America, were the modern countries that occupied the major geographic areas in the Inca Empire. They incorporate all of the extremes of the world's climatic conditions from the steaming jungles of the Amazon basin to the frigid arctic climates of the high Andean ranges, many of which have permanent snow and an average daily temperature of 0 °C. The western limits border the Pacific Ocean with a meeting of the sixty mile wide Humboldt current hugging the coast from Antarctica to Punto Pariñas where it mixes with the warm Equatorial current to form one of the richest ocean fauna in the world. Unfortunately the coastal shore is cut by a series of short rivers running east to west that in many cases only have water a few months out of the year. This results in a narrow strip of coastal desert extending for about 3,000 km along the coast from Tumbes in Peru to Copiapo in Chile, that varies in width from a few kilometers to perhaps 200 at the widest part. The best known geographic area is the Atacama Desert, one of the driest spots on earth. This coastal strip is backed by two to three chains of very high mountains in which peaks reach a height well over 6,000 m, the highest Aconcagua is 6,960 m. Liberally sprinkled among these are a series of active volcanos that produced environmental contamination of much of the water supply in coastal valleys with arsenic carried down by melted snow. Between the chains of the Andes are relatively large valleys many running north/south which are suitable for agriculture, particularly tubers such as potatoes which were early domesticated here.

On the coast there is a shortage of arable land, the few green valleys cutting through the desert were initially inhabited by people attracted by the maritime resources. They eventually domesticated cotton and gourds for use in fishing; later they developed food agriculture to feed relatively large populations through the use of irrigation extending into desert areas with large hydraulic projects, but many small communities continued to live off of the rich maritime resources with minimal sources of fresh water or even managed to devise means of trapping water from dense fogs that often covered the land. Many areas of the desert had a very high water table and methods of agriculture were devised using garden holes sunk several feet into the ground that could feed limited populations with agricultural products often brought to the coast from other areas such as the Amazon basin through an extensive trade network.

Methods of storing and preserving food were developed including freeze drying of the potato. The native indigenous people taking advantage of the varied native flora soon established a rich system of folk medicine with several thousand medicinal plants many of which have been found useful in modern therapy.

While the fauna was not as rich as other parts of the world, the horse, initially a small creature, was an original inhabitant of the Americas that became extinct about 14,000 years ago as did the giant ground sloth. The Andean area had four species of cameloids two of which were eventually domesticated along with the guinea pig. Deer and bears were present as well as the rhea a South American ostrich. There were numerous sea birds along the Pacific coast that served as a source of food along with their eggs, and their skins were used to make clothing. The bird feces (huano) was early a source of fertilizer for agriculture along with inedible fish and nitrates hacked out of desert deposits. The principle red meat of the coastal inhabitants was the sea lion whose bones were made into tools, and whose hide was early used as a type of tent shelter supported by whale bone ribs. The hide was later made air tight and inflated to use as a type of pontoon on a primitive boat. An extensive sea trade had been established by pre-Inca times and large sailing rafts as big as many European boats plied a brisk trade all along the Pacific as far north as present day Mexico.

Our interests began in 1969 with the establishment of a team of medical scientists who working with Peruvian archaeologists began to excavate 30–50 graves from one cemetery each year for the purpose of studying the history of man's health conditions as a product of their society and environment. Thus over the years we covered a series of different coastal valleys extending from Casma in Northern Peru to Tarapaca in Northern Chile, and while we excavated on the coast desert, the populations were a mixture of coastal natives probably of jungle origin, and highland invaders. The periods in time that we covered were from approximately 10,000 years before present to modern republican times. The majority of these people were autopsied or their skeletons were studied and each was more or less well defined as to his position in his society. Modern laboratory methods were applied in attempts to elucidate their genetic make-up as well as the effects of their diet and social structure on their health. The limits on time will only enable us to give a select sample of our work on over 3,000 individuals studied over the past 24 years.

The earliest material studied in Peru was from a site excavated by Dr. Rosa Fung at Bandurria Beach in Huacho. These individuals were salvaged from a cemetery opened up by the rupture of an irrigation canal. They belonged to a fishing community that also took advantage of gathering from the lomas (fog meadows). Among the finds was a hand of wild tobacco probably one of the earliest known to be used by man. The mummies were wrapped in a layer of reed matting tied with rope, and were held in the grave by stakes driven through the body and often weighed down with heavy stones. On opening the bundles the individuals were lying on their side with their hands under their head and legs flexed as if sleeping, wrapped in a cotton twinned blanket, and accompanied by artifacts such as small gourds, often inlaid with decorations. The men wore a string loin cloth and often had several feathers in their hair; the women were generally naked as they were during life. A gill net was recovered in one bundle that had the stone weights attached at the bottom and the gourd floats still in place at the top. The skeleton was the most useful part for study although some skin and muscle was present and was used to determine blood groups and HL-A genetic markers. Harrises lines in the tibia were examined for childhood health and a number of diseases affecting bones were found including what appeared to be treponematosis.

A second cemetery had been opened in this area by the water, and this had individuals buried seated in large baskets, a common form of burial from around 2,500–3,000 years B.P. Such burials were seen in the Department of Ica, Peru as well as the so called Alto Ramirez culture of the Azapa valley in Chile. Some went farther south in Chile a variant of this was seen in which the children were buried in a covered basket and the adults had a second large basket that covered their upper body. Most of these burials were found with the mummy having a string turban on his head and the presence of loomed cloth on the body. Crude ceramics were found in Chile by 2,800 B.P. and metal became more common in graves. Head deformation which had been seen earlier was by now a well established custom among most cultural groups and by the year 1,000 before present a deformer was found among the Mitas Chiribaya that was used to flatten the face by pressure on the cheek bones of infants. The use of hallucinogens was confirmed by the presence of elaborate kits found buried with the body that contained seeds of the hallucinogenic snuff known as wirca (huillca) from the Piptadenia colubrina. There is a broad usage of hallucinogens among the native peoples in the Americas that was first reported by Fray Ramón Pané on Hispaniola in 1498 and in the Andean area this goes back nearly 10,000 years. The Mochica and Chimu cultures of northern Peru used the cactus San Pedro, a source of mescaline, while in the jungle the vine ayahuasca a Banisteriopsis was the source of hallucinogens. All of these are incorporated into modern regional folk medicine.

Some of the larger valleys became hubs for villages, and later cities with an urban population and a cult system that eventually developed a priest caste and later the formation of empires. Special buildings housed the Gods and a special hierarchy was developed based on a theocratic state. In Northern coastal Peru architecture was often associated with artificial pyramidal structures, but in the highlands these sites were often high mountain peaks at times with small offertories where several sacrificial pre-Columbian frozen bodies have been found. In the Arica area around 1,000 B.P. a small temple was found that belonged to the Maitas-Chiribaya culture that was associated with a number of priests who wore special ear rings made of rolled animal skin containing sacred items. They were probably serving the moon goddess and the temple functioned as a fertility site. These priests differed genetically from the general population. More elaborate and elegant architectural examples are to be found in Cuzco and Chavin or the city of Chan Chan in Trujillo in Northern Peru.

Tattooing

Tattooing, the art of making a permanent record on the skin of a living animal or human is an ancient practice. The oldest tattoo that we have found was from the Chinchorros culture and was a thin pencil mustache tattooed on the upper lip of a male adult probably about 6,000 B.P. Generally the area of the mountains and coast south of present day Lima had only occasional forms of this art before 1,000 B.P. but it found a new expression with the migration of large numbers of mountain people moving to the coast about this time. In the Department of Ica there was geometric tattooing of the face, the legs, wrists and arms in both the Huari and Ica cultures. In the Arica area the San Miguel culture had a limited tattooing of certain designs that were also found on cloth.

The area that had the greatest numbers tattooed, and most elegant form of the art was among the coastal fishermen around Casma, Peru (1,000 B.P.) where as many as thirty percent of the population had tattooing in the form of geometric figures, fish, birds, and what appear to be landscapes. The method used, at least in some cases, was shown to be done by sewing the design into the skin with a neeedle and thread dipped in carbon. The designs were on the back, chest, arms and legs as well as the forehead and face.

Non-invasive techniques

The radiograph has many uses in paleopathology. Where there is an outer wrapping to the mummy bundle, a radiograph may serve to decide whether to open the bundle for further study or simply use it for museum display. Harrises lines, best seen on the tibia, are formed during the period of skeletal growth and are due to a variety of health problems that interfere with the normal growth pattern; thus in essence they are a record of childhood health at specific age periods. While they show some remodeling I have always found them a useful record of potential health problems that should be anticipated during the autopsy of children and young adultes, and if they are divided into periods of, birth to 3 years, 4 to 8 years, etc. up until around eighteen they are useful for comparative health studies of different cultural groups. Similar health problems that produced lines across the marrow cavity will also produce lines on the teeth during their period of development. A second area of useful radiographs determines existing pathology that affected the bones, these may be due to infections, tumors, metabolic problems, etc. Again the radiograph serves as a guide for an invasive autopsy. The measurement of bone mass of specific bones such as the tibia, femur and select lumbar vertebrae may also be done radiographically. This is accomplished by using a standard calibrated aluminum wedge when taking the X-ray, and then measuring the density of the trabecular bone using a small inexpensive handheld densitometer with the wedge serving as a technique control standard. The cortical bone width can be measured using a small millimeter rule. Individual bones can be weighed in grams and the length and diameter of the bone measured and several indices may be calculated to measure osteoporosis.

Pre- and post-partum skeletal changes

Working with skeletal material most of us have become familiar with the so called scars of parturition seen at the symphysis pubis and/or the sacral iliac joints. When the skeleton is disjointed, these are simply a record of a past event without really knowing when this occurred during the individuals life. The skeletal material that we worked with in the Andean region is often held together with some bits of skin and even though the organs are long gone the position of the pelvis is such that evidence of a recent delivery can be noted. When one works with mummies that have organs, about forty percent of our material, it is possible to see pregnant females and those recently delivered. We have also found females, probably post-partum, with their sanitary napkins in place and their breasts swollen with what was once milk, others have the cord and placenta still in place having died during delivery, while others died of some intercurrent disease during pregnancy and the fetus is still in utero. This has given us a baseline that we might possibly use for the evaluation of pregnancies and deliveries among individuals with only skeletal remains. Using chemical techniques of atomic absorption analysis we evaluated a wide series of different elements from mummy tissues, initially for diet, but then we noted that certain females had a reduced cortical/trabecular bone iron ratio and these were females that had died with a fetus in the uterus or had evidence for recent delivery. While in all probability there will be other reasons appearing from time to time for these abnormal iron changes but at present in more than 40 individuals most of these changes have been noted in problems related with child bearing and pregnancy. The ratio range for adult males was 3.1–6.2, for adult normal women with no signs of recent childbirth 2.7–10.6, and for women with recent signs of pregnancy or delivery 1.0–2.5.

Table 1. Iron rib/femur rations

Rib/Femur Ratio Male Adult	Rib/Femur Ratio Normal Female	Rib/Femur Ratio Pregnant Females
3.1	2.7	1.0
3.3	3.1	1.0
3.3	3.4	1.0
3.8	4.1	1.4
4.0	4.3	1.4
4.1	4.5	1.5
4.5	5.3	1.6
4.6	5.8	1.9
5.4	10.6	2.5
6.2	–	–
Average 4.23	Average 4.64	Average 1.47

Table 2. Iron values in normal males and females and pregnant or post-partum females

Group	Rib Fe	Femur Fe	Ratio
Normal Males	304	72	4.22
Normal Females	335	75	4.46
Pregnant Females	144	92	1.56

Chronic arsenic poisoning

Many areas of volcanic activity are plagued by the presence of large quantities of arsenic in the environment which with the water run-off results in contamination of the rivers and streams used for drinking, irrigation and farming. As a result there is a serial accumulation of arsenic in living plants and animals that results in eventually chronic arsenic poisoning in the humans who eat these or drink water from contaminated water supplies. Such is the case in rural inhabitants of certain valleys in Peru, Chile and Argentina today, and studies of pre-Columbian peoples who also lived in these valleys shows essentially the same picture thousands of years before present.

This disease leaves certain stigmata on the skin after ingestion of arsenic over a relatively long period, for example two years. There are also a large number of symptoms and eventually lesions related with the gastrointestinal tract, heart and other organs that are difficult to visualize in a mummy, but after an extended period of ingestion, ten years or more, there is the beginning of different types of neoplasms of the skin and organs. Since all geographic areas are not affected it is also possible to use this as a marker of recent invaders from other valleys into areas with arsenic contamination of the environment. Such was the case in the valley of Camerones around 3,000 years B.P. this period was represented by a period of increased violence with skull and parry fractures and presence of projectile point wounds. A group of people were excavated that showed high levels of arsenic, but no skin lesions of the disease. This led us to believe that there had been an invasion by people from an area where arsenic was not present in their environment; when they moved into Camerones they had conflict with earlier inhabitants, but they had not been there long enough to develop the stigmata of arsenic poisoning (perhaps 18 months) although they had been there long enough to develop high tissue levels (perhaps 6–8 months).

Summary

The Andean area is one of the richest sources in the world for the study of early man. Artificial mummification began nearly 10,000 years ago and this was followed by millennia of natural mummification due to the desert-like environment. It is possible to trace man's development in this area from a hunter, fisherman, gatherer, within small kinship groups to that of a farmer with domesticated plants, and animals such as the guinea pig, and cameloids living with the formation of elaborate cults and theocratic governments and eventually of large empires. Belief in the afterlife led to burials which gave a great deal of information on the social and political structure of the societies and autopsies of the mummies that often contained all of their organs enable us to relate the society and environment with the health of the different individuals. Thus we determined that the great majority of diseases of man were universal, but at the same time it was seen that there were certain geographic diseases that in some cases chronic arsenic poisoning for example were associated with the environment or a particular vector such as that of Bartonellosis, seen only in the Andean area. More recent studies have had an emphasis on chemical analysis and these have been useful in reconstruction of diet as well as locating women who have recently delivered children.

Zusammenfassung

Die Andenregion ist eines der ergiebigsten Gebiete der Welt für die Erforschung des frühzeitlichen Menschen. Künstliche Mumifizierung gab es hier schon vor rund 10.000 Jahren, und in den folgenden Jahrtausenden finden wir natürliche Mumien als Ergebnis der wüstenähnlichen Umweltbedingungen. Hier läßt sich die ganze Entwicklung nachvollziehen – vom Jäger, Fischer und Sammler zum Bauern, der Nutzpflanzen anbaute und Haustiere wie Meerschweinchen oder Lamas hielt; von kleinen Familienverbänden zu Theokratien mit hochentwickelten Kulten und schließlich zu großen Reichen. Der Glaube an ein Leben nach dem Tod führte zu Bestattungen, die Aufschluß über die soziale und politische Struktur der Gesellschaft geben. Autopsien der Mumien, die oft noch alle Organe enthalten, erlauben Rückschlüsse auf den Gesundheitszustand der damals lebenden Menschen sowie auf ihre Gesellschaft und Umweltbedingungen. Neben bestimmten – regional spezifischen – Erkrankungen, wie der *Bartonellose,* deren Überträger nur in der Andenregion vorkommt, konnten auch umweltbedingte Krankheiten nachgewiesen werden, wie z.B. die chronische Arsenvergiftung. Es konnten freilich auch Belege für jene Erkrankungen gefunden werden, die weltweite Verbreitung aufweisen. Neuere Untersuchungen haben sich vor allem auf chemische Analysen konzentriert, die Aussagen über die Ernährung zulassen und die die Identifikation von Frauen ermöglicht, die kurz vor ihrem Tod Kinder geboren haben.

Résumé

Les Andes constituent l'une des régions les plus instructives quant aux recherches faites sur l'homme préhistorique. La momification artificielle avait été connue dans cette région il y a déjà 10.000 ans, et on y trouve des momies naturelles datant des millénaires suivants, et formées en raison des conditions environnementales semblables à celles existant dans le désert. Ces restes nous permettent de reconstruire l'évolution entière de l'homme préhistorique – le passage de l'homme chasseur, pêcheur et cueilleur à l'homme producteur qui cultivait des plantes et élevait des animaux domestiques, tels que des cochons d'Inde ou des chameaux; le passage aussi de petites communautés de familles aux théocraties caractérisées par des cultes hautement développés, et ensuite finalement aux grands royaumes. La foi en l'immortalité de l'âme se traduisait dans des funérailles qui permettent de tirer des conclusions sur la structure sociale et politique de ces sociétés. Sur la base des autopsies de momies, dont les organes sont souvent entièrement conservés, il est possible de déduire l'état de santé des hommes de cette époque ainsi que la structure de leur société et leur environnement. Ainsi a-t-on pu constater que la plupart des maladies étaient répandues dans le monde entier, mais qu'il existait toutefois certaines maladies, dont la diffusion était limitée à des régions précises – soit parce qu'elles étaient conditionnées par l'environnement, comme l'arsénicisme chronique, soit parce qu'elles étaient transmises par des agents pathogènes, qui existaient uniquement dans les Andes, ce qui était le cas pour la bartonellose. De récents examens se sont concentrés sur les analyses chimiques permettant et de déduire les habitudes alimentaires de l'homme préhistorique et d'identifier les femmes qui avaient mis au monde un enfant peu avant de mourir.

Riassunto

Le Ande costituiscono una delle regioni piú ricche di testimonianze quanto alle ricerche fatte sull'uomo preistorico. In questa zona la mummificazione artificiale ha avuto inizio ben 10.000 anni fa, e si sono trovate delle mummie naturali che risalgono ai millenni successivi in seguito a condizioni ambientali normalmente proprie del deserto. Questi resti ci permettono di ricostruire la completa evoluzione dell'uomo preistorico: dall'uomo cacciatore, pescatore e raccoglitore all'uomo agricoltore, che coltiva piante e alleva maialini d'India, cammelli o altri animali domestici; dalle piccole comunitá familiari alle teocrazie con culture altamente sviluppate e ed infine ai grandi imperi. La fede nell'immortalitá dell'anima si manifestava nei funerali, che permettono di trarre delle conclusioni sulla struttura sociale e politica di queste societá. Sulla base delle autopsie delle mummie, i cui organi sono spesso ben conservati, è possibile dedurre lo stato di salute dell'uomo di quell'epoca cosí come la struttura della sua societá e del suo ambiente. Inoltre si è potuto constatare che la maggior parte delle malattie erano diffuse in tutto il mondo, ma che tuttavia ne esistevano alcune, la cui diffusione si limitava a particolari regioni, sia per ragioni ambientali, come nel caso dell'arcenismo cronico, che per la presenza di vettori patogeni che esistevano unicamente nella regione andina, come nel caso della bartonellosi. Le ultime ricerche si sono concentrate su analisi chimiche che permettono die stabilire le abitudini alimentari dell'uomo preistorico e anche di identificare le donne che avevano dato alla luce un figlio poco prima di morire.

Correspondence: Dr. M. J. Allison, Department of Pathology, Medical College of Virginia, Virginia Commonwealth University, P.O. Box 662, MCV Station, Richmond, Virginia 23298-0662, U.S.A.

Preparation of the dead in coastal Andean preceramic populations

B. Arriaza

Department of Anthropology, University of Nevada, Las Vegas, Nevada, U.S.A.

Introduction

This paper addresses the earliest known artificial mummification in the world which was practiced by the Chinchorro Culture of South America 8,000 years ago (Uhle, 1919; Alvarez, 1969; Nuñez, 1969; Allison et al., 1984; Arriaza, 1993). Mortuary techniques used by the preceramic Chinchorro Culture will be reviewed and it will be argued that less complex socio-political organizations cannot necessarily be equated with simple mortuary practices. Two fundamental questions will be discussed: 1) why did the Chinchorro develop their laborious mortuary practices? and 2) what is the anthropological relevance of studying preparations of the dead? These are basic questions which do not have simple answers, but are essential for understanding mummification customs and the utility of preceramic mortuary studies in anthropology.

Several basic terms need to be clarified before focusing on early Andean artificial mummification practices. Mummification is used here to mean preservation of the soft and hard tissues of a corpse from an archaeological context. In the Atacama desert of Peru and Chile, natural mummification can be unintentional or intentional and is aided by the extreme environmental conditions of this region, which dehydrate the corpse minimizing decomposition. Unintentional natural mummification is accidental and quite common in the Andes, as the many Peruvian and Chilean mummies found in coastal valleys testify. On the other hand, intentional natural mummification is planned and implies a deliberate action to preserve the body, such as placing the corpse in an area known for its capacity to quickly dehydrate and naturally mummified a cadaver.

Artificial mummification, or embalming as it is sometimes called, does not leave preservation to chance. It requires an additional social dimension since skillful morticians with anatomical knowledge are needed to open the body to extract the internal organs, to refill body cavities, and to restore the body if necessary. This removal of internal organs and subsequent stuffing is done to decrease the likelihood of decomposition and to keep the form intact as much as possible. In brief, in artificial mummification internal and external treatment of the body commonly takes place.

The study of artificial mummification in prehistoric populations is important for many anthropological reasons, and three such reasons will be explored here. First, the mummification process requires high energy expenditure. Second, its undertaking requires specialization and a social infrastructure that facilitates the production of the desired mummy. And third, it denotes that the group has strong spiritual beliefs in the hereafter, along with temporal continuity of mortuary rituals.

Methodology

The Chinchorro literature was reviewed to understand the spread of Chinchorro artificial mummification practices, the radiocarbon dates, and the types of mummies found at different cemeteries. The approach of this essay also considers Chinchorro mummification techniques through time, grave goods, environmental factors, and synchronic and asynchronic comparisons with other Andean mortuary styles.

Chinchorro burial practices

The Chinchorro Culture (7,000 to 1,500 B.C.) extended along the arid coast of southern Peru and northern Chile, from Ilo to Antofagasta, covering a distance of about 900 km (Nuñez, 1969; Bittmann, 1982; Schiappacasse and Niemeyer, 1984, Rivera, 1991; Wise, 1991; Arriaza, 1993). The Chinchorro people were fishermen who also practiced extremely complex techniques for mummifying and caring for their dead between approximately 6,000 to 2,000 B.C. The Chinchorros lacked cotton, ceramic, metal, and woven cloth.

The best evidence of Chinchorro mummies comes from the area of Arica, in northern Chile. These mummies were originally described by Uhle (1917, 1919) in his work "The Arica Aborigines." He classified the mummies into three categories: Simple, Complex, and Mud-

Coated. All of the mummies were buried in an extended position. The simple type corresponds to natural mummification, the complex to rearticulated stuffed bodies, and the mud-coated style to bodies embedded in a cement-like substance. I consider Uhle's Complex and Mud-Coated mummies as two different systems of artificial mummification (Figure 1).

Today, a total of 208 Chinchorro mummies can be accounted for from the literature (Arriaza, 1993). Of these, seventy-one percent (147/208) are artificially mummified (Complex and Mud-Coated) and come from eleven Chilean cemeteries. Most of these sites with artificial mummification cluster in the Arica area, which appears to have been the cultural epicenter for Chinchorro development (Figure 2). The minimum number of artificially prepared mummies found at a given site is one while the maximum is sixty. The total of 208 Chinchorro mummies is low considering the vast amount of time and cultural development noted in the Chinchorro archaeological record. This low figure is likely a result of a lack of systematic analysis of the Chinchorro Culture.

Through their millennial cultural existence the Chinchorro's developed several sophisticated mummification techniques, as reflected in a diversity of asynchronic mummy styles. To make these variations easily comprehensible, I have modified Uhle's (1919) and Allison and co-workers' (1984) classifications as follows: Black, Bandage, Red, Mud-Coated, and Natural mummies (Figure 1). The Black and Red mummies are subdivisions of Uhle's complex types, and the Bandage mummies are basically a variation of the Red ones. A brief outline of the Black, Bandage, Red, Mud-Coated, and Natural mummy types is as follows. The reader is encouraged to consult the literature for further details since a few archaeological inconsistencies, such as the existence of two red stages, are not addressed here (Arriaza, 1993).

Of all the Chinchorro mummies, the Black are the most complex (Figure 3). They are extraordinarily artistic and can be described as a materialization of the spiritual manifestations of the Chinchorro people. This black mummy practice lasted for 2,180 years – from about 4,980 B.C. to 2,800 B.C. and entailed the complete disarticulation of the body, the removal of organs and the defleshing of the bones. The skeleton was then reassembled (Figure 4). A framework made of sticks, twigs, and wrapped cords (securing the bones and sticks) was used to reinforce the skeleton. Longitudinal sticks about the length of the body, one for each leg, coming from the ankle and ascending through the chest, were inserted into the base of the skull. Another stick reinforced the spine. The three sticks joined at the neck and served to reattach the skull to the body. The brain was also replaced with a stuffing of grasses and ashes and the trunk cavity was filled with ashes as well. The whole body was then modelled with a white ash-paste, restoring some of the lost volume.

Sexual characteristics were also modeled. Often human skin or animal skin was placed on top of the white paste including the face. In some bodies the morticians did such a remarkable "embalming" job that the skin of the face and arms appears to be the original, rather than animal skin for example. Sometimes the proportions of the body were maintained, but other times these mummies possess a rather rectangular trunk. A wig made of short human hair was often added to the head. Other times it appears as if the occipital area was not scalped. Later, the completely remodeled body was painted black with a manganese paste (hence its name), with the exception of the hair, which remained its natural color.

The sophistication and retouching of the effigy, or statue-like body, implies the mummies were kept for worship and after an unknown amount of time buried, and perhaps subsequent exhumation took place for social display. Often the mummies were nude, as probably people were in daily life. A few mummies, however, wore a fringe skirt or a waist cord belt. The bodies were wrapped in mat shrouds and buried in groups of about six individuals. Adults of both sexes and children received this Black treatment. Of the scarce grave goods added to the grave, the atlatl was most common.

The Red mummies, although less complex than the Black, were also elaborate (Figures 4 and 5). This late Red treatment lasted about 480 years from 2,570 B.C. to 2,090 B.C. The morticians first decapitated the body, then proceeded to make incisions in the abdomen, groin, knees, and ankles to facilitate the removal of major organs and muscles, which were not replaced. In some cases, the skin appears intact, as if it was rolled back on the arms and legs to allow for muscle removal and then unrolled during the filling process. The head was cleaned of all soft tissue including the brain, which was extracted through the foramen magnum. As in the Black mummies, long sticks were used to reinforce the body. In the Red mummies, however, the sticks were sharpened and slid underneath the skin of the legs, vertebral column, and arms, rather than being wrapped with vegetal cords along with the bone, as was the practice in Black mummies. After the body cavities were dried using hot coals, as evidenced by the burned bone and tissue, they were filled with various elements such as ashes, soils, grasses, camelid wool, and feathers. The skull was also filled with materials similar to those used for the trunk. The face was modelled with black, or reddish clay. It appears that the facial skin of the Red mummy was not replaced as often as in the Black mummies. A wig made of tufts of long unbraided black human hair, up to 60 cm, was attached to the head. This wig was secured to the occipital using a black manganese paste that followed the contour of the cranium. The head, with long hair, appears as if it has on a round black helmet. The black paste was also added to the skull to model or insinuate the facial fea-

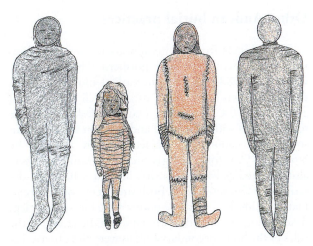

Fig. 1 (top). Artificially prepared Chinchorro mummies. From left to right: Black, Bandage, Red, and Mud-Coated mummies

Fig. 2 (right). Illustration of Chinchorro sites along the Pacific coast of South America. Darkened arrows indicate the spread of Chinchorro artificial mummification characteristics. The dashed arrow indicates the probable spread

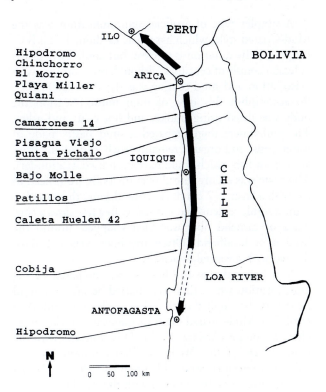

Fig. 3. Close-up of the trunk of a Black Chinchorro mummy. The postmortem damage of the shoulder shows the inner structure of reeds and ashes. Adult female

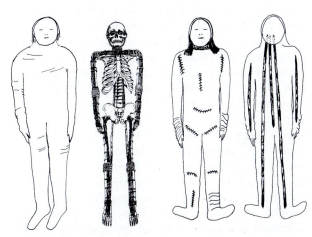

Fig. 4. External and internal representations of a Black mummy (the two figures on the left side) and of a Red mummy (the two figures on the right side)

tures. After the head was reattached, the body was sometimes externally secured with camelid fiber cords at the neck, waist, and ankles. A loincloth also served as reinforcement. The body was then completely painted red with ocher, except the face and the hair, which were left black. The black paste (helmet-like) holding the wig in place was also carefully painted red. The result was a very elaborate red mummy with long black hair and a black face. In at least one case a grass skirt was placed on a female child after the painting took place. Adults of both sexes and children received this type of red treatment and continued to be buried in groups of about six people. Grave goods were also few. They included fishing lines, sinkers, grass skirts, and fire sticks.

The treatment of the Bandage mummy is similar to that of Red mummies described above, except the skin was removed and replaced in strips (Figure 6). Bandages thought to be of sea lion and pelican skin were used also to wrap the bodies. This treatment was noted in children, however, a variation of this technique was seen in one adult Red mummy, which was found with its legs treated in a bandage style.

A simpler type of artificial mummification was the Mud-Coated style, which has a date of about 1,720 B.C. (Figure 7). Its duration is unclear, but could have lasted at least a couple of centuries, although one mummy with evisceration and mud-coating was dated to 2,600 B.C. To accomplish this type of mud mummification the body was dried using smoke, and hot coals and ashes. The embalmers then prepared a mud paste made of water, soils, and organic remains (such as proteins from animals or fish which could have been used as a binder). The paste was evenly spread over the body from head to toe with a layer 1–2 cm thick. After drying, the paste became a solid, cement-like substance. In a few cases a cord made of camelid fiber was then wrapped around the head as a headband. Often the body was naked or dressed with a vegetal loincloth.

It is believed that the bodies were prepared in the same location where they were buried because the mud paste was also applied to the edges of the grave pit. As a result, the Mud-Coated mummies appear as if they were glued to the pit (Arriaza, 1993). This contrasts with the manufacture of the Black and Red mummies, which could have been transportable upon completion. Adults of both sexes have been found with this mud treatment. Often the Mud-Coated mummies were buried alone, but with a few more fishing-related artifacts than the Black or Red mummies (Standen, 1991).

The last type of Chinchorro mummies were naturally desiccated bodies, simply wrapped in a mat shroud (Figure 8). These bodies have been associated with two periods: 7,000 B.C. and about 1,500 B.C. It is speculated that the first period lasted approximately one thousand years, from 7,000 B.C. to 6,000 B.C. and the second 380 years from 1,880 B.C. to 1,500 B.C. (Arriaza, 1993). In both periods the bodies were extended, wrapped in a mat shroud and buried in the sand in a shallow grave pit. There is no evidence of exhumation for social display as could have been the case for some complex artificial mummies. Thus, this natural mummification is a consequence of the aridity of the environment. Although most anatomical features are still present on these mummies, they are primarily composed of skin and bones with few internal organs preserved. These mat bundles of adults of both sexes and children were frequently buried alone, but with more grave goods than the Black and Red mummies (Standen, 1991). They included fishing nets, harpoon heads, fishhooks made from cactus needles, lithic points, fishing lines, sinkers, and camelid hair fringe skirts (Figure 9).

In summary, Chinchorro artificial mummification (Complex and Mud-Coated) practices lasted 4,140 years (5,860 B.C. to 1,720 B.C.). While the complex artificial mummies, that is Black and Red types, persisted for 3,770 years (5,860 B.C. to 2,090 B.C.; Figure 10).

Other Andean burial practices

The Chinchorro Culture appears to have been limited to southern Peru and northern Chile. In Peru, Chauchat (1988) reported the earliest inhumations for the Andes from a site called Paijan which has a date of about 8,000 B.C. At Paijan a couple of skeletons were found buried with legs flexed and lying on their sides. In southern Ecuador, large cemeteries were reported by Stothert (1988) for the Vegas site. Large cemeteries were also reported at La Paloma site, in Peru, which was described by Quilter (1989). Both sites were from approximately 6,000 B.C. and the bodies were buried with legs flexed, lying on one side with hands on the face or at the sides of the body. Secondary burials were also observed at both sites, but there was no evidence of artificial mummification.

In northern Chile, in Arica, the earliest evidence of a burial comes from the Acha site which marks the beginning of what is known as the Chinchorro Culture. The burial comprised a desiccated body, buried in extended position and dated to 7,000 B.C. (Muñoz and Chacama, 1993:28; Aufderheide, Muñoz, and Arriaza, 1993). By about 6,000 B.C. in Arica artificial mummification had developed. Near Lima, Peru for the Chilca site, Donnan (1964) described a burial pattern from 3,420 B.C. somewhat similar to Chinchorro. Bodies were wrapped in reed mats in an extended position and buried in the hut floors, lying on their backs or toward one side. However, artificial mummification and formal cemeteries were not observed there either. In Arica, the artificially prepared Black mummies were in full development at this time, thus, Chilca mortuary practices likely corresponded to a different Andean mortuary tradition.

In the coastal site of Huaca Prieta, at the mouth of the Chicama valley (Peru), Bird (1985) dated human flexed skeletons to 3,000 B.C. Three of the bodies were covered with solidified ashes. A similar pattern was observed at La Paloma (Quilter, 1989: 76). At sites such as El Paraiso, about 2,500 B.C., complex socio-political organization developed allowing for the construction of public monumental architecture (Quilter, 1989; Moseley, 1992).

For later Peruvian mummies, Tello (1929: 131–135 in Cockburn and Cockburn, 1980: 140) stated that Paracas mummies (ca. A.D. 200) were artificially mummified, but subsequent studies have failed to confirm this claim (Allison and Pezzia, 1973 and 1974). The same claim also has been made for the bodies of the reigning Incas, but this has not yet been substantiated.

Returning to the coast of Arica, by 1,500 to 1,300 B.C., the Chinchorro custom of extended burials shifted to burying the dead without artificial mummification, in a flexed horizontal style (lying on one side). This period is known as the Quiani phase (Dauelsberg, 1974). In Quiani the dead were dressed with headbands or turbans

(Figure 11) and an increase in technological development is noted (e.g. dyed yarns and coiled decorated basketry). Later in Arica, agropastoral societies beginning circa 1,000 B.C. created burial mounds in the inland valleys (Santoro and Ulloa, 1985). By about A.D. 500 both maritime and agropastoral societies buried the dead flexed, but now in a vertical seated or fetal position. With the arrival of the Spanish Conquistadors, the converted Indians were given Christian burials and in Arica we have found colonial Indians and European bodies buried in extended positions with hands crossed on the chest embracing a wooden cross with grapes placed on the face.

Reviewing the present archaeological evidence indicates Chinchorro mortuary practices were a unique cultural phenomenon in southern Peru and northern Chile. Although contemporary and even earlier cemeteries have been reported for Ecuador and central Peru they do not possess evidence of artificial mummification.

Why did the Chinchorros artificially mummify their dead?

Care for the dead is a universal cultural phenomenon that as far as we know started with the Neanderthals about 100 thousand years ago (Leakey, 1981: 152). Artificial mummification is uncommon, however, and it can be seen as a negation of death. The group seeks a form of immortality whereby the preserved body is tangible proof that death and decomposition have been somehow defeated. The individual is still among the living and participates with the living in social ceremonies.

Avoidance of cadaver decomposition also appears to be related to beliefs in the afterlife, for instance, the body needs to be preserved, otherwise the soul will perish. Life in the hereafter maybe seen as homologous to worldly existence and burial practices then satisfy these needs. For example, well known are the cases of the Egyptian kings, in addition to having their body preserved, have servants and a vast array of everyday artifacts buried with them. A similar situation took place in the Andes, particularly during Inca times. Wives and servants were buried at the entrance of the palace where the mummy of the Inca King was housed. Sometimes these servants were buried alive, or drugged and strangled before burial (Salomon, 1991).

The Chinchorros, though, were a far less technologically complex society than the Incas or Egyptians. The Chinchorros did not have monumental architecture nor burials with large amounts of grave goods that would indicate differential social rank. In addition, their mummification techniques crosscut sex and age, even including fetuses. So why did the Chinchorro people develop artificial mummification? Few scholars have considered this question.

Mostny (1964, in Bittmann, 1982) believed that artificial mummification began as a consequence of community members dying far from the main camp and the body had to be prepared to bring it back for burial. Although her hypothesis is possible, she was working under the assumption that Chinchorros represented highly mobile hunter-gatherer groups. But today Chinchorro is seen as a sedentary fishing society.

Wise (1991) postulated that Chinchorros developed cemeteries and artificial mummification as a way to secure territory, thus, the Chinchorro people could claim land rights, decreasing competition and gaining unlimited access to natural resources. This is certainly possible since the Atacama coast has few oases, a factor that could have increased competition in the area. Northern Chile, especially Arica is an extremely fertile zone and competition between small groups of people would have been less likely and less intense. Also cemeteries normally develop after people have become settled in an area and not vice versa.

Arriaza (1993) argued that the development of artificial mummification was tied to religious ideology and environmental factors, rather than socio-economic needs. The Arica area is a zone that has always been affected by natural disasters. For example devastating earthquakes occurred in Arica in 1604, 1868, 1877, and 1987 to name a few. *Maremotos* or *Tsunamis* (large tidal waves) nearly destroyed the city of Arica in 1868 and 1877. Recurrent oceanic changes such as red tides, or the warm tropical current, called El Niño, have always caused great concern. Moreover, most beaches at the heart of the Arican Chinchorro community have permanent strong underwater currents that have taken the lives of many bathers. Also, the Pacific coast in this area has a chain of rocky hills that can easily crumble during earthquakes killing its dwellers. Recently, in 1993, winds of up to 60 km/h suddenly hit Arica lifting loose sand from the bare hills and creating a devastating dust storm that clouded the city, causing panic and destruction.

Working with the assumption that these natural disasters also occurred in antiquity, the Chinchorro people must have experienced them. The drowning of several people due to a tidal wave, or any of the above mentioned natural disasters, which caused the sudden loss of human life would have produced enormous social grief and a concerted search for answers. Painting a dead body with earth and keeping the body in the village, perhaps waiting for distant relatives to arrive, could have led to an increase sophistication of mortuary practices, leading the way to artificial mummification. The absolute absence of rain and the salts and sands of the Atacama desert certainly contributed to the excellent preservation of the Chinchorro mummies. Mummification then, may have developed not as an economic strategy, but as a humanistic search for answers to the nature of life and death.

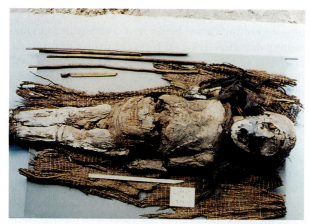

Fig. 5. Subadult Red Chinchorro mummy. Note the clay "helmet" covering the head and the long hair wig coming from under it

Fig. 6. Close-up of a Bandage-style Chinchorro child mummy

Fig. 7 (top). Adult Mud-Coated Chinchorro mummy

Fig. 8 (right). Close-up of a Chinchorro body with natural mummification

The sophistication of the mortuary treatment, the retouching of the mummies, and the millennial duration of their practices, all indicate that mummification was central to the social life of Chinchorro people. In the village, the mummies were probably seen as a shrine, a sacred place where a shaman or others could have communicated with the ancestors petitioning for blessings in their daily lives.

The need for spiritual contacts is well documented in the Andean past. In Inca and colonial times, Andean people always prayed to their ancestral mummies. They brought coca leaves, chicha (an alcoholic beverage), and even sacrificed animals to honor their ancestor long after death. The mummies were fed, and chants and libations were said. The shaman or sometimes the whole family petitioned them for good crops or successful herding (Salomon, 1991). As these Andean people died they were naturally mummified and added to the shrines of ancestral mummies, which were arranged in a hierarchical order. Salomon also said Andean people understood life as a continuous transformation from soft tissue, represented by childhood, to a solid unchangeable state, symbolized by the mummy. Within the mummy a spiritual life arose and after crossing a thin bridge made of human hair, the spirit traveled to its place of mythological origin. There the spirit joined with all its ancestors. A shaman was likely able to bring back the spirit of the mummy, making it alive, warm and caring. A similar situation of public display and worship may have existed for the Chinchorros.

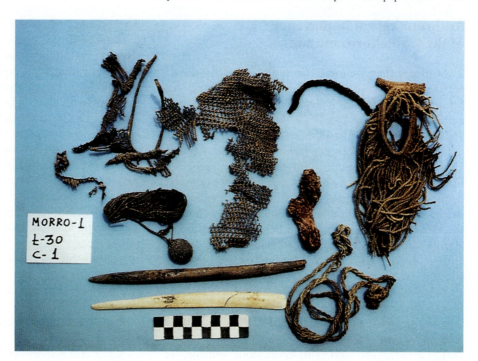

Fig. 9. Chinchorros tool kit showing from left to right fragments of net and cords, fringe skirt, round sinker and string, head of a detachable harpoon, and bone spatula

CHINCHORRO CHRONOLOGICAL SEQUENCE

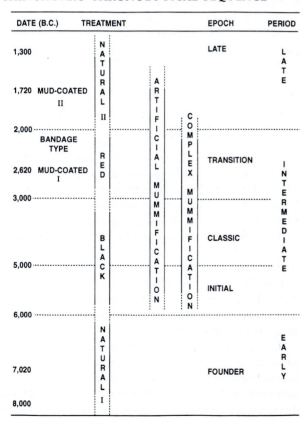

Fig. 10. Chinchorro chronological sequence based on radiocarbon dates of the mummies

Fig. 11. An adult male mummy from the Quiani 7 site (Quiani Phase)

Anthropological importance of the Chinchorro mummies

Mortuary theory correlates complex mortuary rituals with socio-economic gain (Binford, 1971; Tainter, 1978; Salomon, 1991). Death is seen as a rite of passage and as a process leading to future life, either human, animal, or vegetal. Death rituals also can be manipulated to maintain the status quo. A prince making an elaborate burial for his father is promoting his own future reign. Although the centrality of the dead in Inca times cannot be denied, numerous scholars conclude that care for the dead was a socio-political and economic affair to benefit the living (Rowe, 1991; Salomon, 1991). The nature of afterlife, human suffering, and religious ideology often appears to have been dwarfed by immediate economic strategies or benefits.

Archaeological mortuary theory also states that complexity of the treatment of the dead normally correlates with the rank of the deceased individual (Tainter, 1978). Elite individuals were buried or preserved with much energy expenditure, and with continuous worship, as was the case of the ruling Incas. Complex treatment of the dead is also said to be primarily influenced by sex and age in simple societies (Binford, 1971).

None of the above socio-economic theories seem to explain the perplexing Chinchorro spiritual manifestations of caring for the dead. In a cross-cultural study of mortuary practices directed by Carr (Arizona State University), it was found that treatment of the dead in less complex societies is primarily motivated by the character of the individual, responsibility toward ancestors, and by their mythology of a universal order (Arriaza, 1988).

Regardless of age and sex, the Chinchorro's received similar burial treatments. Even fetuses were mummified. In most societies only after a name is given to a child does he or she become a full member of the group, deserving proper burial. The Chinchorro, therefore, seem to be a non-ranked society where their religious ideology entitled everyone to have the same mortuary treatment and burial. Despite the paucity of technological achievements related to subsistence, the study of Chinchorro mummies also reveals that the simplicity of technology does not match the sophistication of religious ideology, as reflected in mummification practices. In other words, paucity of elaborate artifacts cannot be used to infer that people lived a simple social existence, or that early non-ranked societies did not have religious complexity. In brief, it appears the Chinchorro Culture was technologically simple, but ideologically they were highly sophisticated people.

Sedentism is another important factor that must be addressed in relation to Chinchorro mortuary practices. Early populations settling the coastal Andean area have generally been described as nomadic or semisedentary. The presence of cemeteries in many coastal sites from Ecuador to Chile beginning about 8,000 B.C. is indicative of a transition to early sedentism. Moreover, all of these early sites (Las Vegas, La Paloma, Morro 1, and Camarones 14) reveal more than 4,000 years of continuous human occupation in the same locations. In the case of the Chinchorros, artificial mummification did not lead to sedentism; on the contrary, it was sedentism that allowed for the development of artificial mummification. Unfortunately, few Chinchorro coastal dwellings have been found, however, there are a number of clues pointing to a permanent residence at the coast. They include artificial mummification, display of the mummies, subsequent reburial, few temporal variations in subsistence technology, and pathologies associated with a coastal subsistence (such as the development of external auditory exostosis). The large amount of energy invested in mortuary rites and ancestor worship, and the long duration of the mummification practices implies the Chinchorro lived permanently, not temporarily, at the coast.

Craft and activity specializations are attributes commonly used to characterize complex sedentary societies. The sophistication of the Chinchorro mummies indicates they were prepared by specialists with great manual and aesthetic skills. In this case, however, specialization cannot be related to the existence of rank and political complexity, or economic stratification, but rather to sedentism and socio-religious needs.

Moseley (1992) suggested that Chinchorro mortuary practices contributed to the concept of *Kuraka* (an Andean chief). It seems more plausible that Chinchorro ancestral worship formed the basis for the Andean concept of *Huacas* – so common in colonial and present-day Andean religions. *Huacas* are places or things where tutelary spirits and ancestors reside. Many sanctuaries in the Andean hills are still called *Huacas*. The concept of *Huaca* may have an ancient origin that first developed with the worship of the Chinchorro mummies and the shrines where they were kept. To a lesser extent, perhaps the Chinchorro's emphasis on ancestor worship can be seen as the earliest form of Andean group identity, today called *Ayllus*. An *Ayllu* is a community that shares a common ancestry, with a tendency to be endogamous and having access to differential resources. Although artificial mummification of the Chinchorros disappeared about 2,000 B.C. elements of their religious ideology could have transcended time and political complexity to other later coastal populations.

In summary treatment of the dead is influenced by socio-cultural, political and economic variables, but religious beliefs appear to be the social quintessence determining how the Chinchorro bodies were prepared and disposed. This is in contradiction to other archaeological mortuary studies which emphasize socio-economic explanations at the expense of religious-ideological needs. The case of the Chinchorro mummies acts as a reminder that mortuary behavior should be studied as a dynamic interaction of three components: the dead, the living, and the spiritual world, and as a continuous socio-religious process that is generated by the mourners who are attempting to seek harmony within this trilogy and within their technological-environmental limits. In the case of the Chinchorro and other ancient cultures, sociocultural complexity needs to be assessed in terms of ideology and techno-economics, not one necessarily revealed or at the expense of the other.

Summary

The preceramic Chinchorro Culture from coastal South America has the oldest known system of artificial mummification. The Chinchorros practiced artificial mummification from about 6,000 B.C. to 2,000 B.C. and they had a variety of mummification techniques which have been divided into Black, Red, Bandage, and Mud-Coated mummy categories. It appears the modern Chilean city of Arica was the cultural epicenter where the Chinchorro Culture originated. It is hypothesized here that the Chinchorros developed artificial mummification to assuage communal adversity caused by natural cataclysms since these phenomena are common to Arica and surrounding areas. The study of the complex Chinchorro mortuary practices reveals that: 1) the treatment of the dead is a complex process that goes beyond socio-economic explanations, 2) specialization can also be related to religious ideology, not only to socio-economic needs, and 3) non-ranked societies with simple tool kits may have complex religious beliefs and ideology which cannot be inferred by the paucity of artifacts deposited as grave goods.

Zusammenfassung

Die präkeramische Chinchorrokultur des südamerikanischen Küstengebietes besitzt das älteste bekannte System künstlicher Mumifizierung. Die Chinchorros praktizierten die künstliche Mumifizierung von ca. 6000 bis 2000 v. Chr. und hatten verschiedene Mumifizierungstechniken, die in vier Kategorien eingeteilt werden können: schwarze, rote, bandagierte und lehmüberzogene Mumien. Die heutige moderne chilenische Stadt Arica war das damalige kulturelle Zentrum, in welchem die Chinchorrokultur ihren Ursprung hatte. An dieser Stelle wird die Hypothese aufgestellt, daß die Chinchorros eine künstliche Mumifizierung entwickelt hatten, um die allgemeine Not zu lindern, die durch Naturkatastrophen entstanden war, da sich dieses Phänomen in Arica wie auch in der Umgebung der Stadt findet. Die Studie der komplexen Bestattungspraktiken der Chinchorros zeigt folgendes: 1. Das Umgehen mit dem Tod ist ein komplexer Prozeß, der über sozio-ökonomische Erklärungen hinausgeht. 2. Diese Spezialisierung kann auch mit religiöser Ideologie und nicht nur mit sozio-ökonomischen Bedürfnissen verbunden sein. 3. Nichtstrukturierte Gesellschaften mit einfacher Werkzeugausstattung können komplexe religiöse Überzeugungen und Ideologien besitzen, welche nicht aus der geringen Zahl von Gebrauchsgegenständen, die als Grabbeigaben dienten, gefolgert werden können.

Résumé

Le néolithique "acéramique" de la culture de Chinchorro des côtes de l'Amérique du Sud possède un système de momification artificielle connu comme le plus ancien, qui date d'environ 6000–2000 av. J.-C. Cette culture disposait d'une variété de techniques de momification qui à été divisée en quatre catégories: la momie noire, rouge, enroulée de bandages et couverte de glaise. C'est la ville moderne chilienne, nommée Arica, qui apparaît comme centre culturel où se trouvent les origines de la culture de Chinchorro. On établi ici l'hypothèse que les Chinchorro développaient une momification artificielle dans le but d'apaiser la misère commune causée par des cataclysmes naturels, car ce phénomène se trouve à Arica ainsi que dans ses environs. L'étude des pratiques funéraires en leur complexité de la culture de Chinchorro révèle le suivant: 1. Le traitement de la mort es un processus complexe qui va au-delà des explications socio-économiques. 2. La spécialisation peut aussi être liée à une idéologie religieuse et non-seulement à des nécessités socio-économiques. 3. Des sociétés non-structurées avec un équipement d'outils simples peuvent avoir des croyances et idéologies religieuses assez complexes, ce qu'on ne peut pas déduire du petit nombre d'objets d'usage courant, déposé dans la tombe comme mobilier funéraire.

Riassunto

Il sistema piú antico di mummificazione artificiale noto è quello praticato nella cultura dei Chinchorro lungo le coste dell'America Latina, nel Neolitico preceramico. I Chinchorro praticavano la mummificazione artificiale tra il 6000 ed il 2000 a.C. e possedevano varie tecniche di mummificazione che si possono suddividire in quattro categorie. Esistevano le mummie nere, rosse, quelle avvolte in fasce e quelle ricoperte di argilla. L'odierna città cilena conobbe le origini ed il massimo sviluppo della civiltà dei Cinciorro. L'autore esprime l'ipotesi che i Chinchorro avrebbero sviluppato la mummificazione artificiale in seguito a cataclismi naturali, visto che questa pratica è presente sia ad Arica che nei suoi dintorni. Lo studio delle complesse pratiche funerarie rivela quanto segue: 1. le pràssi legate alla morte rappresentano un processo molto complesso che va ben oltre le consuete spiegazioni socio-economiche. 2. La specificità del fenomeno può essere legata anche ad ideologie religiose che non si possono ricostruire sulla base dell'esiguo numero di arredi funerari ritrovati nelle tombe.

References

Allison, M., Focacci, G., Arriaza, B., Standen, V., Rivera, M. and Lowenstein, L. (1984) Chinchorro momias de preparación complicada: métodos de momificación. *Chungará* 13: 155–173.

Allison, M. and Pezzia, A. (1973–1974) Preparation of the Dead in Pre-Columbian Coastal Peru. *Paleopathology Newsletter* 4: 10–12; 5: 7–9.

Arriaza, B. (1988) Modelo bioarqueológico para la búsqueda y acercamiento al individuo social. *Chungará* 21: 9–32.

Arriaza, B. (1993) A Synthesis of the Chinchorro Culture. Paper Presented at the 58th Annual Meeting of the Society for American Archaeology. St. Louis. April.

Aufderheide, A., Muñoz, I. and Arriaza, B. (1993) Seven Chinchorro Mummies and the Prehistory of Northern Chile. *American Journal of Physical Anthropology* 91: 189–201.

Bittmann, B. (1982) Revisión del problema Chinchorro. *Chungará* 9: 46–79.

Binford, L. (1971) Mortuary Practices: Their Study and Their Potential. In J. A. Brown (ed.) *Approaches to the Social Dimensions of Mortuary Practices*. Memoirs of the Society for American Archaeology 35: 6–29.

Bird, J. (1985) The Preceramic Excavations at the Huaca Prieta, Chicama Valley, Peru. In J. Hyslop (ed.). *Anthropological Papers of the American Museum of Natural History*. Vol 62. Part I.

Chauchat, C. (1988) Early Hunter-Gatherers on the Peruvian Coast. In W. Keatinge (ed.) *Peruvian Prehistory*. Cambridge: Cambridge University Press., pp. 41–66.

Cockburn A. and Cockburn E. (1980) (eds.) *Mummies, Disease and Ancient Cultures*. Cambridge: Cambridge University Press.

Dauelsberg, P. (1974) Excavaciones arqueológicas en Quiani, provincia de Tarapacá, depto. de Arica, Chile. *Chungará* 4: 7–38.

Donnan, C (1964) An Early House from Chilca, Peru. *American Antiquity* 30: 137–144.

Leakey, R. (1981) *The Making of Mankind*. London: Abacus.

Llagostera, A. (1992) Early Occupations and the Emergence of Fishermen on the Pacific Coast of South America. *Andean Past* 3: 87–109.

Moseley, M. (1992) Maritime Foundations and Multilinear Evolution: Retrospect and Prospect. *Andean Past* 3: 5–42.

Mostny, G. (1944) Excavaciones en Arica. *Boletín del Museo Nacional de Historia Natural* 22: 135–145. Santiago, Chile.

Muñoz, I. and Chacama, J. (1993) Patrón de asentamiento y cronología de Acha-2. In I. Muñoz, B. Arriaza, and A. Aufderheide (eds.). *Acha-2 y los Orígenes del Poblamiento Humano en Arica*. Arica: Universidad de Tarapacá. pp. 21–46.

Nuñez, L. (1969) Sobre los complejos culturales Chinchorro y Faldas del Morro del norte de Chile. *Rehue* 2: 111–142.

Quilter, J. (1989) *Life and Death at La Paloma*. Society and Mortuary Practices in a Preceramic Peruvian Village. Iowa: University of Iowa Press.

Rivera, M. (1991) The Prehistory of Northern Chile. A Synthesis. *Journal of World Archaeology* 5(1): 1–47.

Rowe, J. (1991) Behavior and Belief in Ancient Peruvian Mortuary Practice. Paper Presented at the Conference "Tombs for the Living: Andean Mortuary Practices." Dumbarton Oaks. Washington, D.C.

Salomon, F. (1991) The Beautiful Grandparents: Andean Ancestor Shrines and Mortuary Rituals as Seen Through Colonial Records. Paper Presented at the Conference "Tombs for the Living: Andean Mortuary Practices". Dumbarton Oaks. Washington, D.C.

Santoro, C. and Ulloa, L. (1985) (eds.) Culturas de Arica. Arica: Universidad de Tarapacá.

Schiappacasse, V. and Niemeyer, H. (1984) *Descripción y Análisis Interpretativo de un Sitio Arcaico Temprano en la Quebrada de Camarones*. Publicación ocasional 41, Museo Nacional de Historia Nacional. Chile.

Stothert, K. (1988) *La Prehistoria Temprana de la Península de Santa Elena, Ecuador: Cultura Las Vegas*. Miscelánea Antropológica Ecuatoriana. Serie Monográfica 10, Banco Central del Ecuador. Guayaquil.

Standen, V. (1991) *El Cementerio Morro 1: Nuevas Evidencias de la Tradición Funeraria Chinchorro (Período Arcaico, Norte de Chile)*. Thesis, Pontificia Universidad Católica del Peru.

Tainter, J. (1978) Mortuary Practices and the Study of Prehistoric Social Systems. In M. G. Schiffer (ed.) *Advances in Archaeological Method and Theory*. No 1, pp. 105–146. New York: Academic Press.

Tello, J. (1929) *Antiguo Peru: Primera Epoca*. Parte I. Excelsior, Lima.

Uhle, M. (1917) Los Aborígenes de Arica. Publicaciones del Museo de Etnología y Antropología. Tomo 1, pp. 151–176. Santiago: Imprenta Universitaria.

Uhle, M. (1919) La arqueología de Arica y Tacna. *Boletín de la Sociedad Ecuatoriana de Estudios Históricos Americanos*. Vol. III, No. 7–8, pp. 1–48. Quito. Ecuador.

Wise, K. (1991) Complexity and Variation in Mortuary Practices During Preceramic Period in the South Central Andes. Paper Presented at the Annual Meeting of the American Anthropological Association. Chicago.

Correspondence: Dr. Bernardo Arriaza, Department of Anthropology, University of Nevada, Las Vergas, Nevada 89154–5012, U.S.A.

Secondary applications of bioanthropological studies on South American Andean mummies

A. C. Aufderheide

Paleobiology Laboratory, Department of Pathology, University of Minnesota, Duluth, Minnesota, U.S.A.

Introduction

Most bioanthropological studies are initially performed in pursuit of the answer to a specific question. Integration of the findings into an archaeological database, however, often provides new and unanticipated applications to other questions, broadening the range of information that can be extracted by the study of an ancient culture. This publication demonstrates how the results of an otherwise unrelated group of studies, when focused secondarily on a specific question, made a significant contribution to the nature of the interactions between two ancient Andean populations.

Geography and climate of the south Andean area

Topography

The high south Andean cordillera follows the contours of South America's southwestern coast and closely approximates it. Only about 125 km separate the Pacific shore from Andean peaks nearly 6,000 meters high in our study area. Westward from the high Andean range lies the 4,000 meter high precordillera that supports the highlands (altiplano; puna), a high plain lying between these two ranges (Figure 1). While their involvement with the geological ice ages is not as well studied as is that of North America or Europe, we know the Andes participated in at least the more recent of the Pleistocene glaciations. During the last glacial recession, meltwater descended the precordillera's western slopes with such violence it lacerated the landscape into a series of steep-walled and v-shaped valleys (quebradas) above a level of about 2,000 meters. From this point to the coast, a gently sloping, broad, coastal plain is intersected by these valleys, today partly-filled with alluvial sandy soil. Thus, these four zones are well defined in our study area, extending inland from the Pacific shore: the broad coastal plain, the strongly-sloped western shoulder of the precordillera with its steep-walled quebradas, the high mountain plain (highlands) and, finally, the high Andes cordillera. Each of these zones manifests its unique topography and climate (Santoro and Nuñez, 1987).

As the flow of moist Atlantic air moves westward across Brazil it impacts the *eastern* high Andean slopes, cools and loses most of its water content there, creating the Amazonian rain forest. Warm, moist Pacific air moving toward the coast from the *west* is similarly cooled by the cold, northward-flowing Peru (formerly called Humboldt) offshore current. The small amount of moisture remaining when the cooled air reaches shore generates daily coastal fog in winter. This extends inland only a few kilometers at which point the warm desert air disperses the moisture again. At occasional sites within this zone, this fog is concentrated by microclimatic factors with a seasonal regularity and intensity to nourish sufficient and predictable vegetation capable of supporting small populations. Such "lomas" are few, small and occur primarily in the central and northern Peruvian coast. Our study area has few of significance today, though several probably existed in antiquity (e.g., near Pisagua, Chile?). Immediately beyond the narrow zone of winter coastal fog, moisture seldom condenses sufficiently to produce rain. The bulk of the coastal plain, therefore (including the zone of steep valleys on the precordillera's western

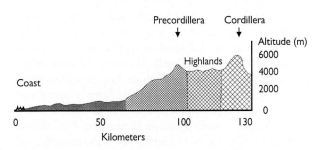

Fig. 1. Schematic profile of Andean topography at the latitude of Arica, Chile in the Azapa Valley (18°30′ south latitude)

slopes), is profoundly arid and has developed the assignation of one of the world's driest areas – the Atacama Desert.

The inland area beyond these two zones contains the more than 4,000 meter high altiplano. This highland plain, flanked by the precordillera and the high Andes range, lies at an altitude cool enough to condense much of the remaining moisture. That escaping condensation at this altitude still reaches the highland plains surface eventually by precipitating as snow on the still higher Andean peaks and flowing down to the altiplano surface as meltwater in summer. At the latitude of our study area, 30–35 cm of rain falls annually in the highlands (Santoro and Nuñez, 1987). While capable of presenting sundrenched vistas of awesome beauty, these highland plains more frequently tax the resolve of even its hardiest residents with nearly perpetually cool temperatures and often with blustery, cold rain, sleet and even snow with freezing winter temperatures.

To understand the relationships between coastal and highland people it is necessary to focus on the available water (and therefore vegetation) lying between these populations. Following disappearance of the glacial ice about 8,000 B.C., the coastal and western precordillera zones have received little or no rainfall for reasons outlined above. Since then and even today most of these glacially water-sculpted valleys persist as dry canyons, incapable of supporting human populations. An occasional valley, however, serves to conduct accumulated rainfall and snowmelt from the highlands down to the sea. Most of these are diminutive streams only a few meters wide, but some, like the San Jose River of the Azapa Valley, may measure as much as 20 or more meters wide. Furthermore, in their initial course down the steep mountain slopes, the valleys contain little soil and the streams there are bordered by a narrow strip of vegetation often only a few meters wide that marks the course of the brook at the very apex of the valley's v-shape. Not until the valley emerges on the coastal plain below 1,000 meters does it contain enough soil to permit irrigation agriculture. These only occasional, water-containing valleys could have lent themselves as travel routes between the coast and the highlands for small trader groups, though passage through the upper valley portions would have required a physically taxing traverse.

This description of the geography of northern Chile between about 18–23° south is applicable to the areas as far as northern Peru as well. However, latitude effects produce variations, some of which have considerable effect on human residence potential. Already mentioned are the fog-supported vegetation areas called lomas. These are much more common in northern Peru than in northern Chile. Of great importance is the recurring climatic phenomenon known as "el Niño" during which, for reasons related to the earth's rotational ("Coriolis") effect but not completely understood, the cold, north-flowing Peru current is diverted westward, away from the shore. The resulting effects are calamitous: torrential, destructive rains can occur in otherwise desert areas; and a dramatic reduction in marine resource (and, secondarily, avifauna) productivity found in the warm water replacing that of the diverted current (Moseley, 1975). These effects are usually more severe on the Peruvian than the Chilean coast. Additionally the winter highland temperatures at latitudes of the southern Atacama Desert (at about the latitude of modern Antafogasta) are too low to permit year-round highland pastoralism, forcing movement of herds to a lower level (and the human cooperative arrangements necessary to achieve that). The Peruvian portion of that desert also receives a little more moisture, blurring the sharply-defined junctional portions of the described zones and, in the more northern parts of the desert, mitigating the effects of aridity.

Settlement history

The present status of attempts to reconstruct the populating pattern of the New World are fraught with uncertainty. A small number of sites in South America bear radiocarbon dates of 25,000 years or more (Cruxent, 1968). While these are controversial (Lynch, 1990), it is probable that at least some of them will prove to be accurate. Currently, convincing archaeological continuity between these older sites and those less than 10,000 years old is often not clear. Many, therefore, regard these simply as dating errors or as occasional (accidental?) arrivals of small, perhaps, groups that probably did not survive to contribute significantly to South America's present population.

The traditional explanation of the New World's native population origin attributes it to trans-Beringeal migrations from northeast Asia about 11,000 years ago. After passing through Central America these migrants are believed to have entered South America, perhaps in the region of the Caribbean coast of Columbia where some drifted into the jungle areas of Brazil while others moved to the highland plains and followed them southward (Bryan, 1987). Setting aside the very ancient dates mentioned above, a vexing problem haunting efforts to reconstruct these migrations lies in the fact that some of the oldest dates in southern South America are not very different from those in northern North America. A substantial amount of additional archaeological data will need to be generated to resolve these problems (Jennings, 1978).

We do have a modest amount of information for both highlands and coastal regions in our study areas (Santoro and Nuñez, 1987). Evidence for highland occupation prior to 8,000 B.C. (Paleo-Indian period) at the latitude of modern Arica, Chile (18° 30′ south) is lacking. Between 8,000 and 6,000 B.C. (early Archaic period) the scattered lithic remains of paleolithic hunters suggest their permanent, year-round occupation at marshy highland sites, surviving on camelids and deer that were attracted there by the grasses and plants such boggy soils supported. The only artifacts related to the coast are a very few mussel shells (*Choromytilus chorus*). Farther south (Salar de Atacama) winter highland temperatures forced hunters to locate their bases at lower levels.

Warmer, drier climate between 6,000 and 3,000 B.C. (Middle Archaic) and extensive volcanic activity forced abandonment of these year-round highland bases. During this interval, summer hunting visits to the altiplano occurred, but base camps must have been at lower altitudes. Numerous authors (Santoro and Nuñez, 1987; Muñoz and Chacama, 1982; Bird, 1943) call attention to the development of new coastal sites at this time, and Santoro and Nuñez (1987) suggest these may represent highland groups who were driven down by the highlands' climatic and environmental changes to exploit coastal resources, though no genetic and only slender archaeological data exist to support that speculation. Others (Moseley, 1992) suggest this is an artifact of rising sea levels.

By the Late Archaic period (3,000–2,000 B.C.) the highlands were again occupied year-round in extreme northern Chile and at the oases and lower valleys in the Salar de Atacama area. Some pottery, beads, pigments (used for rock art) and also some mussel shells appear together with camelid and rodent bones at the more northern, permanent highland sites. In the region's more southern parts permanent residence was established at intermediate altitudes from which summer hunting expeditions to the highlands were carried out as well as winter retreats to the oases for fruit and local game. Important to this period, also, was evidence of the initiation of pastoralism (Santoro and Chacama, 1982) with camelid breeding and a wool industry that had probably been initiated a little earlier (4,000 B.C.) at more northerly latitudes (Cáceres, 1985; Brotherston, 1989).

Thus, the Archaic period is characterized by permanent occupation of highland (in northern section) or intermediate (southern section) altitudes by people whose economy focused primarily on hunting with lesser amount of gathering. Highland bases were abandoned for climatic and environmental reasons in the Middle Archaic period. This was followed by reoccupation and the development of both incipient agriculture and early pastoralism during the Late Archaic periods. Evidence of coastal contact is limited to small numbers of mussel shells present in significant numbers only during the latter part of the Late Archaic period.

Coastal residence, like that of the highlands at these latitudes (17°40′–23°40′ south) was initiated during the Early Archaic period. A shell midden at Quebrada Las Conchas near modern Antofagasta on Chile's northern coast is radiocarbon dated at about 7,500 B.C. (Llagostera, 1979). The oldest human remains (about 7,000 B.C.) from that coast is a spontaneously mummified body from the Quebrada Acha, a tributary to the lower Azapa valley (Muñoz, et al., 1993). The next-oldest is an anthropogenically mummified body from about 5,810 B.C. (Allison et al., 1984), followed by other bodies similarly treated and dated at about 4,900 to 4,000 B.C. (Aufderheide et al., 1993) and a host of both spontaneously and anthropogenically mummified bodies down to about 1,700 B.C. (Allison et al., 1984). On the basis of archaeological features all of these bodies are identified as members of the Chinchorro cultural group. These burial sites are scattered along about 500 km of northern Chile's littoral zone.

Problem statement

The question on which we focused the results of our studies is: What was the relationship between the coastal populations of what is now northern Chile and the residents of the highlands at the same latitude during the 7,500 B.C.–1,500 B.C. interval of the Archaic Period. Specifically, is there evidence of food and other material goods exchange? cultural exchange? population movement? amicable or hostile relationship?

Study material

The answers to the questions stated above were pursued primarily through an intensive study of the coastal populations represented by human mummified remains excavated from burial sites on the coast and the low valleys of northern Chile between 18°30′ and 20°30′ south latitude. Most of them were located in or near the Azapa Valley at Arica, Chile. They were excavated by the professional staff at the Archaeological Research Institute of the University of Tarapaca in that community and the remains were and are curated there. Data from the highland occupations are abstracted from reports of sites in those areas, most of which were examined and reported by archaeologists from that same institute. The coastal populations studied by the author are listed in Table 1 and characterized below.

Table 1. Coastal cultures from North Chile studied for chemical dietary reconstruction

Culture	Time period	N[a]
Chinchorro	7,500 B.C.–1,500 B.C.	67
Alto Ramirez	1,000 B.C.–A.D. 350	30
Cabuza	A.D. 400–A.D. 1,000	32
Chiribaya	A.D. 1,100–A.D. 1,300	48
San Miguel	A.D. 1,300–A.D. 1,450	27

[a] Number of adult samples studied

Chinchorro (7,500–1,500 B.C.)

This earliest group to occupy the west coast of South America between 17°40′ and 24° south latitude first appears at Quebrada Las Conchas at about 23°40′ south, a midden dated to about 7,500 B.C. (Llagostera, 1979). Their origin remains unknown. The oldest and spontaneously mummified body from this group, dated to 7000 B.C. was found in a tributary of the Azapa Valley at 18°30′ south (Aufderheide et al., 1993; Muñoz, Arriaza and Aufderheide, 1993). The mummified bodies of more than 150 other members of this group have been excavated from coastal sites of northern Chile. Some of these have been mummified by elaborate, intentional (anthropogenic) mummified procedures (Allison et al., 1984) while in others the mummification was the spontaneous, unintended result of local climatic effects. This perceramic group survived the area's environment by living in small groups, probably at the band level, at river mouths, carrying out a highly adaptive and almost exclusively maritime subsistence economy. The type site at Chinchorro Beach is located at Arica, Chile, having been excavated and reported by Max Uhle (1919). A major group of nearly 100 anthropogenically mummified bodies was found within the community of Arica (Mo1 site) in 1984 (Allison et al., 1984) and three years later 69 spontaneously mummified bodies were studied from an immediately adjacent location (Mo1–6 site). Most, but not all, of the other bodies studied are curated currently at the Arturo Prat University in Iquique (Nielson collection: Munizaga, 1961) and at the National Archaeological Museum in Santiago. The majority of our bioanthropological studies of the Chinchorro people were carried out on bodies from the Mo1 and Mo1–6 sites. Their bodies were buried in the sand or clay of the site in extended positions without tombs. The only grave goods were articles included within the mostly totora reed mats that enveloped their bodies or in the sand immediately adjacent to the cadavers.

Quiani (1,500–1,000 A.D.)

The Quiani people can be viewed as "transitional Chinchorros" in that they have most of the features of Chinchorros but without intentional (anthropogenic) mummification practices, have a modified burial position (right side down with partially flexed lower extremities) and give evidence of early horticultural experimentation (but principally with industrial products like cotton and gourds). Their economy remains almost exclusively marine-oriented.

Alto Ramirez (1,000 B.C.–A.D. 350)

While these bodies are usually classified as a single group, based principally on unique ceramic and textile designs, varying burial modes (shaft, tumulus, supine) imply possible heterogeneity. Such grave goods suggest a highland relationship, as does their agropastoral economy, and so this group is viewed as among the first highland groups that migrated to and settled in the low valley sites near the coast. Most of the bodies on whom we carried out studies came from sites within or near the Azapa Valley in the vicinity of Arica, Chile, but a few were excavated from a coastal site at Pisagua about 85 km farther south.

Cabuza (A.D. 400–1,000)

These are Tiwanaku-related, agropastoral migrants from the expanding empire centered in the circum-Titicaca area in the highlands. They also settled in the lower Azapa Valley.

Chiribaya (A.D. 1,000–1,250)

This principally agricultural (with additional pastoralism) group appears to have evolved from indigenous origins following collapse of the Tiwanaku empire. They farmed the irrigated, flat alluvial soil of the lower Azapa Valley.

"Late" cultures: San Miguel, Gentilar (A.D. 1,250–1,500)

These groups are similar to the Chiribaya in economy and location, separated on the basis of ceramics and introduction of higher technology (Rivera, 1991).

Study methods

Anatomic dissection

Following careful unwrapping, specimens of hair, nails, skin, bone and muscle were acquired. Internal dissection of body cavities and their viscera was then performed with examination of all viscera, both *in situ* and of extirpated organs. Coprolites were collected for study. Samples were rehydrated and sectioned for microscopic study by the method reported by Sandison (1963). After gross study of all soft tissues a complete osteological examination was performed.

Biochemical studies

The trace element studies were performed by atomic absorption spectrometry, the details of which have been reported previously (Aufderheide, 1989; Aufderheide and Allison, 1992a). Stable isotope ratios were carried out by ion ratio mass spectrometry as reported by Tieszen et al. (1992). Strontium isotope ratios were performed in a similar manner by Harold Krueger at Geochron Laboratories in Cambridge, Massachusetts. Cocaine detection in hair samples was performed as reported in Cartmell et al. (1991). Parasite ova and dietary analysis of rehydrated coprolites was studied by Karl Reinhard as reported (Reinhard and Bryant, 1992a). Marc Kelley et al. (1991) carried out dentition studies.

The evidence

While the time period (7,500–1,500 B.C.) for the question of interest includes only the coastal populations of Chinchorro and Alto Ramirez, data from the chemical dietary and cocaine studies will be included for the other populations also to provide context and perspective.

The artifacts

Since the Chinchorro mummies were not buried in tombs but merely wrapped and interred directly in the soil, grave goods were limited to those found within the wrappings and in adjacent sand. These clearly reflected their maritime subsistence economy. Fishhooks were abundant, fashioned from cactus spines. Fishlines were made of vegetal material. Harpoons were constructed with a detachable forepiece that was connected to the shaft, whose flanged distal end accepted the forepiece held into position only by friction. The forepiece was fitted with a lanceolate, flaked point and bone barbs. A small (about 20×15×10 cm) net bag was made of vegetal cords, probably used for shellfish collecting. Hafted sea lion ribs (chopes) were used to pry shellfish off rocks. Prior to about 2,000 B.C. no ceramics or agricultural tools were identifiable. All these items emphasize that their subsistence efforts focused almost entirely on marine resources. The grave goods found with the burials of the Alto Ramirez bodies, however, included ceramics, abundant camelid bones, wool textiles, spindles and a number of agricultural items (maize, potatoes, pepper). No precedence for such items exists among the Chinchorros prior to about 2,000 B.C. (Rivera, 1991).

The clothing

The Chinchorro bodies were almost all wrapped in totora reed (*Scirpus californicus*) mats. Some had inner, more finely woven mats surrounded by an outer, more coarsely woven one. Several infants, however, were wrapped in small (about 25×25 cm) hides with yellowish, attached camelid fur suggestive of guanaco nature (though we have not completed immunodiffusion confirmation). In the occasional adult with a similar piece of camelid hide included, the segment was also of a size about equal to that enveloping the infants. In the adults such hides were found beneath the inner totora reed envelope, lying flat on the abdomen. Alto Ramirez textiles were principally composed of woven camelid wool, and heads were sometimes encircled by "turbans" made of wool yarn (Rivera, 1991).

Anatomic findings

External auditory canal exostoses are found commonly in individuals chronically exposed to cold water, such as marine divers. These were present frequently among not only the Chinchorros (Standen et al., 1984) including the oldest body from 7,000 B.C. (Aufderheide et al., 1993), but also among the Alto Ramirez population (Standen et al., 1985), where even one female demonstrated the lesion. Dentition differences between Chinchorros and Alto Ramirez populations were also significant. Less than one percent of Chinchorro teeth revealed evidence of dental caries while 11.5 % of Alto Ramirez teeth were carious (chi square test = p<.05) (Kelley et al., 1991).

Evidence of violence was sought in these populations, in an effort to evaluate the presence of group warfare. The Chinchorros revealed no evidence of skull fractures characterized by geometric shapes associated with war weapons, nor were "parry fractures" of the left ulnar diaphysis identified. Only one Chinchorro body demonstrated evidence of interpersonal violence – an adult male with unhealed fractures of the right mandible and zygoma as well as two large, penetrating chest wounds probably produced by harpoon thrusts.

Diet

The diet of an ancient population may be assessed using a variety of methods:

Environmental nature: The flora and fauna normally expected to be found in an environment of the type known to have existed at the population's residence site can provide a general idea of the possible nature of their diet (Styles, 1985). For the Chinchorros, living at river mouths and beach sites, this includes marine resources, a few edible plants normally found at estuaries in this region, such as the common totora plant (*Typha angustifolia*) and algarrobo (*Prosopis tamarugo*) at a few sites, as well as a few rodents. With such an approach, it would, however, not be known whether coastal residents normally ascended the highlands to hunt camelids and other altiplano animals.

Subsistence artifacts: For the Chinchorros, these would predict an almost exclusively marine diet.

Quantitative analysis of floral and faunal remains: Except for a few fish vertebrae, no such remains were present in association with Chinchorro bodies at our studied site.

Coprolite analysis: Reinhard and Bryant (1992b) examined coprolites from 21 Chinchorro bodies and found about 50 percent of the recovered components identified were of marine animal origin while the remainder were of plant nature. However, interpretation of these results must include the recognition that ingestion of shellfish meat will be recognizable in coprolites only by inclusion of the occasional shell of the smallest of these, that meat from large sea mammals like sea lions will escape detection and that the nutritional fraction of some ingested plants is low. These considerations could underestimate the marine meat fraction substantially.

Dental analysis: As noted above, the remarkably low prevalence of caries (Kelley et al., 1991) implies a low carbohydrate (vegetal) dietary fraction.

Principles of chemical dietary reconstruction

The principle involved in this approach lies in the fact that certain major dietary components, such as the vegetal fraction, normally contain some chemicals not shared by other components. If these accumulate in body tissues they may be measured and compared with values in tissues of animals with known, controlled diets. The principal chemicals measured for this purpose are total strontium and the stable isotope ratios of carbon, nitrogen, sulfur and strontium (Aufderheide, 1989).

Strontium is a mineral commonly present in many soils. Its salts are usually sufficiently soluble to appear in ground water, and most plant roots readily absorb it and distribute it to other parts of the plant. When plants are ingested, mammals absorb the strontium, and the blood distributes it throughout the body. However, the body treats it much like calcium, with the result that it quite promptly becomes incorporated into the hydroxyapatite crystal and undergoes long-term storage in bone mineral. Hence, human ingestion of plants results in substantial absorption of the plants' strontium content with subsequent deposition of the strontium in the human bone. If an animal eats such a plant it, too, will store the strontium in its bones. However, if a human eats the *meat* from such an herbivore, the meat contains essentially no strontium, so that the meat fraction of a human diet contributes no strontium to that human's bone strontium content. For this reason the human bone strontium content reflects the vegetal dietary fraction. This can be quantitated by comparing it to that of an herbivore, because in such a mammal 100 percent of the diet is composed of plants. The bone strontium content is often related to the calcium content of a bone and expressed as a fraction: $Sr/Ca \times 1000$ (Aufderheide, 1989).

Restrictions on the application of this approach include the fact that ground water percolating through buried bone may, depending on local conditions, alter the skeletal strontium content by leaching it from or depositing additional strontium into the bone and thus obscuring its antemortem value pattern. In addition, certain plants have a much lower strontium content than most plants do; if the control herbivore did not consume the same proportion of such plants, then the vegetal fraction of the human may be substantially underestimated. Maize kernels are an example of this phenomenon. In addition, because the strontium of soils varies with geography the strontium contents of control animal and plant values are site-specific.

Dietary applications of isotope ratio analysis is based on similar assumptions. Plants producing three-carbon (C3) sugars use different enzymes to produce these sugars from carbon dioxide in air than do plants producing four-carbon (C4) sugars. But about one percent of the carbon in carbon dioxide of the earth's atmosphere is the isotope ^{13}C while the remainder is ^{12}C. Both enzyme systems that plants employ for this photosynthetic process discriminate against ^{13}C, but those of C3 plants do so to a greater extent than those of C4 plants. Since the absorbed sugars (with their varying $^{13}C/^{12}C$ ratios) are incorporated into proteins that are synthesized within the body, that protein's final $^{13}C/^{12}C$ ratio will reflect the proportion ingested, and therefore also will reflect the ratio of C3/C4 plants ingested. Comparison of the carbon isotope ratio of human protein with that of potential vegetal dietary resources available at a site thus can predict ratios of C3/C4 plants ingested (Tieszen and Chapman, 1992). The carbon isotope ratios are generally expressed in relation to that of a limestone standard (Peedee belemnite ore: PDB) according to the formula:

$$\delta^{13}C \text{ ‰} = \frac{^{13}C/^{12}C \text{ sample} - ^{13}C/^{12}C \text{ PDB standard}}{^{13}C/^{12}C \text{ PDB standard}} \times 1000.$$

C3 plants usually generate a protein signal of about $\delta^{13}C = -27$ while the value for that of maize and C4 plants is about -12. C3 plants prefer temperate or cool environments while most C4 plants flourish best under warmer conditions (Tieszen and Chapman, 1992).

In addition, the nitrogen ($^{15}N/^{14}N$), sulfur ($^{34}S/^{35}S$) and strontium ($^{87}Sr/^{86}Sr$) ratios of terrestrial sources differ from those of marine environment food sources to a degree sufficient to be exploited for such source predictions. Furthermore, nitrogen ratios undergo enhancement at each trophic level (Tieszen et al., 1992).

Applications of chemical dietary reconstruction to mummies

Three different approaches were used to detect possible diagenetic effects. The first was based on the assumption that hydroxyapatite dissolution by ground water would remove both calcium (Ca) and strontium (Sr) at equal rates, leaving the Sr/Ca unchanged while deposition of Ca and Sr ions from ground water could be expected to alter this ratio. When the Sr/Ca ratio was related to the absolute Ca concentration for individual groups it became clear that the Sr/Ca did not change until the absolute Ca concentration deviated from normal to a degree greater than 20 percent (Aufderheide and Allison, 1992a: Table 2). About eight percent of the analyzed adult specimens had values exceeding 20 percent, so these were eliminated from the database. In addition a few "outlying" Sr/Ca values that exceeded the group mean values by more than four standard deviations were also eliminated. The remainder of the values (88 percent of adult specimens analyzed) were found to conform to a Poisson distribution. Finally, the pattern of values for infants up to two years of age was examined. For physiological reasons detailed elsewhere (Aufderheide and Allison, 1992a) the strontium values in this age group undergo unique fluctuations, averaging about one-half of that of their mothers at birth, declining to a nadir about 35–50 percent of birth values at about 18 months and then rising to adult levels before age ten years. Because these chronological changes are based on normal, physiological events involving bone mineral turnover rates, the pace of skeletal growth and weaning, the demonstrated persistence of this pattern in an archaeological population can be employed to ensure that diagenesis has not obliterated the antemortem strontium pattern in

Table 2. Trace element and isotope values for North Chile coastal populations

	Total Sr/Ca	$^{87}Sr/^{86}Sr$	% Vegetal Sr/Ca	% Vegetal index[a]	Adjusted % vegetal index	Marine index	Terrestrial meat index	$\delta^{15}N$
Chinchorro	.49	.7088	.19	.06	.06	.89	.05	22.1
Herbivore	1.54	.7071	1.00	1.00	1.00	.00	.00	6.4
Sea Lion	0.97	.7088	0.00	0.00	0.00	1.00	.00	24.9

[a] Index = dietary component value. See text for detailed explanation of this table

a studied skeletal sample (Aufderheide and Allison, 1992a). After removal of the aberrant values from the database as indicated above, the database demonstrated retention of the characteristic pattern of infancy providing reassurance for preservation of the antemortem strontium values in the database, permitting their interpretation by the usual criteria.

The Chile site presented a formidable challenge to the methodology of chemical dietary reconstruction. It involved multiple populations living in a single area over a 9,000 year interval. The diets of these could be expected to include patterns of preagricultural and postagricultural, terrestrial and marine, wild and domesticated plants and animals, C3 and C4 plants and marine trophic effects. Identification of the influence of these many variables on the diets of the seven studied groups needed to be accomplished using only the chemical values of bone calcium and strontium concentration as well as the bone or hair protein stable isotope ratios of strontium, carbon, nitrogen and sulfur. The latter was an innovative contribution by Dr. Roy Krouse of the physics laboratory at the University of Calgary, Canada (Krouse and Herbert, 1988).

Table 2 demonstrates an example of these applications. The initial Sr/Ca values showed no statistically significant differences between the various studied groups. However, the $^{87}Sr/^{86}Sr$ value in soil, reflected in the herbivore value of 0.7071 was found to be substantially different from the universal marine value (.7092). The .0021 range of values between these two was great enough to reflect the contribution made to the *total* strontium in the bone by each of the two individual possible alternative food sources at this site: terrestrial plants and marine foods. This could be done by comparing the human bone strontium isotope ratio to that of the herbivore. By so doing it was found that, for the Chinchorros, only 19 percent of the bone strontium was contributed by strontium in terrestrial plants. Correction of the originally determined Chinchorro total Sr/Ca value (.49) by 19 percent revealed a Chinchorro vegetal Sr/Ca of .09 and comparison of this value to that of the herbivore bone control (1.54) showed that terrestrial plants constituted only six percent of the Chinchorro diet.

We had expected to be able to estimate the marine fraction of the total strontium concentration by comparing the human value to that of the purely marine animal – the sea lion. However, we found that, while human consumption involved primarily the low-strontium meat of sea mammals, shellfish and fish, sea lions ingested not only the fish meat but also most of the high-strontium bones and many smaller mussel shells, elevating their bone strontium to a much greater degree than would consumption by humans of the same amount of only the meat from such marine food sources. Since the sea lion bone strontium, then, proved to be an inappropriate "100 percent marine diet" comparison control, we were fortunate to be able to employ the same animal's nitrogen or sulfur isotope ratio for that purpose, since these were unaltered by ingestion of fish bone or mussel shell. On that basis, the Chinchorro nitrogen isotope value is about 89 percent of the sea lion protein value. Since all of that is of marine origin, this would represent the marine fraction of the Chinchorro's Sr/Ca (.40). Thus, if .40 represents 89 percent of a 100 percent marine diet, the Sr/Ca value for a 100 percent marine meat diet would be .45. The total marine component of the diet, therefore, can be obtained for any of the groups by comparing its marine Sr/Ca fraction with .45. Having established the marine and vegetal dietary fractions, the terrestrial meat fraction represents the remaining component. In addition, we needed to deal with the problem of maize as a vegetal component. While Chinchorros consumed no maize, all of the subsequent groups did. This required an adjustment for calculation of the vegetal fraction estimate (adjusted vegetal index), because maize contributed only about half as much strontium as most other plants in this area (Aufderheide and Allison, 1992b). That adjustment was made based on the assumption that about half of the vegetals in the diet (other than that of the preagricultural Chinchorros) was maize. With the above approach the three principal dietary components for all groups studied is shown in Table 3.

Table 3. Principal dietary components of North Chile coastal populations (in percent)

	Meat		Vegetal
	Terrestrial	Marine	Terrestrial
Early Chinchorro	12	80	8
Late Chinchorro	5	89	6
Alto Ramirez	24	42	34
Cabuza	29	31	40
Chiribaya	27	31	42
San Miguel	30	31	39

Cocaine in mummy hair

Chewing of the raw coca leaf has been a south Andean cultural tradition for centuries (Plowman, 1986). During the post-Conquest period, motives for this practice have been attributed to its alleged beneficial effects for various illnesses, hunger and fatigue. Ethnohistorical reports suggest it may have played a central role in religious ritual among the Incas (Garcilaso de la Vega, 1989 [1608]). Modern anthropologists emphasize its value as an expression of cultural identity (Allen, 1988). The amount of cocaine absorbed by chewing the unprocessed leaf of the coca plant is sufficient to be detected in the blood by modern chemical instrumentation (gas chromatography mass spectrometry: GCMS, or radioimmunoassay: RIA) (Paly et al., 1980), though probably much too little is absorbed to effect a major mood alteration. It is an intensely tenacious cultural practice that has resisted both clerical and governmental efforts to extinguish it (Allen, 1988).

In practice a half-dozen leaves are rolled into a quid that is placed between the teeth and the cheek where it is allowed to become saturated with saliva (it is not actually chewed). Alkalinization by addition of powered lime or ashes results in the extraction of the leaf's alkaloids, including cocaine, which is then absorbed. At least some of the cocaine absorbed into the blood is deposited in cells of the hair follicle and becomes incorporated into the hair shaft. In graves of cultural groups among whom this practice is common, both characteristic lime containers may be found as well as the coca leaves themselves within specialized cloth bags or placed in an intraoral location as a quid (Cartmell et al., 1991).

In an effort to establish the demography and possibly the purpose as well as the antiquity of coca leaf chewing practices, we used both GCMS and RIA methods for detection of cocaine and its principal metabolite benzoylecgonine on samples of mummy hair obtained from six different pre-Conquest groups that occupied the Azapa Valley and coast in northern Chile over a period of more than 5,000 years. Results are tabulated in Table 4. It will be apparent that none of the Chinchorro hair samples gave a positive reaction to these tests for detection of cocaine. The earliest cultural group among whose members we found positive results was Alto Ramirez (beginning 1,000 B.C.). Between one-half and two-thirds of adults in later groups demonstrated evidence of having chewed coca leaves. Both males and females did so with equal frequency. The fact that infants under age two years had the same frequency of positive reactions as did adult females suggests the cocaine reached the infants by transplacental passage or through the breast milk. The significance of these results for our purpose in this report lies in the observation that there is no evidence of coca leaf chewing practices among the coastal populations of northern Chile until the arrival of the earliest migrants from the highlands – the Alto Ramirez people.

Table 4. Results of cocaine and benzoylecgonine studies in mummy hair

	Number positive	Number tested	% positive
Chinchorro	0	23	0
Quiani	0	3	0
Alto Ramirez	2	13	15
Cabuza	10	16	63
Chiribaya	54	97	56
San Miguel	2	8	25
Chiribaya			
0–2 yr	12	19	63
3–14 yr	7	17	41
15+yr	35	61	57
Female adults	17	28	61
Male adults	16	30	53

Synthesis

Both the highlands and the coastal areas were populated during the Early Archaic period (8,000–6,000 B.C.) at the latitude of our study area. The origin of the settlers in neither of these areas is known with certainty. Though one hundred horizontal kilometers and 4,000 meters of altitude of the world's driest desert separated these two populations, an occasional stream-containing valley penetrated the arid land between them and conceivably these might have been employable as communication channels. To answer the question whether or not they actually did so, we must focus our attention on the evidence available from studies of the Chinchorro people, because the poor preservation conditions in the highlands have left us few human remains or perishable artifacts for study. Yet the highland sites contain sufficient clues so that we can recognize that the original settlers there were primarily hunters of camelids, deer, rodents, and birds with little opportunity to gather edible plants. In spite of the climatic and altitudinal hardships, they survived, though climatic and environmental conditions drove them to base their camps at intermediate altitudes during the Middle Archaic period. Their return to the permanent highland sites in the Late Archaic period is distinguished by the evolution of agricultural and pastoral practices. Except for some mussel shells at highland sites (few in the Early Archaic and many in the latter part of the Late Archaic), there is little archaeologic evidence there of contact with coastal groups.

The evidence from the Chinchorro group, that occupied the coastal area for more than five millennia during all three segments of the Archaic period, includes artifacts, clothing, anatomic changes in their bodies, parasite acquisition related to their subsistence activities, cultural practices (coca leaf chewing) and chemical reconstruction of their diets. Of course, not all of these items are of equal importance. Nor are the bodies we have available to study equally distributed among those nearly 6,000 years of coastal occupation: we have a single body from about 7,000 B.C., a second a millennium later, increasing numbers after 5,000 B.C. and nearly 150 mummies between 3,000 and 1,500 B.C. Nevertheless, all the clues we have seem to point in the same direction, emphasizing the minimal contact these two populations had, and the dramatic differences they present in the evolution of their subsistence economies.

Their clothing, for example, is composed almost entirely of vegetal material (split totora reeds) available at their habitation sites near river mouths. However, some of the children were wrapped in small sections of camelid hides and a few adults had similar sections included within their totora wrappings. Because the normal residence area for camelids is the highlands, use of camelid items by the Chinchorros could be viewed as evidence of at least contact with highland traders, even if the Chinchorros did not actually ascend to highland areas for participation in camelid hunts. Yet such a deduction is not necessarily compelling. The hide fur suggests they may be from guanacos, and even during historic times guanacos are known to descend seasonally from the highlands via valley streams all the way to the coast. Thus the few camelid hides found on these bodies could have been obtained by the Chinchorros during the winter using their atlatl and dart weapons when the occasional guanaco appeared on the coast near their habitation sites. In fact, the small number of these useful hides found among Chinchorro bodies strengthens that implication.

Another group of items relates to the nature of the Chinchorro subsistence economy during that very prolonged coastal occupation. The Peru current along the northern Chile coast is frigid water. A principally maritime subsistence inevitably involves frequent and prolonged exposure to that cold water. Such exposure is known to correlate with external auditory canal exostoses (Kennedy, 1986). Such lesions were found in the oldest of these bodies (Aufderheide et al., 1993) as well as in many of the others at all periods (Standen et al., 1984). Artifacts, too, are almost entirely marine related, including fishhooks, fishlines with sinkers, lures, fishnets, harpoons and similar items. No digging or other agriculture-related items appear earlier than 2,000 B.C., nor are ceramics present before that time. Additionally, coprolite studies identified fish tapeworm ova (*Diphyllobothrium pacificum*) in several Chinchorros. Since this parasite cycles normally between the local sea lions and fish, consumption of the infested fish can make man an accidental host.

While such artifacts, bone lesions and parasitism emphasize the Chinchorros' marine exposure, reconstruction of their diet provides even more specific and impressive evidence. If contact with highland residents were common, one would anticipate that the agricultural items developed there at least during the Late Archaic period and the terrestrial meat abundantly available through their hunting and pastoral activities would be reflected also in the Chinchorros' diet. Yet the chemically reconstructed diet pattern demonstrates an absolutely uninterrupted continuum of a diet overwhelmingly dominated by foods of marine origin beginning with the oldest body studied and continuing, unchanged, to the most recent groups about 2,000 B.C. (Table 3). True, Reinhard's study of coprolite content suggests a lower fraction was of marine origin, but for reasons documented above, that approach to dietary reconstruction underestimates fish consumption, and sea lion meat (probably a major Chinchorro dietary item) may be missed altogether.

Another cultural feature (coca leaf chewing) seems to be sending the same message. Coca plants are native to the jungle of the Andes' eastern slopes. To reach the western coast of Chile the cultural diffusion path of this practice must traverse the highlands. The results of our studies indicate that the Chinchorros did not indulge in this practice. It does not appear on the coast of northern Chile until the arrival there of the Alto Ramirez people, migrants from the highlands of the Lake Titicaca area.

Finally the absence of "military-type" fractures and the finding of only one example of interpersonal violence argues for the absence of intergroup warfare, further evidence of social separation of these populations.

Interpretation

These bioanthropological studies help paint a most unique picture of two different populations occupying northern Chile during the nearly 6,000 year interval of the Archaic period between about 7,500 B.C. to 1,500 B.C. Separated by only about 100 km, the highland occupants, probably under duress from a demanding climate, evolved culturally from hunting to a pastoral and finally agropastoral subsistence economy. During that same period the coastal populations persisted in an uncontested and largely unchanging maritime subsistence until the arrival of migrants from the highlands. Throughout these almost 6,000 years the results of the bioanthropological studies suggest that, in contrast to probable practices in more northern contemporary South American indigenes (Feldman, 1992), the degree of contact between occupants of the coast and the highlands was so minimal that it escapes detection by application of these methods.

These results indicate the potential value of "secondary" applications of bioanthropological studies (those other than the original purpose for their performance), and also emphasize how highly efficient was the Chinchorros' chosen form of cultural adaptation to this otherwise formidable terrain.

Summary

Bioanthropological studies normally are targeted at very specific questions. Like archaeological artifacts, however, their results may be applied to questions other than those for which they were originally performed ("secondary applications"). In this publication the results of a variety of studies, including chemical dietary reconstruction and chemical detection of cocaine in mummy hair, were focused on the secondary question of the relationship and degree of social interchange between the coastal and highland populations of northern Chile during the Archaic period. Although living within 100 kilometers of each other, they were separated by a desert of formidable but not impassable terrain and aridity. Study results suggest that, while the highland populations evolved culturally from hunting into a diversified hunting-agropastoral subsistence economy, the coastal residents persisted in a marine hunting economy over the entire 6,000 year interval without evidence of significant contact between these neighboring populations.

Zusammenfassung

Bioanthropologische Untersuchungen zielen gewöhnlich auf ganz bestimmte Fragen. Wie archäologische Artefakte können ihre Ergebnisse jedoch auf andere Fragen angewandt werden, als für solche, welche ursprünglich durchgeführt worden waren (sogenannte „sekundäre Anwendungen"). In dieser Publikation werden die Ergebnisse einer Reihe verschiedener Untersuchungen einschließlich der chemischen Rekonstruktion der Ernährungsweise und der chemischen Ermittlung von Kokain im Mumienhaar auf die sekundäre Frage nach dem Verhältnis und dem Grad des sozialen Austausches zwischen der Küsten- und der Hochlandbevölkerung Nordchiles in der archaischen Epoche gerichtet. Obwohl sie nicht weiter als 100 Kilometer voneinander entfernt lebten, trennte diese Populationen eine Wüste enormer, aber nicht unüberwindlicher Trokkenheit. Untersuchungsergebnisse deuten darauf hin, daß, während die Hochlandbewohner ihre Jägerkultur zu einer breitgefächerten Jagd- und Agropastoralgesellschaft entwickelten, die Küstenbevölkerung ihre Fischwirtschaft über den gesamten Zeitraum von 6000 Jahren ohne Anzeichen eines bedeutenden Kontaktes zwischen diesen benachbarten Populationen beibehielten.

Résumé

Les études de bioanthropologie ont généralement pour sujet des questions très spécifiques. Cependant, comme pour les artéfacts archéologiques, les résultats de telles recherches sont susceptibles de s'appliquer à des questions autres que celles auxquelles ils s'agissaient initialement d'exécuter (applications secondaires). Dans cette publication, les résultats de diverses études, dont la reconstruction chimique de régimes alimentaires et la détection chimique de la cocaïne dans les cheveux des momies, sont appliqués à la question secondaire des relations et du degré d'interaction sociale des peuplades côtières et montagnardes du nord du Chili à l'époque archaïque. Bien qu'ils n'aient vécu qu'à une centaine de kilomètres de distance, ces peuples étaient séparés par un désert d'un relief et d'une aridité considerables, sans être toutefois infranchissable. Les résultats de cette étude suggèrent que bien que les populations des montagnes aient évolué d'une culture basée sur la chasse à une économie diversifiée de subsistance agropastorale et de chasse, les résidents de la zone côtière ont conservé une économie basée uniquement sur la chasse d'animaux marins durant une période ininterrompue de 6000 ans sans preuve de contact significatif entre les populations voisines.

Riassunto

Gli studi bioantropologici hanno normalmente come obiettivo la soluzione di problemi molto specifici. Tuttavia, come per i manufatti archeologici, i risultati possono trovare applicazione nella soluzione di problemi diversi da quelli per i quali tali studi erano stati eseguiti ("applicazioni secondarie"). Nella presente pubblicazione i risultati di diversi studi, inclusa la ricostruzione chimica del regime dietetico e il rilevamento chimico di cocaina nei capelli delle mummie, sono stati focalizzati sulla questione secondaria delle relazioni e del grado di interscambio sociale tra le popolazioni costiere e quelle montane del Cile settentrionale durante il periodo arcaico. Sebbene viventi a 100 km l'una dall'altra, esse furono separate da un deserto costituito da terreno improduttivo, arduo ma non invalicabile. I risultati degli studi suggeriscono che, mentre le popolazioni montane manifestarono un'evoluzione culturale passando da un'economia di sussistenza basata sulla caccia ad un'economia diversificata venatorio-pastorale, le popolazioni costiere persistettero in una economia basata sulla pesca per l'intero arco di 6000 anni, senza evidenza di contatti significativi tra queste popolazioni contigue.

Acknowledgements

The author expresses his appreciation for the guidance and instruction in mummy dissection provided by Dr. Marvin Allison, to Larry Cartmell for the cocaine studies, to the National Science Foundation Research Experiences for Undergraduates 1987 grant (award #BBS-8713295) for assistance in specimen collection, to Mary Aufderheide for field assistance and interpretive suggestions and to Sara Hammer for her invaluable editing and clerical services.

References

Allen, C. J. (1988): The Hold Life Has: Coca and Cultural Identity in An Andean Community. Smithsonian Institution Press. Washington, D.C.

Allison, M. J., Foccaci, G., Arriaza, B., Standen, V., Rivera, M., Lowenstein, J. M. (1984): Chinchorro, momias de preparación complicada: métodos de momificatión (Chinchorro, mummies of complicated preparation: mummification methods). Chungará 13, 155–173.

Aufderheide, A. C. (1989): Chemical analysis of skeletal tissues. pp. 237–260 in M. Y. Iscan, K. A. R. Kennedy (eds.): Reconstruction of Life from the Skeleton. Wiley-Liss. New York.

Aufderheide, A. C., Allison, M. J. (1992a): Strontium patterns in infancy can validate retention of biogenic signal in human archaeological bone. Paper presented at the First World Congress on Mummy Studies held at Puerto de la Cruz, Tenerife, Canary Islands, February 3–6, 1992.

Aufderheide, A. C., Allison, M. J. (1992b): Chemical dietary reconstruction of North Chile prehistoric populations by trace mineral analysis. Paper presented at the First World Congress on Mummy Studies held at Puerto de la Cruz, Tenerife, Canary Islands, February 3–6, 1992.

Aufderheide, A. C., Muñoz, I., Arriaza, B. (1993): Seven Chinchorro mummies and the prehistory of northern Chile. American Journal of Physical Anthropology 91, 189–201.

Bird, J. (1943): Excavations in North Chile. Anthropological Papers of the American Museum of Natural History xxxviii, Part iv. New York.

Brotherston, G. (1989): Andean pastoralism and Inca ideology. pp. 240–255 in J. Clutton-Brock (ed.): The Walking Larder. Unwin-Hyman. London.

Bryan, A. L. (1987): Points of order. Natural History 6/87, 6–11.

Cácares M. J. (1985): The Prehispanic Cultures of Peru. National Archaeological Museum. Translated by Daniel Sandweiss. Lima, Peru.

Cartmell, L. W., Aufderheide, A. C., Springfield, A., Weems, C., Arriaza, B. (1991): The frequency and antiquity of prehistoric coca-leaf-chewing practices in northern Chile: Radioimmunoassay of a cocaine metabolite in human-mummy hair. Latin American Antiquity 2(3), 260–268.

Cruxent, J. M. (1968): Theses for meditation on the origin and distribution of man in South America. pp. 11–16 in Biomedical Challenges Presented by the American Indian. Pan American Health Organization Scientific Publication No. 165.

Feldman, R. A. (1992): Preceramic architectural and subsistence traditions. pp. 67–86 in D. H. Sandweiss (ed.): Andean Past 3. Cornell University Latin American Studies Program. Ithaca. New York.

Garcilaso de la Vega, El Inca (1987 [1609]): Royal Commentaries of the Incas and General History of Peru, Part 1. Translated by H. V. Livermore. Reissued. University of Texas Press. Austin. Originally published 1965.

Jennings, J. D. (1978): Origins. pp. 25–27 in J. D. Jennings (ed.): Ancient North Americans. W. H. Freeman. San Francisco.

Kelley, M. A., Levesque, D. R., Weidl, E. (1991): Contrasting pattern of dental disease in five early northern Chile groups. pp. 203–213 in M. A. Kelley, C. S. Larsen (eds.): Advances in Dental Anthropology. Wiley-Liss. New York.

Kennedy, G. E. (1986): The relationship between auditory exostosis and cold water: a latitudinal analysis. American Journal of Physical Anthropology 71, 401–415.

Krouse, H. R., Herbert, M. K. (1988): Sulphur and carbon isotope studies of food webs. pp. 315–325 in B. V. Kennedy, G. M. LeMoine (eds.): Diet and Subsistence: Current Archaeological Perspectives. The University of Calgary Archaeological Association. Calgary.

Llagostera M., A. (1979): 9,700 years of maritime subsistence on the Pacific: An analysis by means of bioindicators in the north of Chile. American Antiquity 44(2), 309–324.

Lynch, T. F. (1990): Glacial-Age man in South America? A critical review. American Antiquity 55(1), 12–36.

Moseley, M. E. (1975): The Maritime Foundations of Andean Civilization. Cummings. Palo Alto, California.

Munizaga, J. (1961): La collecion arqueológica A. Nielson, de Iquique (The Nielson Archaeological Collection at Iquique). Revista Chilena de História y Geográfia 129, 232–246.

Muñoz, I., Arriaza, B., Aufderheide, A. C. (1993): Archaeological Studies of the North Chile Coastal Area. University of Tarapaca. Arica, Chile.

Muñoz, I., Chacama, J. (1982): Investigaciones arqueológicas en las poblaciones preceramicas de la costa de Arica (Archaeological investigations of the preceramic coastal populations of Arica). Documentos de Trabajo 2, 3–97. Universidad de Tarapacá. Arica, Chile.

Paly, D., Jatlow, P., Van Dyke, C., Cabieses, F., Byck, R. (1980): Plasma levels of cocaine in native Peruvian coca chewers. pp. 86–89 in F. R. Jeri (ed.): Cocaine 1980. Proceedings of the Interamerican Seminar on Coca and Cocaine. Lima, Peru.

Plowman, T. (1986): Coca chewing and the botanical origins of coca (*Erythroxylum spp.*) in South America. pp. 5–34 in D. Pacini, C. Franquemont (eds.): Coca and Cocaine: Effects of Policy in Latin America. Cultural Survival Report No. 23, Cultural Survival Inc. Cambridge.

Reinhard, K. J., Bryant, Jr., V. M. (1992a): Coprolite analysis: A biological perspective on archaeology. pp. 245–288 in M. B. Schiffer (ed.): Advances in Archaeological Method and Theory 4. University of Arizona Press. Tucson.

Reinhard, K. J., Bryant, Vaughn M. (1992b): Investigating mummified intestinal contents: Reconstructing diet and parasitic disease. Paper presented at the First World Congress on Mummy Studies held at Puerto de la Cruz, Tenerife, Canary Islands, February 3–6, 1992.

Rivera, M. (1991): Prehistory of northern Chile: A synthesis. Journal of World Prehistory 5(1), 1–47.

Sandison, A. T. (1969): The study of mummified and dried human tissues. pp. 490–502 in D. Brothwell, E. Higgs, G. Clark (eds.): Science in Archaeology: A Survey of Progress and Research. Thames and Hudson. London.

Santoro, C., Chacama, J. (1982): Secuencia cultural de las tierras altas del area centro sur Andina (The cultural sequence of the south central Andean area). Chungara 9, 22–45.

Santoro, C. M., Nuñez, L. (1987): Hunters of the dry puna and the salt puna in northern Chile. pp. 57–109 in D. H. Sandweiss (ed.): Andean Past, Vol 1. Anthropology Department, Cornell University Press. Ithaca, New York.

Standen, V., Allison, M. J., Arriaza, B. (1984): Patologías óseas de la población Morro-1, asociada al complejo Chinchorro: Norte de Chile (Skeletal pathology of the Morro-1 population: northern Chile). Chungará 13, 175–182.

Standen, V., Allison, M. J., Arriaza, B. (1985): Osteoma del conducto auditivo externo: Hipótesis en torno a una posible patologia laboral prehispanica (Osteoma of the external auditory canal: hypothesis of a possible pre-Hispanic occupationally-related pathological lesion). Chungará 15, 197–209.

Styles, B. W. (1985): Reconstruction of availability and utilization of food resources. pp. 21–59 in Robert I. Gilbert, Jr. and James H. Mielke (eds.): The Analysis of Prehistoric Diets. Academic Press. New York.

Tieszen, L. L., Chapman, M. (1992): Carbon and nitrogen isotopic status of the major marine and terrestrial resources in the Atacama Desert of northern Chile. Paper presented at the First World Congress on Mummy Studies held at Puerto de la Cruz, Tenerife, Canary Islands, February 3–6, 1992.

Tieszen, L. L., Iversen, E., Matzner, S. (1992): Dietary reconstruction based on carbon, nitrogen and sulfur stable isotopes in the Atacama Desert, northern Chile. Paper presented at the First World Congress on Mummy Studies held at Puerto de la Cruz, Tenerife, Canary Islands, February 3–6, 1992.

Uhle, M. (1919): La Arqueología de Arica y Tacna (The Archaeology of Arica and Tacna). Boletin de la Sociedad Ecuatoriana de Estudios Histórics Americanos. Quito.

Correspondence: Dr. Arthur C. Aufderheide, Paleobiology Laboratory, Department of Pathology, University of Minnesota, Duluth, MN 55812, U.S.A.

The Prince of El Plomo: a frozen treasure

P. D. Horne

Department of Pathology, York County Hospital, Newmarket, Ontario, Canada

Introduction

Until the discovery of the "Iceman" in 1991, the only region where ancient human remains were preserved by high altitude cold was the Andes between Argentina and Chile. The first such frozen individual was discovered in 1905. In 1954 the Prince of El Plomo, the topic of this paper, was discovered (Mostny 1957, Horne & Quevedo K. 1984). In 1964 the mummy of Cerro El Toro was discovered and well documented (Schobinger 1966). In recent years more than 115 high altitude sanctuaries and several other mummies have been found and documented. All of these human remains and sanctuaries lie between 5,200 and 6,700 meters above sea level. All are the remains of the short-lived, yet very powerful Inca civilization and date from approximately 1,475 A.D. to 1,536 A.D. – the latter date coinciding with the arrival of the Spanish.

The great Inca civilization did not arise in isolation but was the ultimate in a continuum of several thousand years. The dominance of the Quechua-speaking Inca hinged not only on the power and abilities of the rulers but on a highly organized economic, religious and political system. Ethnic groups in conquered territories were often allowed to keep their customs and sometimes even local chiefs so long as tribute was paid to the supreme Inca.

By about 1,100 A.D. the Inca began to move into the Valley of Cuzco, Peru and for approximately the next five hundred years began to occupy adjacent territory. In the middle of the 15th century imperialist expansion occurred. By about 1,475 A.D. the Empire reached its maximum. By the time of the Spanish conquest the Empire stretched more than 3,000 km north/south and about 800 km east/west linked by 17,000 km of roads.

The high peaks of the Andes were sacred to the Inca. High altitude sanctuaries are strung out along the Andes chain from Ecuador to Cerro el Plomo peak at the southern end of Inca territory. Cerro el Plomo is a dominant peak some 45 km east of present day Santiago, Chile.

In 1954 three mountain climbers discovered an intact ceremonial complex with tomb at the 5,400 meter level on El Plomo. The complex consisted of four low structures, three rectangular, one circular, each with a stone retaining wall and filled with stone rubble. Under one of these structures covered by a cap stone, was found the tomb.

The body and accompanying grave goods were purchased by the National Museum of Natural History, Santiago, Chile. Following careful and detailed studies the body, along with its artifacts, was placed in a glass fronted freezer and put on display.

Recent El Plomo mummy studies

Almost thirty years later, in 1982, I was invited by the United Nations Ecomonic Scientific and Cultural Organization (UNESCO), Paris and the National Museum, Santiago to study the body in detail from both medical and conservational points of view.

First let us look at the mummy as it was found in 1954. The preservation, due to constant high altitude low temperatures and low relative humidity, was exquisite. The male child, as was shown later, had been sacrificed and buried alive. He was found fully clothed in the sitting position. Approaching death he had pulled his knees to his chin, clasped folded hands around his knees on which he laid his head (Fig. 1).

The child's face was covered by a thin layer of red paint composed of iron ochre. Four broad yellow bands on each side of the face completed the decoration. The yellow pigment was shown to be composed of arsenic sulphide mixed with animal fats. The facial design, all but obscured during the thirty years since recovery, was brought back to sharp focus through infrared photography. The shoulder length hair was divided into more than 200 fine braids and held in place by a *llatu* or headband woven from human hair. His head was crowned by a llama wool head-dress topped by condor feathers.

The child's principle garment was a sleeveless tunic or *uncu* woven of black llama wool and measuring 94 cm by 47 cm with a central slit for the head to pass through. The tunic was trimmed with dyed llama fur. A large grey alpaca wool shawl or *yacolla* was folded over his

Fig. 1. The "Prince of El Plomo" with accompanying grave goods

shoulders. Unused soft leather moccasins each made from a single leather strip and trimmed with a red embroidered band adorned the feet. A heavy silver bracelet was found on the left wrist while a silver H-shaped pectoral ornament completed the attire; both silver ornaments attesting to his high social status. Grave goods consisted of a laminated gold llama 6.5 cm in height and a smaller red *Spondylus* shell llama. A female figurine made of soldered and laminated silver is exquisitely clothed in the attire of the day even to imported parrot feather decoration. A woven bag, magnificently worked in feathers was found to be tightly packed with still aromatic coca leaves. Three hollow balls made of animal intestine were found to be filled with human hair, fingernail clippings and deciduous teeth.

The significance of each of these articles of clothing and grave goods was well documented by the Spanish at the time of conquest not long after the child's death but will not be detailed here. However, studies of these chronicles and modern archaeological studies show the child to be from Qolla Suyu the southeastern quarter of the Empire near Lake Titicaca. This means the child walked or was carried some 2,000 km to reach his destination of Cerro El Plomo.

These same chronicles tell us that human sacrifice among the Inca was not a common occurrence but reserved only for auspicious occasions. In this case, perhaps to appease El Plomo, the source of life giving water.

Children for sacrifice were chosen from the social elite. Four methods of sacrifice were in place at time of conquest: 1) strangulation with a piece of cord, 2) placing the child's neck over an anvil and breaking the cervical vertebrae with a stone, 3) opening the rib cage with an obsidian blade and removing the still beating heart and 4) the method chosen in this case – burial alive. In each case the child was first numbed with several drafts of *chica,* a maize beer, and coca was chewed as well. Evidence for *chica* drinking in this case may be the vomit stain on front of the child's tunic. Orientation of ceremonial structures and information gleaned from the chronicles would indicate this child was sacrificed at the summer solstice (December 23) the great feast of Capac Racmi.

In July 1982, a multidisciplinary team of local experts which I was able to put together, began a six week study of the remains. The tedious task of clothes removal and their conservation was carried out by the Museum's curator of textiles. The body was then sealed in a transparent plastic bag with silica gel to prevent moisture buildup. The body was removed from refrigeration and the bag for only short periods at a time.

The body was first examined for any superficial microbial contaminants. None were observed visually and numerous skin swabs failed to grow microorganisms on any of the various culture media employed at 0°, 22°, and 37 °C. Appropriately stained sections of skin confirmed the absence of microbial contamination of the superficial tissues.

The body was re-xrayed and submitted to CAT scan studies. These studies showed the child to be 8 to 9 years of age. No pathology, trauma or Harris lines were observed. It was noted however, that the right first metatarsal was shorter and thicker than normal. All internal organs were intact and in a fine state of preservation.

The child's anus was widely dilated. The fecal remnant studied in the 1950's was no longer available for examination. Studies of the faeces at that time revealed numerous ova of *Trichuris trichiura* and some specifically unidentifiable cysts of amoeba (Pizzi & Schenone 1954). The investigators at that time suggested that since no report of *Trichuris* came from Europe prior to the 1740's this nematode was introduced by the conquistadors. However, since then, there have been numerous reports of *Trichuris* in the ancient Old World. These range from the intestinal contents of a Han Dynasty mummy of China (Wei 1973) to excrement from prehistoric salt miners in Austria dating to 800–350 B.C. (Aspock et al. 1973).

While cleaning and rearranging the childs hair I found numerous eggs or nits of the common head louse. No adult lice were found. Both myself and others have found evidence of human head lice in mummies of both Old and New Worlds from the most ancient of times.

A tiny skin biopsy was submitted to micro elution methods for blood grouping and was shown to be group A confirming earlier studies (Hart et al. 1983).

Due to the unique nature of this mummy, declared a world treasure by the United Nations, I decided not to open the body to sample internal organs for fear of allowing the entrance of opportunistic organisms. Attempts to biopsy the liver with a sterile biopsy needle proved unsuccessful. The tiny puncture was sealed with sterile wax.

A complete examination of the surfaces of the mummy proved impossible due to its flexed position. The child appeared well nourished, if not obese, and weighed 10.62 kg. The plantar portions of his feet showed marked hyperkeratosis with edema in the lower legs especially the lower right leg. It is of interest to note that the nail beds of four fingers of the left hand exposed to cold were of a blue/black color while the nail beds of the clasped right hand protected from the cold were of a normal color. This would indicate the child reached his final destination alive.

An especially exciting find was noted on the child's left hand. At the base of the thumb were two small (2–3 mm) slightly raised lesions with the appearance of common verrucae.

I immediately removed one of these and prepared it for examination by transmission electron microscopy at the nearby medical school. Sections showed the typical

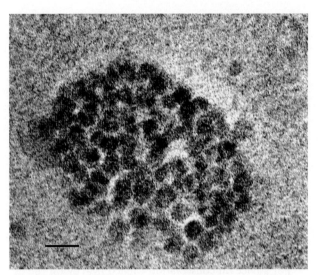

Fig. 2. Papilloma virus from wart on child's hand, ×300,000 (bar: 0.03 μm)

papilloma virus the causative agent of the common wart (Fig. 2). This was a most exciting find being the first time viruses were demonstrated in ancient tissue.

As in the previous study eight ulcerated lesions were noted on the lower limbs. These had not been biopsied in the previous study. The lesions ranged in size from 0.5 to 1 cm with well defined borders and were covered by a fibrinous exudate. I biopsied one of these lesions and prepared it for histological examination. In the area of the lesion microscopic examination revealed numerous dilated vascular channels (Fig. 3). These were filled by a homogeneous eosinophilic material which may represent hemolysed blood. These vessels appear to be capillary in origin although the possibility that they may represent lymphatics cannot be entirely excluded. Similar dilated vessels are seen in the subcutaneous fat, some being considerably larger than those in the dermis. The epidermis

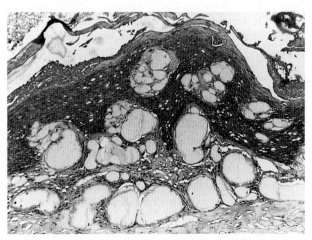

Fig. 3. Acanthotic vascular lesion. The dilated capillaries extend from the epidermis, through the dermis, into subcutaneous fat, ×100

is thickened, hyperkeratotic and assumes a papillary configuration where the capillaries are most abundant. Numerous polymorphonuclear leukocytes and possibly other types of white blood cells and abundant fibrin are observed in the ulcerated areas. Various attempts to demonstrate tissue antigens, eg. vimentin and Factor VIII, both by myself and colleagues, proved unsuccessful. However, the histologic features are consistent with an angiokeratoma, probably angiokeratoma circumscriptum which is secondarily inflamed and ulcerated (Horne & Quevedo K. 1984).

Conservation

For advice on conservation I relied heavily on the extensive expertise of the Canadian Conservation Institute (CCI), Ottawa. During the thirty years the mummified remains were in the Museum freezer, some deterioration had occurred as was evidenced from some superficial cracking of the epidermis.

Upon my arrival I found the mummy sitting unprotected in a morgue refrigerator at 4 °C. My first reaction was to enclose the body in a plastic bag with silica gel desiccant to lower humidity.

The recommendations of CCI were rather lengthy and will not be detailed here. In summary it was the CCI's contention and mine that we were dealing with a freeze-dried or almost completely freeze-dried body. It has been estimated that the dry weight of humans is 23–30 % of live weight. The average weight of an 8.5 year old male American is 37 kg (Behrman 1992). It would follow that for an obese 8–9 year child 10.6 kg would be close to its dry weight.

It is probable that the body was freeze-dried by time of recovery. This process could also have occurred or been enhanced during the thirty years in the museum freezer. It was felt, however, that the following factors should be taken into account for its continued preservation:

1) Low and constant relative humidity in the range of 30 %.
2) Low and constant temperature in the range of 4–5 °C.
3) A dust free atmosphere composed of an inert gas such as argon or at least an oxygen reduced atmosphere.
4) Low, indirect lighting well below 100 lux to avoid color changes in the textiles and the child's skin color. These recommendations were put forward as the minimum requirements for continued preservation.

The body is now being kept at optimum conditions in the conservation area of the Museum while a replica of the body is on display.

Summary

The freeze-dried remains of an 8 to 9 year old Inca child were found at 5,400 m level on Cerro El Plomo peak in the Andes, near present day Santiago, Chile. The child had been sacrificed and buried alive shortly before the Spanish conquest of the mid 16th century. A careful examination of the well preserved remains showed no major pathology. Skin lesions on the lower limbs proved to be a rather rare paediatric condition: *angiokeratoma circumscriptum*. Electron microscopic studies of a lesion at the base of the left thumb showed papilloma virus – the causative agent of *verruca vulgaris*. A summary of recommendations regarding future conservation is presented.

Zusammenfassung

Auf dem Cerro El Plomo in den chilenischen Anden, in der Nähe des heutigen Santiago, wurden auf 5.400 m ü. M. die gefriergetrockneten Überreste eines 8–9 Jahre alten Inka-Kindes gefunden. Das Kind war kurz vor der spanischen Eroberung um die Mitte des 16. Jahrhunderts geopfert und lebendig begraben worden. Eine sorgfältige Untersuchung der gut erhaltenen Überreste zeigte keine größeren Pathologien. Hautläsionen an den unteren Gliedmaßen erwiesen sich als eine eher seltene Kinderkrankheit, *Angiokeratoma circumscriptum*. Elektronenmikroskopische Studien der Läsion am Ansatz des linken Daumens zeigten Papillomaviren, die Verursacher der *Verruca vulgaris*. Empfehlungen für die weitere Aufbewahrung werden in einer Zusammenfassung präsentiert.

Résumé

Sur le Cerro El Plomo dans les Andes Chiliennes, près du Santiago d'aujourd'hui, on a trouvé à 5.400 m d'altitude les restes lyophilisés d'un enfant de 8–9 ans appartenant à la tribu des Incas. Il avait été sacrifié et enterré vif peu avant la conquête espagnole au milieu du XVIème siècle. Au cours d'examens détaillés des restes parfaitement conservés, aucune pathologie majeure n'a été détectée. La présence de lésions de la peau aux extrémités inférieures s'est révélée comme le symptome d'une maladie infantile plutôt rare, dite *angiokératome circomscriptum*. Les examens sous microscope électronique de la lésion á l'insertion du pouce gauche ont permis de détecter des virus du papillome, qui provoquent la *verrue vulgaire*. Dans un résume sont proposées des recommandations pour une conservation ultérieure de ces découvertes.

Riassunto

Sul Cerro El Plomo nelle Ande cilene nelle vicinanze della Santiago odierna, a 5.400 m s. l. m. furono ritrovati i resti essicati e gelati di un bambino tra gli otto ed i nove anni, appartenente agli Inca. Il bambino era stato sacrificato prima della conquista spagnola verso la metà del XVI sec. ed era stato sepolto vivo. Uno studio accurato dei resti ben conservati non presenta nessuna patologia di rilievo. Delle lesioni cutanee alle estremità inferiori risultano essere dovute ad una patologia infantile piuttosto rara, *l'angiocheratoma circumscriptum*. Le analisi al microscopio elettronico della lesione all'origine del pollice sinistro hanno evidenziato la presenza del virus papilloma che provoca la *verruca vulgaris*.

Nel riassunto vengono formulate raccomandazioni per l'ulteriore conservazione della mummia.

References

Aspöck H., Flamm H., Picker O. (1973) Darmparasiten in menschlichen Exkrementen aus prähistorischen Salzbergwerken der Hallstatt-Kultur (800–350 v. Chr.). Zbl Bakt Hyg 223: 549–558.

Behrman R. E., ed. (1992) Nelson Textbook of Pediatrics, 14th ed., W. B. Saunders Co., Toronto, p. 23.

Hart G. D., Kvas I., Soots M., Badaway G. (1983) Blood group testing of ancient material. In: Hart G. D. (ed.) Disease in Ancient Man. Toronto: Clark Irwin 159–171.

Horne P. D., Quevedo Kawasaki S. (1984) The prince of El Plomo: A paleopathological study. Bulletin of the New York Academy of Medicine 60: 925–931.

Mostny G., ed. (1957) La Momia Del Cerro El Plomo. Boletín del Museo National de Historia Natural. Tomo XXVII, No. 1, Santiago de Chile.

Pizzi T., Schenone H. (1954) Hallazgo de huevos de *Trichuris trichiura* en contenido intestinal de un cuerpo arquelógico incaico. Boletín Chileno Parasitologia 9: 73–75.

Schobinger J. (1966) La "Momia" Del Cerro El Toro. Suplemento al Tomo XXI de Annales de Arqueologia y Ethnologia, Mendoza, Argentina.

Wei O. (1973) Internal organs of a 2,100 year old female corpse. Lancet 2: 1198.

Correspondence: Mr. Patrick D. Horne, Department of Pathology, York County Hospital, 596 Davis Drive, Newmarket, Ontario, L3Y 2P9, Canada.

European mummies

European memories

European bog bodies: current state of research and preservation

D. Brothwell

Department of Archaelogy, The University of York, York, U. K.

Introduction

Of the various well preserved ancient bodies which have been discovered in different world environments, none have produced as many human remains as peat bogs. These have mainly been derived from peat deposits in North-West Europe, although some peat-preserved remains have appeared elsewhere, including in America. Altogether, it has been estimated that well over 1400 bodies, or parts of bodies, have thus been found. However, it is not usually stated that the majority of finds are of bones only, or in some cases only clothing with perhaps human skin and hair. So it is difficult to estimate with precision the number of well preserved bodies which have been found, although for Britain and Ireland I would estimate that there have been about 30 well preserved bodies out of around 200 human finds in peat bogs (15 %).

In date, the bog bodies range widely from 900 BC to 1,800 AD, but those of special archaeological interest concentrate between about 600 BC to 300 AD and appear to be mainly linked to the cultural practices of the early Germanic peoples of northern Europe. While some bodies may represent evidence of severe punishment, the deaths of others may indicate ritualistic sacrifice. A few may have been normal burials into bogs (Gebühr, 1979).

In contrast, bodies which have been buried in bogs within the last few centuries are likely to have missed burial in a Christian cemetery for one of two reasons: either they are bodies hidden for criminal reasons, or they are individuals rejected by their societies and have thus been denied normal Christian burial. For instance a female body from the Great Bog of Ardee, currently being re-investigated at the Royal College of Surgeons in Ireland, is known to have been buried about 1775. Disappointed in love, the woman is known to have committed suicide by poison. Normal burial of her body was refused and she was thus disposed of in the bog.

Most bodies are from raised bogs, formed from limited plant varieties dominated by sphagnum moss. The antibiotic properties of sphagnan, an organic component derived from this species of moss (Painter, 1991), has probably been more important in the preservation of bog bodies, than tannin or the wet anaerobic acid conditions (pH 3.6–4.0). Depth of burial into peat does not seem to be an important variable, and while the Ardee woman was only about 60 cm down in peat, the Tollund man was 2.5 m below the surface, and some bodies have been deeper.

Records of bog body discoveries extend back to 1640, this date referring to a find in Shalkholz Fen in Schleswig Holstein (Glob, 1969). In 1773, in the Ravholt district of Denmark, a naked peat body was found which provided early evidence of the violence which was even-

Fig. 1. General view of Tollund man, showing variation in the degree of preservation

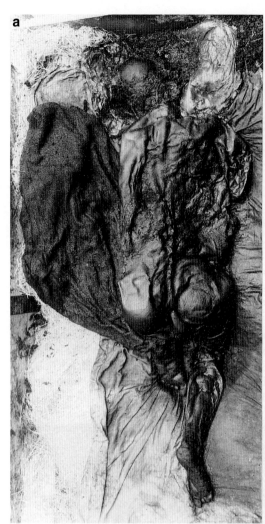

Fig. 2. General views, back (a) and front (b), of the Borre Fen III woman. Note the severely damaged and decayed head region

tually to be seen in later bodies. Under the chin was a deep cut, forensic evidence of a kind which was to be revealed again in the Grauballe man and Lindow II individual (Fischer, 1980; Bourke, 1986). During the nineteenth century, a number of the bodies were submitted to preliminary examination, and even tentative conservation. The Rendswühren body from near Kiel was studied in 1873, and its surprising brain preservation was explained as being due to the presence of cholesterol in the tissues.

Over the years various other bog bodies received preliminary examination, in order to satisfy judicial enquiries. Some bodies were reburied and others were handed on to museums, especially in Denmark, Germany, Holland and Ireland. but discoveries at Tollund, Grauballe and Borre Fen (Figures 1, 2, 3a and d) in Denmark between 1946 and 1952 established a new level of investigation of such finds. In particular, the various research studies on Grauballe man (reported in *Kuml* for 1956) established criteria for the detailed study of such bodies (Krebs and Ratjen, 1956; Helbeck, 1958). Recently, further research on bog bodies has been initiated in Britain, Denmark, Ireland and Holland. This work has been partly on newly discovered bodies, as well as on earlier finds in need of re-investigation.

Taphonomic problems

In considering the preservation of European bog bodies over the past eight years, I have been impressed by the great range of variation in the survival of tissues. Generally, it appears that tissue survival depends on such factors as how fast a body is immersed in water, temperature and time of year, presence of insect predators, and the extent and types of internal micro-organisms. Calcified and keratinous structures (bones, teeth, skin, hair and nails) are the most resistant to decay, with muscles and ligaments showing variable survival. Surprisingly, the intestinal tract does not self-digest after death and can be very well preserved, as seen in Tollund man (Fig-

European bog bodies: current state of research and preservation   163

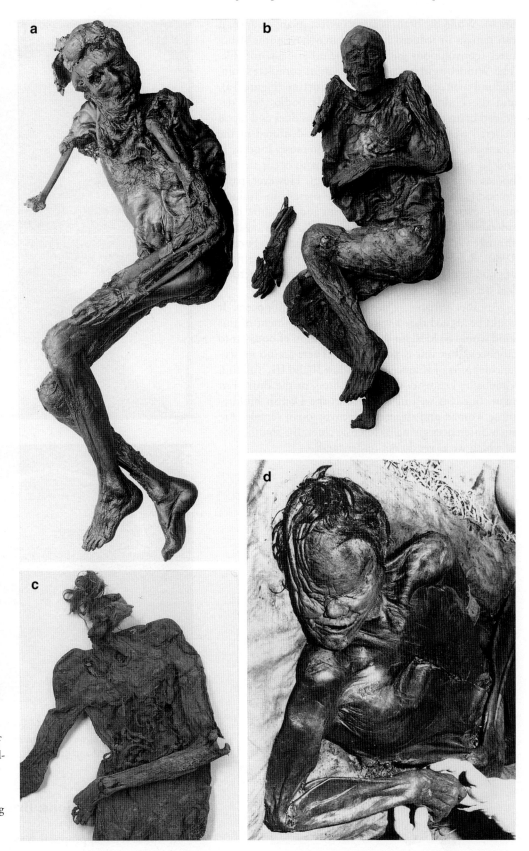

Fig. 3. General view of four bog bodies from Denmark and the Netherlands. (a) The Borre Fen I man. (b) The Huldremose woman. Note the right arm which was hacked into pieces (the fingers were removed by the peat cutter). (c) One of the two Weerdinge bodies from northern Holland. (d) The face and chest region of the Grauballe man, showing the long autopsy incision in midline

ure 4). While the brain is rarely perfect in appearance, it can remain in reduced form and may have some structures (convolutions and hemispheres) still identifiable. The Windeby girl has perhaps the most complete brain, Tollund man a somewhat reduced structure (Figure 5a), and the Huldremose and Lindow II bodies very degraded brain tissue (Figure 5b). Food residues and faecal material were preserved and could be identified in the Grauballe, Tollund, Borremose, Huldremose, Lindow II and III bodies. The least likely organs to occur in such bodies are the kidneys, liver and lungs. However, paradoxically, these soft structures were present in Tollund man – even though parts of the arms and legs show considerable decomposition. In Grauballe man also, a shrunken liver and lung tissue were identified (Fischer, 1980). Preservation of the Lindow II body was externally as good as these two bodies, yet most internal structures had decomposed. Why?

Another variable is the occurrence of adipocere. It could be argued that a waterlogged environment of this kind would be ideal for the transformation of body fat to adipocere, but this is not the case. There are no noticeable deposits of adipocere in Lindow II and III, Tollund, Grauballe, Damendorf, Rendswühren, or the Borre Fen man. But it is well in evidence in the Danish Huldremose woman and the Irish Meenybraddan female. Are such differences giving us forensic clues, the significance of which we as yet do not understand?

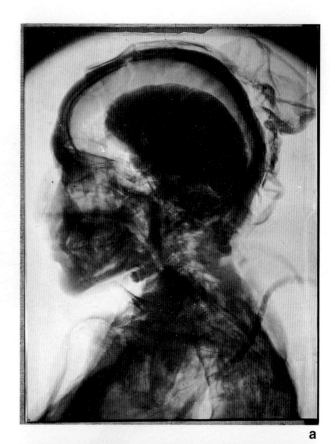

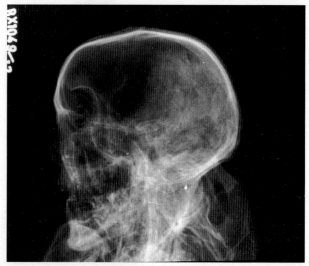

Fig. 5. Examples of the variable preservation of the brain in bog bodies. (a) The well formed but shrunken brain in Tollund man. (b) The more decayed and posteriorly placed brain of the Huldremose woman

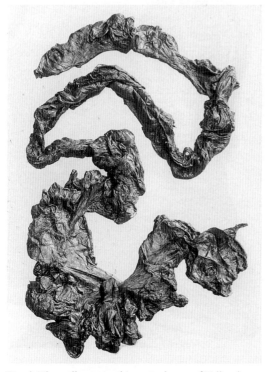

Fig. 4. The well preserved intestinal tract of Tollund man

Current investigations: Britain and Ireland

It would seem appropriate here to mention work currently being undertaken in Britain and Ireland. The discovery of parts of two bodies at Lindow Moss, near Manchester in 1986 is now well known (Stead, Bourke and Brothwell, 1986), but in fact additional finds subsequently came to light, unfortunately cut to pieces by the industrial peat cutting machinery. Detailed studies on these and the previous finds will be published in 1995, but some comment may be briefly made here. In the hope of discovering further, more intact, bodies an excavation at the site was carried out in 1987, but without further discoveries of human remains.

Considering the first and subsequent finds, the Lindow II man is now fairly complete (body A in Figure 6). Lindow III, a male, is represented by parts of the thorax, abdominal area and both arms and legs – but not the head (B and C in Figure 6). Lindow I is represented by a head only, probably from a female, but the possibility that Lindow I and III are the same individual is still being considered. A surprising find in Lindow III are two abnormally small hand phalanges, which probably indicate a minor congenital abnormality associated with the thumb. Could this have been the reason why the individual was selected for ritual death in the bog? – if indeed it was a ritual death.

New research in Ireland has been stimulated by the Lindow finds, so that three bodies have been under investigation. Initial studies are now completed on the Meenybraddan body, found in 1978 but placed in cold storage for some years (Delaney and O'Floinn, 1995). The woman was of medieval date (AD 1,050 – AD 1,410) and is one of the best preserved bodies of that period. Following conventional radiography, CT scanning, and endoscopy, it was autopsied and eventually freeze-dried. She was well nourished at death, and much body fat had been converted into adipocere. While there is some evidence of anthracosis, and possibly infection of the lungs and pleura, the actual cause of death could not be determined.

While the most recent bog body count for Ireland is 82 (O'Floinn, 1995), only two others are being currently re-investigated. Of these, the male from Castle Blakeney is of Iron Age date, being found 3 m deep in 1821 and naturally dried to a stable state. On discovery, it was in a good state of preservation, but has deteriorated over time. Nevertheless, it has been CT scanned recently by Professor Max Ryan in Dublin and is the subject of further study at present by Dr. M. Delaney and colleagues. The third body is that of the eighteenth century suicide from Ardee, mentioned already.

Physique

The physique of the individuals from bogs is not easy to establish, although stature established soon after discovery and before shrinkage may be more reliable than dimensions taken later. Where stature was obtained as in

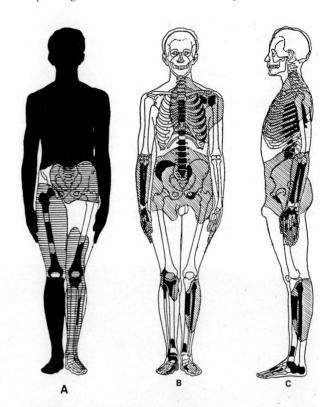

Fig. 6. Diagrammatic representations of the Lindow II body (A) and Lindow III (B and C). The areas present in black are the initial discovery

the Tollund, Borre Fen and Lindow men, they were of moderate stature (range 155–173 cm). Because of the small sample size no significance can be given to such dimensions as evidence of European Iron Age stature. Compression and distortion of the body laterally or antero-posteriorly, makes estimates of robustness highly tentative. The Borre Fen, Tollund and Rendswühren men appear to be gracile, but the Grauballe and Lindow II individuals could well have been more robust. The Huldremose, Borre Fen III and Meenybraddan women and Windeby girl appear to have been plump, and displayed varying amounts of body fat converted to adipocere.

Cranial and facial variation are difficult to establish, owing to the common occurrence of distortion, and in some cases the drying out has resulted in considerable shrinkage. Indeed, we need experimental studies to investigate techniques for the rehydration and expansion of shrunken dried bones and bodies, as well as techniques to correct for the different kinds of distortion.

Evidence of trauma

One of the most interesting aspects of the prehistoric European bog bodies is the varied and common evidence of trauma. Excluding minor superficial incised wounds on the sole of the right foot of Tollund man, the majority of evidence is of forensic interest in that the injuries are not accidental or self-inflicted, but result from acts of punishment or aggression on the part of others. In six instances, there is evidence of more than one form of injury to the person. In the case of the second Lindow individual, there are multiple injuries – presumably of ritual significance – the sequence of trauma still being debatable (Figure 7a). In one of the Weerdinge bodies from the Netherlands, a chest wound (Figure 3c) deserves further investigation, and so too does the head region of the third Borre Fen body, and this woman remains one of the most interesting problems in forensic interpretation (Figure 2a and b).

The well built Borre Fen III woman (530 BC±100) was found lying on her stomach, her body appearing to be compressed but undamaged except for the head region. The posterior part of the head was said to be intact but scalped, and the face was crushed (Fischer, 1980). It seems to me that this is one of a number of bog bodies which deserve careful re-investigation. Like the first discovered Borre Fen body of a man (Figure 3a), the nature of the broken skull bones indicates violence at, or soon after, death. In fact the woman does not simply have a crushed face, for there is clearly damage at the back of the head. Indeed, I suspect that the claim of scalping cannot be supported by the anatomical detail, and the detached area of scalp is simply an indication of the

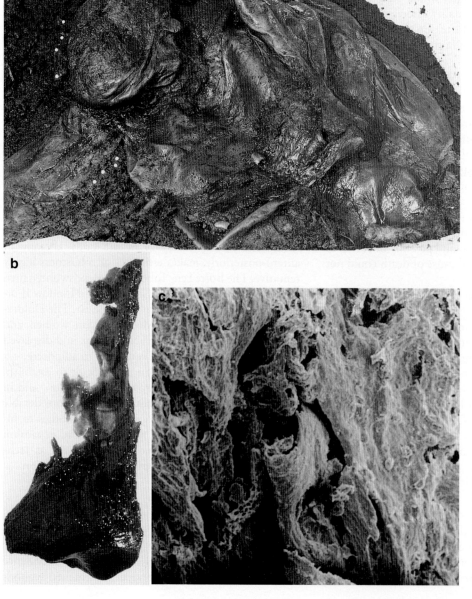

Fig. 7. Preservation aspects of the Lindow bodies. (a) Lindow II partially exposed from the peat. (b) A hand phalanx from Lindow III, displaying considerable erosion. (c) High magnification (SEM) detail of an area of phalangeal surface erosion (Lindow III)

severe beating about the head the woman received about the time of death. This is clearly still a controversial issue (Andersen, 1990), and only further study will resolve the problem.

How worthwhile re-examinations are in the case of some bog bodies is exemplified by the recent study of the Huldremose woman (Figure 3b) from Denmark (Brothwell, Liversage and Gottlieb, 1990). Discovered in 1879, it received only preliminary consideration before a detailed restudy, with conventional radiography and CT scanning, was undertaken in 1986 at The National Museum in Copenhagen. Contrary to previous comments (Glob, 1969), her right arm was not broken off, but was hacked off – possibly into a number of pieces (the fingers were also taken off accidentally by the peat cutter). Other damage with a sharp weapon were noted below the left knee, above the left heel, below the right knee, at the right medial malleolus of the tibia, and on the upper aspect of the right foot (Figures 8 a–d).

False, or pseudo-pathology, was produced on the Huldremose body by ground pressure on the softened bones and the drying of the body in the museum. This resulted in pseudo-biparietal thinning and pseudo-healed-fractures of the left forearm and left thigh (Figures 9 a and b). Abnormal bone shapes of post-mortem origin were also noted in the Tollund man (Fischer, 1980), and the Grauballe individual (Krebs and Ratjen, 1956). It is also possible, but this needs further investigation, that the apparent "dislocations" at the left knee and elbow of Gruballe man, the right carpals from the radius in the Huldremose woman, and the right ankle in the Castle Blakeney man from Ireland, are the result of partial decay in the joints, allowing bone slipping, as a result of rough handling by the original excavators.

Chemical aspects of bog bodies

As regards the chemistry and molecular biology of these bog finds, research progress has been slow. While blood group studies have been undertaken on well preserved ancient bodies over the last sixty years, only a very few bog finds have been tested, and these are only from Britain and the Netherlands (Connolly, Evershed, Embery, Stanbury, Green, Beahan and Shortall, 1986; van der Sanden, 1990). A, B and O substances have been identified, but the overall sample is so small that the gene frequencies must remain unknown until more bodies are considered for their blood groups. In the case of the Lindow II individual, M substance of the MNSs system was also identified.

As yet, soft tissue from only eight bodies have been selected for DNA analysis. Information of this kind is of course potentially of value in confirming the sex of incomplete bodies, of detecting possible family links in bodies from the same site (Lindow, Borre Fen, Weerdinge) and in assembling information on genetic variation in past populations. Unfortunately, even though attempts were made to remove PCR inhibitors, only negative results were achieved (Osinga, Buys and van der Sanden, 1993; Hughes, Jones and Connolly, 1986).

The study of chemical elements in these human remains is as yet at an elementary stage, and limited to a few bodies. Lindow II and III have probably received the most attention so far (Connolly, Evershed, Embery, Stanbury, Green, Beehan and Shortall, 1986; Pyatt, Beaumont, Buckland, Lacy and Storey, 1991; Pyatt, Beaumont, Lacy, Magilton and Buckland, 1991). The elemental content of the Lindow samples was considered from the point of view of interchanges between a body and its burial environment, possible diet related differences between bodies, and any possible cultural factors which might influence the chemistry of the skin surface.

Twelve elements were noted in the skin of Lindow II and ten were found in Lindow III. Nine elements are shared in common, but there are noticeable percentage differences for some of these elements. While it is difficult from the results to suggest dietary contrasts, further consideration of some element differences might still establish this. Interesting speculation has arisen as regards body decoration. While tests for indigotin (the colorant in the ancient body decoration woad) were negative (Taylor, 1986), and infra-red photography did not reveal tattoos or body paint, nevertheless the elemental analysis might suggest body decoration. In particular, Cu values are enhanced, while Al and Si are well represented and may indicate a clay mineral base. Because of these element anomalies, it has therefore been tentatively suggested by some of my colleagues that at least the Lindow III individual may have displayed some body decoration at death.

As regards the preservation of lipids and other organic compounds, studies on the Lindow bodies have been able to establish the presence still of endogenous cholesterol (Evershed and Connolly, 1988). In the case of fatty acid changes, there was no evidence of adipocere formation in the Lindow bodies, but analyses undertaken on the Meenybraddan body showed typical changes, with the predominant formation of palmitic acid (Delaney and O'Floinn, 1994; Evershed, 1992). It would be interesting to have far more studies on the proportions of fatty acids and the extent of conversion to adipocere in the European bog bodies. Indeed, this may help to reveal differences in the events after death and of the physicochemical environments within the bogs.

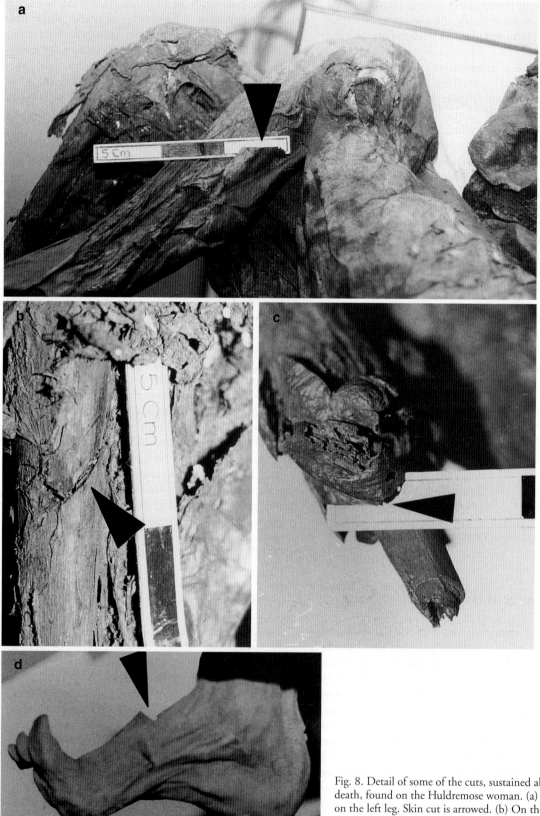

Fig. 8. Detail of some of the cuts, sustained about the time of death, found on the Huldremose woman. (a) Below the knee on the left leg. Skin cut is arrowed. (b) On the upper shaft of the right tibia. (c) On the right upper arm (skin shows straight cut – arrowed). (d) On the upper arch of the right foot

Keratin structures

Anaerobic bog environments permit the survival of hair and nails in a good state of preservation, and there is considerable research potential in these structures. Variation has both forensic and dietary implications. Hair on the head and face tells varied stories in the bog bodies. The beheaded Osterby man (Figure 10 c) had long hair skilfully knotted, while the beheaded Roum Fen woman had roughly-cut short hair (Figure 10 a). The Huldremose woman had long hair which was cut off (Figures 10 b and d) and thrown onto the body after death (Brothwell et al., 1990). The Elling woman from the same large bog as Tollund man, and similarly hanged, had long plaited hair which had presented problems to the executioner, who had wound it up and round itself to remove its contact with the rope. The short beard stubble in the Borre Fen, Tollund and Grauballe men could perhaps suggest captivity and the inability to shave for some days before death. The Lindow II man had a well trimmed beard, with detail at the cut ends showing that scissors and not a knife had been used.

Finger and toe nails are not usually found in place in these bodies. As in the Lindow II body, distance of nails from a body may give clues to the burial history of the individual. High magnification study of nails surfaces can also provide clues to the lifestyle and labour of such individuals. For instance, there is no doubt from the smooth well rounded finger nails of Lindow II, that he did not undertake normal agricultural work, which would have severely marked the nail surfaces.

One final point regarding keratin remains. Until recently the chemistry of this material has not been investigated from bog remains. It certainly has considerable potential value in terms of dietary and other interpretation, and it is good that bog body hair samples are beginning to be investigated.

Contents of the intestinal tract

We do not know how many bog bodies had evidence of food or helminth parasites within the intestinal tract. Some, such as the Meenybradan woman, appear not to have identifiable food residues within the gut. In fact less than ten bodies have so far been found to contain food and to have been studied in detail. Helbaek (1950, 1958) and Brandt (1950) established high standards of analysis in their studies on the Tollund, Grauballe and Borremose individuals. The Huldremose, Lindow II and III studies, are more recent specialised investigations of food remains (Brothwell, Holden, Liversage, Gottlieb, Bennike and Boessen, 1990; Holden, 1986, 1994).

Surprisingly, there is great variation in the plants eaten by these different individuals, and this has given

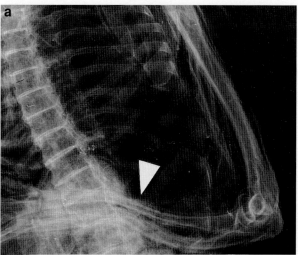
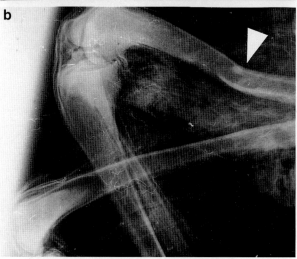

Fig. 9. Bone deformity (pseudo-pathology) in the Huldremose woman, due to pressure onto the decalcified bones long after burial. (a) Left forearm with radius and ulna curvature (arrowhead). (b) Femur shaft deformity

rise to the view that some of these "meals" may have had ritual significance. Tollund Man, for instance, had eaten many species, including barley, willow-herb, linseed and gold of pleasure. Grauballe Man and Lindow II had wheat and barley as major cereals in the intestines, together with some weed seeds, while a Borre Fen individual had no cereal component, just wild species. Corn spurrey *Spergula arvensis* is present in the Huldremose, Grauballe and Borre Fen individuals, and raises the question whether it was collected from a wild or cultivated form of the plant?

It should be mentioned that the histology of the gut and other tissues in Lindow II shows considerable degradation (Bourke, 1986). This was similarly found in the Meenybraddan woman, where no nuclei were seen and tissue structures had been infiltrated by bog plants and fungal hyphae (Delaney and O'Floinn, 1994).

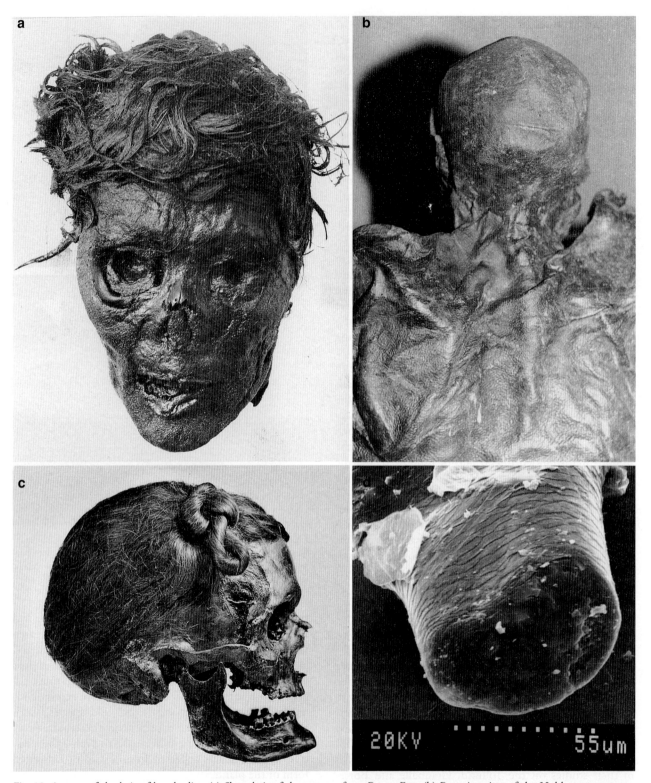

Fig. 10. Aspects of the hair of bog bodies. (a) Short hair of the woman from Roum Fen. (b) Posterior view of the Huldremose woman, showing the "shaved" head. (c) The knotted hair of Osterby man. (d) High magnification (SEM) detail of a cut hair from the scalp of the Huldremose woman

Evidence of intestinal parasites, in the form of numerous distinctive ova, have been found in various archaeological sites, as well as bog bodies. Lindow II had both *Trichuris trichiura* (whip worm) and *Ascaris lumbricoides* (maw worm). Ova counts ranged from 2,700 to 5,700 ova per gram (Jones, 1986). Other bog bodies have produced evidence of helminths too, although the Grauballe and Tollund individuals are reported to have only had *Trichuris*. However, there would seem to be a need to restudy intestinal samples and for instance in a new analysis of the Grauballe parasite ova, the species identified included *Eimeria mira*, a parasite of red squirrel, no doubt ingested when the squirrel was eaten as food (Hill, 1989).

Finally, let me mention the problem we had in obtaining a small food sample from the Huldremose woman – because it may be relevant for the Iceman. In the Huldremose X-rays and scans, it was clear that food debris remained in the gut. The problem was how to get to it. The body was dry and hard, and endoscopy via a natural orifice was impossible. To cut a long story short, we noted an area of intestine which was close to the abdominal wall. We then got permission from the Danish National Museum to perform micro-surgery and lift out a small window over the gut. By this means we successfully removed a food sample and the small area of thoracic wall was replaced by their Conservation Department – without obvious scarring.

Conclusion

The better preserved European bog bodies have presented us, over the years, with a variety of research and conservation problems. Work on them has helped to establish standards for the investigation of human remains, whether from bogs, arid environments or low temperature conditions. But I hope I have also shown that there is still much to be done on bog bodies, not only in applying new techniques, but in the reinvestigation of bodies found long ago.

Summary

Parts of at least 1,400 human bodies have been found in peat bogs, mainly in north-west Europe. Of these, about 300 have been relatively well preserved, and date from 900 B.C. to 1,800 A.D. Detailed studies on a number of Danish bodies have set high standards for the investigation of ancient bodies, including radiography, forensic aspects and the evaluation of food residues in the intestinal tract. New discoveries continue to be made, as for instance at Lindow Moss in England, and studies on these remains suggest that trace elements on the skin surface may be used to indicate body decoration. Bog body studies also suggest that DNA may not survive in some burial environments. From a detailed restudy of the Danish Huldremose woman, it is clear that there is potentially much information to be obtained by the application of new techniques to bodies which may have been excavated many decades ago.

Zusammenfassung

Teile von mindestens 1400 menschlichen Leichen sind in Torfmooren, vor allem in Nordwesteuropa, gefunden worden. 300 davon sind relativ gut erhalten und stammen aus der Zeit von 900 v. Chr. bis 1800 n. Chr. Die detaillierten Studien an einer Reihe von dänischen Leichen, inklusive Radiographie, gerichtsmedizinischer Studien und Analysen von Speiseresten im Darmtrakt haben für derartige Untersuchungen hohe Maßstäbe gesetzt. Es werden immer wieder neue Entdeckungen gemacht, z. B. in England, in Lindow Moss, wobei sich an diesen Funden gezeigt hat, daß sich mit Hilfe von Spurenelementen auf der Hautoberfläche dekorative Hautzeichnungen nachweisen lassen. Aus Analysen von Moorleichen geht auch hervor, daß DNA in manchen Grabstätten möglicherweise nicht überlebt. Eine detaillierte Neuuntersuchung der dänischen Huldremose-Frau zeigt, daß sich durch neue Techniken auch von Leichen, die schon vor vielen Jahrzehnten ausgegraben wurden, noch sehr viele Informationen gewinnen lassen.

Résumé

On a retrouvé des parties d'au moins 1400 corps humains dans des tourbières, surtout dans celles du Nord-Ouest de l'Europe. Environ trente d'entre eux pour lesquels on a obtenu des dates allant de 900 avant J.-C. à 1800 après J.-C. étaient assez bien préservés. Des analyses détaillées portant sur plusieurs corps trouvés au Danemark sont de haut niveau et indiquent une méthodologie pour l'étude de ces corps anciens qui comprend l'usage de la radiographie, celui des méthodes utilisées en médecine légale et l'analyse des restes de nourriture trouvés dans le tube digestif. On continue à faire de nouvelles découvertes comme, par exemple, à Lindow Moss en Angleterre dont l'étude suggère que des traces d'éléments trouvés sur la peau pourraient indiquer la présence de décoration corporelle. Les études des corps provenant de tourbières semblent également montrer que l'ADN ne survit peut-être pas sous les conditions du milieu ambiant de certaines des sépultures. L'étude de la femme d'Huldremose au Danemark, qui a été refaite, démontre que l'on peut obtenir de plus amples informations en appliquant des techniques nouvelles aux corps exhumés il y a des dizaines d'années.

Riassunto

In torbiere, soprattutto nell'Europa nord-occidentale, sono state ritrovate parti di millequattrocento cadaveri umani, trecento dei quali sono relativamente ben conservati e provengono dal 900 a. C. fino al 1800 d. C. Gli studi dettagliati fatti su una serie di cadaveri danesi, avvalendosi di radiografie e studi di medicina leglale, nonchè analisi di resti di cibo nel tratto gastrointestinale, hanno fornito informazioni molto utili. Vengono continuamente fatte nuove scoperte, come per es. in Inghilterra a Lindow Moss dove si è visto che, con l'aiuto di analisi di oligominerali sulla superficie cutanea, si sono potuti dimostrare tatuaggi decorativi. Dalle analisi di cadaveri di questo tipo risulta anche che il DNA in alcuni siti funerari probabilmente non pote sopravivere. Uno studio dettagliato della donna danese di Huldremose ha dimostrato che con tecnologie avanzate si possono ricavare informazioni utili anche da cadaveri dissepolti molti decenni fa.

Acknowledgements

Over the past thirty years, I have visited various museums in Denmark and Germany to view the bog bodies. In particular, I am most grateful to Dr. David Liversage and members of the National Museum in Copenhagen for various assistance in studying some Danish bog bodies in detail. Similarly, Dr. Maire Delaney of University College Dublin and Dr. Raghnall O'Floinn of the National Museum of Dublin have been most helpful in providing information on Irish bog bodies. Dr. W. van der Sanden, and the previous Keeper of Archaeology of Drents Museum in the Netherlands, provided me with information and an opportunity to examine a number of Dutch bog bodies. Last, but not least, I wish to thank Dr. Ian Stead of the British Museum in London, for his invitation to join in the investigation of the partial bog bodies discovered at Lindow Moss, near Manchester.

References

Andersen, S. R., Ophthalmopathologische Befunde bei Moorleichen. *Klinische Monatsblätter für Augenheilkunde*, 197: 187–190, 1990.

Bourke, J. The medical investigation of Lindow Man. In: *Lindow Man: The Body in the Bog*, eds Stead, I., Bourke, J. and Brothwell, D. Brit. Mus. Pub.: London, 1986, pp. 46–51.

Brandt, J. Plant remains in the body of an early Iron Age man from Borre Fen. *Arboger for Nordisk Oldkyndighed og Historie*. 348–350, 1950.

Brothwell, D., Holden, T., Liversage, D., Gottleib, B., Bennike, P. and Boesen, J. Establishing a minimum damage procedure for the gut sampling of intact human bodies: the case of the Huldremose woman. *Antiquity*, 64: 830–35, 1990.

Brothwell D., Liversage, D. and Gottleib, B. Radiographic and forensic aspects of the female Huldremose body. *J. Danish Archaeol.*, 9: 157–178, 1990.

Connolly, R. C., Evershed, R. P., Embery, G., Stanbury, J. B., Green, D., Beahan, P. and Shortall, J. B. The chemical codaposition of some body tissues. In: *Lindow Man The Body in the Bog*, eds: I. Stead, J. Bourke and D. Brothwell, Brit. Mus. Pub., London, 1986, pp. 72–76.

Delaney, M. and O Floinn, R. A bog body from Meenybraddan bog, County Donegal, Ireland. In: *Bog Bodies*, eds Turner, R. and Scaife R, Brit. Mus. Pub., London, 1995, pp. 123–132.

Evershed, R. P. and Connolly, R. C. Lipid preservation in Lindow man. *Naturwissenschaften*, 74: 143–145, 1988.

Evershed, R. P. Chemical composition of a bog body adipocere. *Archaeometry*, 34: 253–265, 1992.

Fischer, C., *Bog Bodies of Denmark*. Pp 177–193. In: *Mummies, Disease and Ancient Cultures*, eds A. and E. Cockburn. Cambridge University Press: London 1980.

Gebühr, M. Das Kindergrab von Windeby: Versuch einer Rehabilitation. *Offa*, 36: 75–107. 1979.

Glob, P. V. *The Bog People. Iron-Age Man Preserved*. Faber and Faber, London, 1969.

Helbaek, H. The Tollund Man's last meal. *Arboger for Nordisk Oldkyndighed og Historie*. 328–41, 1950.

Helbaek, H., The Grauballe Man's last meal. *Kuml*, 111–116. 1958.

Hill, G. Human Helminth Parasites in Archaeology. MSc thesis. London, 1989.

Holden, T. G. Preliminary report on the detailed analyses of the macroscopic remains from the gut of Lindow Man. In: *Lindow Man The Body in the Bog*, eds, Stead, I., Bourke, J. and Brothwell, D. Brit. Mus. Pub., London. 1986, pp. 116–125.

Hughes, M., Jones, D. and Connolly, R. Body in the bog but no DNA. *Nature*, 323: 208, 1986.

Jones, A. K. G. Parasitological investigations on Lindow Man. In: *Lindow Man: The Body in the Bog*, eds, Stead, I., Bourke, J. and Brothwell, D. Brit. Mus. Pub., London. 1986, pp. 136–139.

Krebs, C. and Ratjen, E. Det radiologiske fund hos moseliget fra Grauballe. *Kuml*, 138–150. 1956.

O'Floinn, R. Recent research into Irish bog bodies. In: *Bog Bodies*, eds, Turner, R. and Scaife, R. Brit. Mus. Pub. London, 1995, pp. 137–145.

Osinga, J., Buys, C. H. C. and van der Sanden, W. A. DNA and the Dutch bog bodies. DNA Newsletter, London, 1993.

Painter, T. J. Lindow Man, Tollund Man and other peat-bog bodies: the preservative and antimicrobial action of sphagnan, a reactive glycuronoglycan with tanning and sequestering properties. *Carbohydr. Polymers*, 15: 123–142, 1991.

Pyatt, F. B., Beumont, E. H., Buckland, P. C., Lacy, D. and Storey, D. M. An examination of the mobilisation of elements from the skin and bone of the bog body Lindow II and a comparison with Lindow III. *Environmental Geochemistry and Health*, 13: 153–159, 1991.

Pyatt, F. B., Beumont, E. D., Lacy, D., Magilton, J. R. and Buckland, P. C. Non Isatis sed Vitrum or the colour of Lindow Man. *Oxford J. Archaeol.*, 10: 61–73, 1991.

Stead, I., Bourke, J. and Brothwell, D. Lindow Man, the Body in the Bog. Brit. Mus. Pub., London, 1986

Taylor, G. W. Tests for dyes. In: *Lindow Man The Body in the Bog*, eds, Stead, I, Bourke, J. and Brothwell, D. Brit. Mus. Pub., London, 1986, p. 41.

van der Sanden, W. A. *Mens en Moeras*, Doctoral thesis, University of Leiden (monograph, Drents Museum), 1990.

Correspondence: Prof. Don Brothwell, Department of Archaeology, University of York, King's Manor, York, YO1 2EN, U.K.

Selection of a conservation process for Lindow Man

V. Daniels

Department of Conservation, The British Museum, London, U.K.

Introduction

The normal processes of decomposition start soon after death and, under normal circumstances, it may be only a matter of days before a human body loses recognisable form. However, in a few instances human bodies have retained a recognisable appearance for hundreds, or even thousands, of years by preservation due to natural circumstances. Rapid desiccation produced by burial in dry sand produces preservation by reducing the water content of the body below a point where microorganisms will thrive. Alternatively, the presence of large quantities of salts, e.g. the use of natron by the ancient Egyptians, can produce preservation (1). Extreme cold can halt the deterioration of animal flesh is well known from our domestic freezers, but as soon as temperatures return to normal, biodeterioration sets in. However, natural freeze-drying can occur to bodies in cold regions when water vapour from ice in bodies is removed by sublimation into cold air currents resulting in preservation by desiccation. The majority of examples of body preservation in northern Europe occur due to accidental or intentional burial in peat bogs. Skin and hair are often preserved but internal organs less frequently. There are over 120 recorded sites where human remains have been discovered in bogs in Great Britain and Ireland (1). The conditions are often quoted to be anaerobic, acidic and wet, and until recently these conditions were considered to be those necessary for the preservation of skin. However, Painter (2) suggests that bodies are preserved in sphagnum peat bogs because they are tanned by spagnan, a pectin-like polysaccharide from the cell walls of sphagnum mosses. Additionally, sphagnan aids preservation by reacting with the digestive enzymes secreted by putrefactive bacteria and immobilising them on the surface of the peat.

Fig. 1. Lindow Man being excavated from the surrounding peat, in the laboratory

The discovery of Lindow Man

On 1. August 1984 the remains of the lower part of a leg complete with a foot were found at Lindow Moss, Cheshire, England. The site was a sphagnum peat bog which was being excavated for horticultural purposes. Excavation was halted by the discovery of the human remains and on 6 August the rest of the body was carefully removed *in situ* in its block of peat. The body and peat were wrapped in water-soaked plastic foam and thick plastic sheeting. Post excavation deterioration was minimized by cold storage of the body in a nearby hospital mortuary in Macclesfield. In late September the body was transferred to the British Museum's Department of Conservation and stored in a specially constructed cooled container at 4 °C.

The body excited the attention of the media and the uncovering and scientific examination of Lindow man was well covered by television, radio and the press. However by the time the main conservation treatments were started the media attention had subsided to a reasonable level.

Once most of the peat had been removed from the body (Figs. 1 and 2) an important step was the construction of a mount so that the body could be safely handled and easily viewed from both sides. This was achieved by the construction of a two-part shell which was moulded to the shape of Lindow Man. The construction of the mount was aided by the use of two types of polyester casting tapes. Delta Lite (which ceased to be commercially available during the course of the work) and Scotchcast. These materials are used in hospitals as a substitute for plaster bandages when applying splints to broken limbs (3). The front of the body was covered with Clingfilm as a separator and then strips of polyester bandage, 127 mm in width, were cut into squares and moulded to the body and overlapped to give strength. When all the body was thus covered, the bandage was sprayed with water which caused it to become rigid in a few minutes (Fig. 3). The mould was removed and further strengthened by painting on Tiranti Rigid Laminate, a polyester resin thickened with glass beads, over reinforcing strips of coarse and fine fibreglass.

When the first half of the shell had been made, the body was turned over and the remainder of the peat block excavated away. The back of the body now had Clingfilm applied as a separating layer and the mould making process repeated. A lip round the edges of each half of the shell enabled the two halves to be fixed together by nuts and bolts, allowing the body to be turned over at will.

Large numbers of experts volunteered their services for the scientific examination of Lindow Man, the main part of this work is described by Brothwell (4), and was complete before the interventive conservation was started, however, investigations continue to the present day although on a much smaller scale.

During the scientific examination it was important to keep the body wet and cool; this was achieved by regular spraying of exposed areas with chilled, recently boiled, distilled water. Parts of the body not being worked on were covered with "Clingfilm", a very thin polyvinyl

Fig. 2. Lindow Man, back view, the final traces of peat being removed

chloride based domestic food packaging film. The temperature of the body was monitored using several thermocouples and the body returned to the freezer when the temperature rose above 10–12 °C. This generally gave a working time of two hours. During examinations the body was also kept cool by the use of cold packs of frozen wet paper towels inside polythene bags, and by passing cool air over the exposed area. Temporary support for parts of the body was conveniently obtained by the use of peat wrapped in Clingfilm (Fig. 4).

Test for tannins

Graubelle Man (5) was found to be partially tanned when excavated and it was, therefore, thought necessary to test whether Lindow Man was in a similar condition. Tannins are released from several types of plant material but are not generally associated with peat bogs. Three samples were examined, tissue from the inner thoracic cavity, tissue from the outer lumbar region and peat from near the body.

The test used was based on that described by Reed (6). Samples of tissue weighing 0.02 g were separately refluxed for four hours in either 1:1 water : acetone (8 ml) or equal volumes of 6 M hydrochloric acid, acetone and water. One ml of each solution was adjusted to pH 7 using sodium hydroxide solution, and 1 ml of 1 % ferric potassium sulphate was added. A blue or brown coloration would indicate the presence of tannins. No coloration was detected. The test was repeated using 0.15 g of peat; no tannins were detected. The test gave a strong positive reaction when carried out on 0.02 g of tanned goatskin. It was concluded that the skin of Lindow Man was not tanned to any significant extent.

At the time, the alleged preservative properties of sphagnum moss (2) were unknown and no attempt was made to detect spagnans.

Selection of a conservation treatment

Within a few months of discovery, Lindow Man was showing signs of apparent biodeterioration and unless deep freezing was to be chosen for the permanent storage, some type of conservation treatment would be required to induce stability. Monitoring the biodeterioration of the body was carried out by Ridgeway (3) from September 1984 to March 1985. Cultures were made from the distilled water used to spray the body, from the peat and from swabs taken off various sites on the body. The fungi isolated were of no pathogenic consequence and included a number of *Pseudomonas* spp. *Penicillium*, *Mucor*, *Vertisillium* and *Candida*, but none of these had

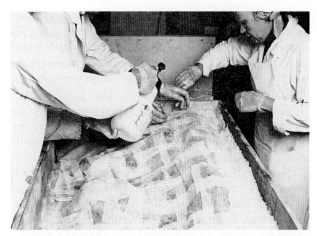

Fig. 3. Casting tape bandages being sprayed with water to create a casing which enabled Lindow Man to be turned over

Fig. 4. Lindow Man's head resting on a cushion of peat wrapped in Clingfilm

caused any detectable deterioration in the state of preservation of the body.

Although the British Museum Department of Conservation had built up a considerable expertise in the treatment of waterlogged wood, leather and other materials found on waterlogged archaeological sites, no waterlogged human remains had previously been encountered. To obtain background information on the conservation of bog bodies, the head of the conservation team, Sherif Omar, visited several of the European museums where bog bodies had been conserved. Tollund Man had been impregnated with wax; Graubelle Man had been partly tanned when excavated and the process was completed by soaking in oak bark and then impregnating with various oils (5). These and many other bodies were examined and conservation treatments compared. Although some excellent results had been obtained none of the conservators visited were completely satisfied with the treatments and thought freeze-drying was an alternative process which might yield superior results. Freeze drying was first used routinely for the conservation of waterlogged material since the 1970s as it avoids shrinkage to a considerable extent. It is often used in conjunction with a pretreatment with a consolidant such as polyethylene glycol (PEG).

Conservation of waterlogged material with PEG is a process which can be carried out in many ways and the PEG is available in a wide range of molecular weights from 200, a liquid, to over 6,000, a waxy solid. The PEG is soluble in water and in various organic solvents. Different PEG grades may be mixed; for waterlogged wood treatment a mix of liquid and solid grades is popular. Objects may be impregnated for various times and then allowed to air dry or be freeze-dried. Thus, the number of possible significantly different PEG treatments is very large. An additional treatment sometimes used for waterlogged leather was glycerol impregnation thus its use was considered at the same time as the use of PEG.

There was no doubt that some kind of stabilisation was necessary for storage of Lindow Man in near ambient conditions, as a small sample allowed to naturally air dry shrank considerably, to 50 % of the original area and became hard and brittle.

Normally the development of a new conservation process proceeds with the treatment of genuinely old materials in an appropriate state of decay which may be sacrificed for the experiments. A sufficiently large quantity of the material is needed so that comparisons of treatments and measurements are possible. Replicate results are needed to avoid variations in the samples and method of application. In this case several hundred square centimetres of uniform bog body skin would have been needed; this was unavailable.

The next best approach is to obtain new materials and degrade them artificially. Discussions with freeze-drying specialists revealed that pigskin had pathological similarities to human skin and was sometimes used as a substitute for experiments in freeze drying. Thus it was decided to use degraded pigskin.

Fresh pigskin does not bear a close resemblance to the skin on a bog body. In an attempt to produce a realistic model material for the experimental work, pigskin was cut into strips and packed into peat for several months. The strips were placed in two sealable glass jars. One of the jars was topped up with distilled water and the other with a slurry of peat from Lindow Moss in distilled water. To produce anaerobic conditions, the distilled water was degassed by boiling and cooling it just before use. Further deoxygenation was brought about by bubbling nitrogen through the filled jars for ten minutes before sealing them. The jars were placed outside the laboratory for four months (January to April 1985). At the end of this period the skins were examined. The samples from the peat were firmer and browner than at the start of the experiment but those from distilled water had swelled and become gelatinous. It was decided to proceed with the experimental work using only the peat-treated skins as these were the best match to the skin on the bog body.

Various factors go to make a successful conservation treatment. Treated material should not undergo any undesirable colour change, it should be of comparable flexibility to the original, it should be stable in near ambient conditions and there should be no distortion or shrinkage. Materials often swell when waterlogged, thus some shrinkage is acceptable but non-uniform shrinkage causes distortion, thus most conservation treatments aim to minimise shrinkage. The experiments were designed to measure shrinkage, but at the same time colour, flexibility and stability were also monitored.

The skin samples were cut into rectangular shapes and their outlines drawn on pieces of cardboard; these shapes were later used to calculate the shrinkage on freeze-drying. The cut skin samples were immersed in the following solutions which reflect the concentrations previously found successful for waterlogged organic materials. A set of untreated samples was retained for comparison. PEG 400 and glycerol are liquids at room temperature, the other PEGs are solids:

10 % v/v PEG 400 in distilled water
10 % w/v PEG 1500 in distilled water
10 % w/v PEG 2000 in distilled water
10 % w/v PEG 2000 in 10 % vv PEG 400
10 % v/v glycerol in distilled water

After eight weeks the samples were removed, blotted dry and were then freeze-dried, sandwiched between stiff board to minimize warping. Pre-freezing was done in a domestic chest freezer at $-26\,°C$ and the freeze-drying carried out with a chamber temperature of $30\,°C$. Samples were removed from the freezer-dryer after three days, covered with polyethylene sheeting and allowed to equilibrate to ambient conditions for a week. Their outlines were retraced onto cardboard.

Although roughly rectangular, the treated skin samples had shrunk in a non-uniform manner, and the area of the sample could not be calculated by direct measurement. Tracing round the sample before and after treatment yielded pieces of card which, when cut out, had weights proportional to their areas. These pieces of card could easily be weighed and the percentage area shrinkage calculated. These shrinkages are shown on Table 1.

It is important to bear in mind that percentage area shrinkage is roughly twice that of linear shrinkage. Similarly, volume shrinkage is roughly three times that of linear shrinkage. For example, if the 10 mm sides of a square of area 100 mm² shrink by 10 %, the new area is 81 mm², 19 % less than the original area. Similarly, a cube with sides shrunk by 10 % has a volume of 729 mm³, 27.1 % less than the original.

Table 1. Shrinkage of freeze-dried pigskin samples

Pretreatment	No. of samples	Average area shrinkage
None	5	22.9 %
10 % PEG 400 in 10 % PEG 2000	6	21.6 %
10 % PEG 400	6	11.5 %
10 % PEG 1500	5	13.1 %
10 % PEG 2000	5	11.8 %

Pigskin samples were cut from areas where the skin was seen to be uniform, ie away from folds in the skin, the backbone, etc. However, some samples shrank in an irregular way. Results from these pieces were used in the assessment of shrinkage. Ideally these experiments should have been repeated but, as time was at a premium, it was not possible to do this.

PEG 2000 left a white waxy deposit on the samples. If used on the body this would later have caused problems when attempts were made to remove it from areas where there was hair; depilation could result. Ellam (7) reports that PEG 400, a liquid, gave good results for freeze-drying bone and notes that higher concentrations than 30 % were not satisfactory. It was decided to use PEG 400 for the conservation of Lindow Man.

The Meenybradan bog body

After the Lindow Man body had arrived at the British Museum, the National Museum of Ireland asked whether it would be possible for us to treat another bog body. It was decided that the treatments chosen for Lindow Man would be first applied to the 'new' body, Meenybradan Woman, to gain more experience in the chosen conservation technique. Meenybradan Woman arrived at the British Museum in July 1985. She had been excavated in 1976 and subsequently stored in a domestic freezer. During transport from Ireland the body was wrapped in a survival blanket inside a wooden box packed with solid carbon dioxide. On arrival at the British Museum it was placed in a domestic freezer at –26 °C. After a few days it was allowed to thaw out for examination and photography. The body was subsequently stored at 4 °C.

Meenybradan Woman's body is much more complete than Lindow Man's and possessed both legs, although these were disconnected from the torso (Figs. 5 and 6). The bones were in good condition. There were deposits of adipocere distributed over the body and under the skin; these are apparent in the Figures as white patches. Adipocere is produced by the prolonged storage of animal fat under anaerobic cool conditions. No information was available on the effect of freeze-drying on adipocere so experiments had to be carried out to investigate this.

Loose fragments of adipocere of about 2 cm³ were used for the experiments. One sample was freeze-dried without any pretreatment and retained its shape and characteristic physical properties. Another sample was soaked in 10 % PEG 400 for five days before freeze-drying; this sample also responded well to treatment. It was concluded that the conservation treatment proposed for Lindow Man could be used on Meenybradan Woman.

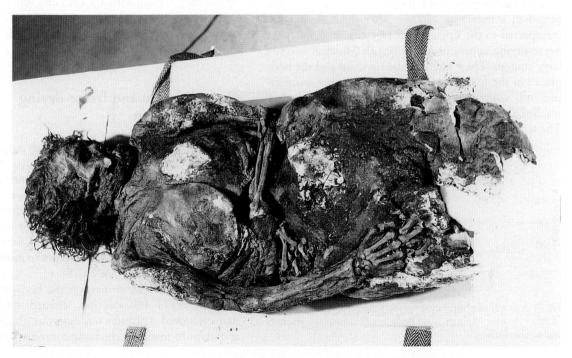

Fig. 5. Meenybradan Woman after conservation

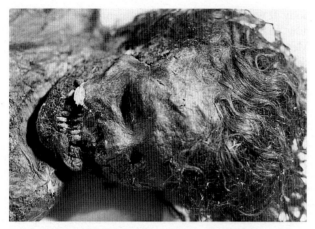

Fig. 6. Meenybradan Woman's head, after conservation

Table 2. Linear shrinkage measurements during conservation of Meenybradan Woman

Position	% shrinkage
Left leg, bone	5.6
Left leg, skin	1.8
Ditto at 90° to the above	1.3
Left hip	1.2
Between hips	0.6
Left hand, bone towards abdomen	1.3
Left hand, bone towards the head	2.3
Left hip, longitudinal	0.0
Left hip 90° to above	0.0

Before the main process of conservation was performed, the stomach contents were removed for examination by a surgeon, because it was believed these would not respond well to freeze-drying.

The body was secured to a Perspex (poly {methyl methacrylate}) sheet using strips of polyester bonded fabric (Tyvek) and cushions made from polyester wadding in plastic bags. The aim was to secure limbs during immersion as they were liable to float. Stainless steel pins were fixed into the skin and bone to provide reference points for shrinkage measurements. Loose fragments were put in a bag of nylon net and the whole body placed in a bath of 15 % w/v PEG 400 in distilled water. Special attention was given to the skull to ensure that no air was trapped there. After soaking for four weeks the PEG was drained off, a thermocouple inserted in the abdomen and the body wrapped in Clingfilm. The body was then placed in a freezer at –26 °C. After three days it was transported to the English Heritage Laboratories as the freeze-drying apparatus at the British Museum was not large enough. The Clingfilm was removed and the body placed in the freeze-dryer, where the chamber temperature was controlled at –32 °C. When the body had reached a temperature of –32 °C a vacuum was applied. The temperature/time graph is shown in Fig. 7.

The process was monitored by periodically weighing one of the legs and measuring the body temperature. The process was deemed to be complete when body temperature rose to –5 °C. After 31 days of freeze-drying the body was removed from the freeze-dryer, wrapped in a survival blanket and placed in a wooden box with dry silica gel (to prevent condensation). The body was returned to the British Museum; ideally it would have been allowed to reach ambient temperature in the freeze-dryer but operational difficulties prevented this.

After a week of acclimatization of the body to ambient conditions the survival blanket was removed. Linear shrinkage of the body was generally small; between 1 and 2 % (see Table 2).

Skin texture and details became much easier to see, eg eyelashes and eyebrows became visible. The skin was supple and the peat that remained on the body was easy to remove. The hair was rather tangled at the end of the treatment and was interspersed with a dried-out scum from the impregnation solution. This was later removed by carefully rinsing with 1:1 IMS and water.

Generally, the result was very good and there was no reason to suppose comparable results would not be obtained on Lindow Man.

Consolidation and freeze-drying of Lindow Man

Lindow Man was cleaned as much as possible and then placed on a Perspex sheet; strips of polyester bonded fabric (Tyvek) were used to secure the body to the mount. Pressure from the strips was evenly distributed over the body by using polyethylene bags filled with polyester wadding between the body and the fabric strips. Ten pairs of stainless steel pins were stuck into various parts of Lindow Man in order to monitor shrinkage after conservation (see Fig. 8).

After four weeks immersion the body was removed from the PEG 400 solution, drained, and the fabric strips removed. Polyester wadding was placed inside the abdomen to maintain the body contours, and Clingfilm

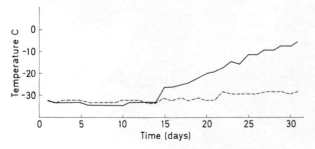

Fig. 7. A graph of temperature versus time for the freeze-drying of Meenybradan Woman. —— Body temperature, ---- chamber wall temperature

wrapped round the body to avoid drying during transport to the English Heritage Laboratories where freeze-drying was to take place.

Thermocouples were placed in the skull and abdomen (Fig. 9). The body, still on its sheet of Perspex and wrapped in Clingfilm, was placed inside the freeze-drying apparatus. For five days the chamber walls were refrigerated at −28 °C; at the end of this period the thermocouples registered a temperature of −24 °C in the skull and −20 °C in the abdomen. The Clingfilm was removed and the vacuum applied. A pressure of between 100 and 200 millitorr was maintained during the freeze-drying. Figure 10 shows how the temperatures varied during the course of the treatment. After 23 days the body was weighed daily. The treatment was considered to be complete when negligible weight loss occurred; this happened on the 29th day and the refrigeration was turned off. After three more days the body had warmed up and the temperature was stable. The body was removed, covered with Tyvek fabric sheets and placed in a wooden box containing dry silica gel.

On its return to the British Museum the body was left to acclimatise to ambient conditions for seven days before any further conservation work was done.

Results of treatment

Generally the skin had become lighter in colour. However, lighter and darker marks could be seen on some areas of the skin which were reminiscent of the folds of Clingfilm. The cause of these marks is unclear but may have been caused by a surface morphology change during freezing or absorption of plasticiser from the film.

Freeze-drying enabled the last traces of peat to be removed easily and the excellent preservation of parts of the body was revealed. The skin texture was still very good over most of the body. The skin had become stronger and more rigid than in the wet state but was still flexible and could be handled with greater ease. There was no smell from the body.

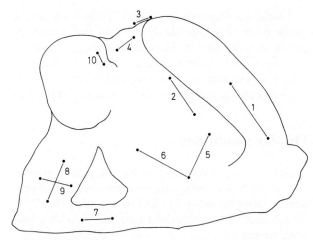

Fig. 8. Pin positions on Lindow Man used to monitor shrinkage

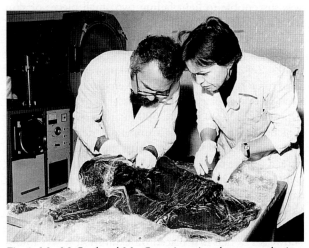

Fig. 9. Ms. McCord and Mr. Omar inserting thermocouples into Lindow Man prior to freeze-drying. The freeze-dryer can be seen on the left of Mr. Omar

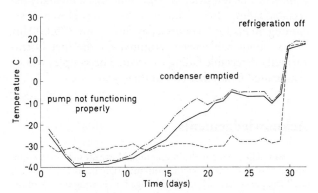

Fig. 10. A graph of temperature versus time for the freeze-drying of Lindow Man. —— Temperature in skull, --- temperature in abdomen, —·—· chamber wall temperature

Some of the stainless steel pins inserted to monitor shrinkage had come out during the freeze-drying and it was impossible to relocate them. Only five shrinkage measurements were possible (Table 3).

Table 3. Linear shrinkage during freeze drying of Lindow Man

Reference no	Position	Distance between pins (in mm)		% linear shrinkage
		Before conservation	After conservation	
1	Left arm	123	117.5	4.47
2	Chest	70	67.6	3.43
8	Right arm	85	83.8	1.41
9	Right arm	107	105	1.61
10	Ear	46	44	4.30

Linear shrinkage was less than 5 % which was similar to the linear shrinkage obtained in the preliminary experiments using pigskin. The conservation treatment was considered very successful. At the time of writing the body has been conserved for over eight years and shows no sign of change except for a slight lightening of the skin.

Display conditions

After conservation, Lindow Man appeared in some special exhibitions, where the environmental conditions were: relative humidity 55±5 %, temperature 19±2 °C, illuminance 50 lux, ultraviolet level, less than 75 microwatts/lumen.

To determine whether the skin colour of Lindow Man was susceptible to light fading two small pieces of skin were taken from the back of the body where light exposure had been minimal. The dark brown colour of the skin was measured instrumentally and found to be $L^*a^*b^*$ 42.05, 2.44, 2.26. Each piece of skin was mounted on a glass microscope slide and half covered with opaque tape. One piece was placed in a Microscal light fastness tester exposed to 13,000 lux and 600 microwatts/lumen ultraviolet content. The first instrumentally detectable fading was seen after 500,000 lux hours and the sample continued to fade until after 10 M lux hours the sample was light brown. The use of a UV absorbing filter reduced the rate of fading by 30 %. Thus it was demonstrated that the skin of Lindow Man was susceptible to light fading.

The other piece of skin was placed in Lindow Man's showcase. When first placed on exhibition it was found that the illuminance was too high, however, in its present position the desired environmental conditions have been achieved. The test piece has shown no instrumentally detectable fading after 25 months of storage. However, assuming 50 lux for 10 hours a day this is only 375,000 lux hours, about half what is required for the first instrumentally detectable fading to occur. The samples will be remeasured in another two to three years.

Acknowledgements

I would like to thank the International Institute of Conservation for permission to reuse material from my original article (8) and Dr. C. Price, previously at English Heritage for allowing the use of the freeze-drying facilities and especially Miss J. Watson for supervising the freeze-drying process, and taking measurements when required.

Summary

Lindow man was found buried in a peat bog in Cheshire, England. The body was kept cool until a conservation process was decided upon. An attempt was made to simulate this type of degraded human-skin by aging pig-skin in oxygen-free peat and water mixture. The skin produced was a better match for the skin on the body than fresh pig-skin. It was decided to freeze-dry the body after impregnation with polyethylene glycol (PEG) solution. Shrinkage measurements were made after freeze-drying the aged pig-skin impregnated with various grades of PEG. As a result of the experiment a 15 % solution of PEG 400 was selected.

Before Lindow Man was treated, another bog body, Meenybradan Woman was conserved using the proposed method. The results were satisfactory and thus Lindow Man was treated in the same way with excellent results. For the conservation, a two part polymer resin shell was made which was moulded to the shape of Lindow Man; this greatly facilitated the handling of the body.

Zusammenfassung

Der Lindowmann wurde in einem Torfmoor in Cheshire, England, gefunden und bis zu einer Entscheidung über die Konservierungsmethode kühl gelagert. Es wurde beschlossen, die Leiche mit Polyethylenglykol (PEG) zu tränken und gefrierzutrocknen. Um die stark beeinträchtigte menschliche Haut zu simulieren, wurde Schweinehaut in einer Mischung aus sauerstofffreiem Torf und Wasser künstlich gealtert. Die so erhaltene Haut entsprach in ihren Eigenschaften besser der Haut der Leiche als frische Schweinehaut. Nach Gefriertrocknung der mit verschiedenen PEG-Konzentrationen imprägnierten, gealterten Schweinehaut wurden Schrumpfungsmessungen durchgeführt. Als Ergebnis dieser Versuche entschied man sich für eine 15%ige Lösung von PEG 400.

Noch vor der Behandlung des Lindowmannes wurde eine andere Moorleiche, die Frau von Meenybradan, mit der vorgeschlagenen Methode konserviert. Das Resultat war zufriedenstellend, und so wurde auch der Lindowmann nach demselben Verfahren behandelt – mit ausgezeichnetem Erfolg. Für die Aufbewahrung wurde ein zweiteiliger Polymerharzmantel in der Form des Lindowmannes angefertigt, was die Handhabung der Leiche wesentlich erleichterte.

Résumé

L'homme de Lindow a été trouvé dans un marais tourbeux à Cheshire en Angleterre et gardé au frais jusqu'à ce qu'une décision concernant la méthode de conservation soit prise. Il a été décidé d'imprégner le cadavre de polyéthylène-glycole (PEG) et de le lyophiliser. De la peau de porc, vieillie artificiellement dans un mélange d'eau et de tourbe exempte d'oxygène, simule la peau humaine, étant celle de l'homme de Lindow fortement atteinte. Par ses caractéristiques, la peau de porc ainsi traitée ressemblait plus que la peau de porc fraiche à la peau d'un cadavre. Après la lyophilisation de la peau de porc, vieillie par imbibition de PEG en différentes concentrations, le retrait de la peau a été mesuré. Sur la base de ces examens il a été décidé d'employer une solution de 15 % de PEG 400.

Bien avant le traitement de l'homme de Lindow, un autre cadavre, celui de la femme de Meenybraden, a été conservé suivant la susdite méthode. Le résultat ayant été satisfaisant, l'homme de Lindow a été, lui aussi, soumis aux mêmes procédés, ce qui a abouti à des resultats excellents. Pour la conservation, un manteau en deux parties de résines de polymérisation, ayant la forme de l'homme de Lindow, a été fabriqué, ce qui a facilité notablement le maniement du cadavre.

Riassunto

L'uomo Lindow è stato rinvenuto nello Cheshire in Inghilterra e tenuto a temperature basse finchè non si è deciso quale metodo adottare per la conservazione. Si è scelto di immergere il corpo in polietilen-glicolo (PEG) e di deidratarlo a freddo. L'elevato grado di decomposizione della pelle umana si è potuto simulare facendo invecchiare della pelle di maiale in una soluzione di acqua e di torba priva di ossigeno. Il risultato corrispose più alla pelle del corpo rinvenuto che non alla pelle fresca di maiale. Dopo la deidrazione a freddo della pelle di maiale invecchiata e impregnata di diverse concentrazioni di PEG, sono state effettuate delle misurazioni sul grado di raggrinzamento della pelle. In base a questa esperimentazione è stato deciso di usare una soluzione del 15 % di PEG 400.

Ancora prima che l'uomo di Lindow fosse sottoposto ai trattamenti era stata messa in conservazione la mummia di una donna rinvenuta nella torbiera di Meenybradan secondo il metodo consigliato. Il risultato fu soddisfacente e quindi si adottò lo stesso procedimento anche per l'uomo di Lindow che portò a degli ottimi risultati. Per la conservazione la mummia è stata ricoperta di un manto in resina polimerica diviso in due parti, il che ne facilita la maneggiabilità.

References

(1) Brothwell, D., The bog man and the archaeology of people. British Museum Publications, London (1986).
(2) Painter, T., Chemical and microbial aspects of the preservation process in *Sphagnum* peat, in, Turner, R. C. and Scaife, R., The Lindow Moss bog bodies. British Museum Press (1995) 88–99.
(3) Newey, H., Dove, S., and Calver, A., Synthetic alternatives to plaster of Paris in excavation, in, Recent advances in the conservation and analysis of artefacts, Black, J. Summer Schools Press, London (1987) 33–36.
(4) Stead, I. M., Bourke, J. B., and Brothwell, D., Lindow Man, The body in the bog. British Museum Publications, London (1986).
(5) Glob, P. V., The bog people. Faber and Faber, London (1977).
(6) Reed, R., Ancient skins parchment and leathers. Seminar Press, London (1972) 266–281.
(7) Ellam, D., Wet bone: the potential for freeze drying in archaeological bone, antler and ivory. Occasional paper No 5 UKIC, London (1987) 34–35.
(8) Omar, S., McCord, M and Daniels, V., The conservation of bog bodies by freeze-drying. Studies in Conservation *34* (1989) 101–109.

Correspondence: Dr. V. Daniels, Department of Conservation, The British Museum, Great Russell Street, London WC1B 3DG, U.K.

Guanche mummies of Tenerife (Canary Islands): conservation and scientific studies in the CRONOS Project

C. Rodríguez-Martín

Instituto Canario de Paleopathologia y Bioantropologia, Santa Cruz de Tenerife, Canary Islands, Spain

Introduction

Mummies are not just objects that reflect isolated elements of a culture. They represent a special link between human biology and the cultural practices of different societies (Rodríguez-Martín, 1992a).

The amalgam between human biology and culture, reflected in mummies, constitutes an important field of research. The long tradition of mummy studies in different scientific disciplines is due to the soft tissue preservation and, in certain cases, the superb persistence of the external forms of the corpses.

Although most people identify the word "mummy" with ancient Egypt, it is true that such specimens have been found in many other regions of the world, including: Peru, Chile, Columbia, Bolivia, Ecuador, Brazil, Venezuela, North and Central America, Alaska, Aleutian Islands, Greenland, some regions of Europe, Siberia, Australia and Melanesia, and, of course, the Canary Islands.

Most authors agree that Guanche mummies have long been famous since the Spaniards discovered them in the burial caves of Tenerife (Hooton, 1925). The conquerors were astonished by the wonderful preservation of those specimens. Therefore, it is not strange that Guanche mummies have been a matter for debate from the XVth century until present times. Reason or reasons for mummification, method or methods of mummification, social status of the mummies, existence of specialists in that work, and many similar topics have been discussed during the last 500 years. But the problem always appeared in the fact that most researchers have always used historical sources lacking direct observation of the corpses, analysis of the specimens, application of modern techniques, and experimentation. Perhaps, this was due to the scarcity of these most valuable specimens in the main museums and centers of our archipelago and the difficulties that this kind of studies implies (Schwidetzky, 1963). Here, as everywhere, spoliation has played its own and disagreeable role. These are the reasons, until the *CRONOS Project, the Bioanthropology of Guanche Mummies*, that research on this branch of Canarian Prehistory has not been so brilliant as we might expect in a land where secular mummies have appeared by hundreds and thousands.

Now, the question appears: Who were the Guanches? Guanche is the traditional name of the ancient inhabitants of Tenerife (the biggest of the seven Canary Islands) deriving from the Berber "wan zenete" which means "... of the Zenetes", but according to the studies of the French Sabin Berthelot during the XIXth century this word was used to designate all the Canarian aborigines.

These people came from the Berber tribes of northwestern Africa and although the exact date of the arrival to the archipelago is not yet known most researchers agree that it was around the I century of our Era (radiocarbon dating of mummies fluctuates between 400 AD and 1,400 AD).

Physically, they were tall individuals for the time they lived, with an average stature of 1.70 m for males and 1.57 m for females (García-Talavera, 1992). In general, we can affirm that they were robust and well adapted to the environment of Tenerife. However, due to the geographic and climatic differences of the island, today we cannot view the Guanche population as an indivisible whole (Aufderheide et al., 1992b; Rodríguez-Martín, 1992b).

As we have previously seen, the Canarian prehistory, and of course the prehistory of Tenerife, show lack of important data. So, the Archaeological Museum of Tenerife decided in 1989 to begin with a major research project: *CRONOS Project, the Bioanthropology of Guanche Mummies* in order to try to elucidate some of these data.

Over a period of 3 years we carried out the different studies, analyses, and researches on the Guanche mummies, and the following institutions took part in them: Archaeological Museum of Tenerife, Natural Science Museum of Tenerife, and University Hospital of the Canaries (these institutions belong to the Cabildo of Tenerife or Government of the island); University of La Laguna (Tenerife); University of Minnesota-Duluth Campus (USA); Augustana College, Sioux Falls, South Dakota (USA); Park Nicollet Medical Center, Minneapolis (USA); and the Manchester Museum (United Kingdom). *CRONOS* was the first interdisciplinary and international research project carried out on the prehispanic cultures of the Canaries, and almost all its goals have succeded.

Discovery and recovery of Guanche mummies

The Guanches placed their deceased in natural caves with the same characteristics of dwelling caves, and these caves were located in the ravines at the ends of the villages (Diego-Cuscoy, 1968). But, as Ruíz-Gómez et al. (1992) point out, archaeology has demonstrated that there are a lot of burial caves at different altitudinal levels that are not related to permanent settlements (Roque Blanco, near 2,000 m, or Llano de Maja or Cueva del Salitre, more than 2,000 m, are three examples). Arco-Aguilar (1976) affirms that the burial caves in Tenerife are located at any altitude in the coast, in the midlands or in the highlands, but always avoid the forest (it may be related, in the opinion of Criado and Clavijo (1992), with the absence of natural caves, hollows or shelters in the forest and with the fact that the Guanches avoid the forest for pastures). On the other hand, there are burials that appear near the dwellings and dwellings that were used as burial caves (this is the case of Los Guanches Cave, Icod, in the northern slope of Tenerife, or Los Auchones, Anaga, Santa Cruz de Tenerife, in the east of the island).

Fig. 1. Map of the island of Tenerife showing burial caves with mummies

One of the characteristic of these burial caves is that they were protected from external hazards with dry stone walls. At the same time, they are located in inaccesible places.

In the absence of appropriate caves the Guanches buried the corpses in small hollows, natural shelters, or, more uncommonly, in stone cracks far from the settlements (Criado and Clavijo, 1992). For Ruíz-Gómez et al. (1992) this fact can indicate that the deceased was buried in the same place where he died.

In caves or in hollows flat stones, vegetal remains or wood were placed on the floor and then the corpse was placed upon these artifacts. So, the Guanches never buried the corpses; they only placed the body in a conditioned space avoiding contact between soil and corpse. As Aufderheide et al. (1992a) affirm, this is a unique feature of the Guanches' mortuary practices. This practice enhanced desiccation and could even, if circumstances were appropriate occasionally, conceivably induce natural mummification. There was no preference for geomorphological environment of the burial.

When a burial cave with mummies appears, the staff of the Archaeological Museum (archaeologists and conservationists) and the Canarian Institute of Paleopathology and Bioanthropology (physical anthropologists) goes to the place with specialists in Physical Geography in order to study the burial.

The first step is to fill a sheet with all the needed data:

1. Name of the burial cave.
2. Municipality.
3. X, Y, Z, coordinates.
4. Exposition of the cave.
5. Orientation of the cave.
6. Type of cavity.
7. Size.
8. Geomorphological environment.
9. Altitude.
10. Climatic level.
11. Annual average temperature.
12. Rain (outside the cave).
13. Insolation and humidity outside the cave.
14. Temperature and humidity of the burial.
15. Vegetation around the burial.

Then maps, drawings and photographs are made in order to complete all the geographic and climatic data. Video-tape is also used.

The range of temperatures inside the burial caves in Tenerife is 21.9 °C–34.1 °C, with an average of 24 °C, and the humidity fluctuates between 36.3 % and 67.5 % with an average of 48 % (similar to the ideal conditions in museums and centers of research: 20 °C and 45 % of humidity).

Cultural artifacts related with mummies

Traditionally, spoliation has made it very difficult to study the Guanche mummies and their funerary offerings. Therefore, it is very difficult to assess if the mummies were always accompanied by cultural artifacts. Only in 20 % of the mummy collection of our museum has it been possible to establish the relation between mummies and offerings (Ruíz-Gómez et al., 1992). It seems clear that we must differentiate the artifacts located upon the corpse and those that surround the mummy, because, in the opinion of the previous authors, the first of them may be personal funerary offerings and the second one may be a collective or, perhaps, an unknown offering.

The artifacts that appear in these funerary offerings are:

1. Necklace beads. These are the more abundant artifacts that appear in relation with the corpses.
2. Fragments of pottery of different typology.
3. Obsidian knifes called by the Guanches "tabonas".
4. Punchs of animal bones.

These artifacts are not different from those used in daily Guanche life. Therefore, we must reject the possibility that the funerary offerings were made in a different way.

Guanche mummification

A long debate has ensued over five centuries about the method or methods that the Guanches used to mummify the corpses. As Hooton (1925) states, for we know that only the upper classes were embalmed, and many burials of the lower class Guanches have been found and may be found today, in which the bones are perfectly clean and devoid of all soft parts. But the question is: what kind of embalming? It is clear now that just as with the chronicles and histories, writen during or shortly later the Conquest, it is impossible to reconstruct the embalming process or processes. But until the *Cronos Project*, almost all the researchers, with the exception of Hooton (1925) and Chil y Naranjo (1877), have used the historical sources as the only way to elucidate the problem.

As Aufderheide et al. (1992a) point out, of the various methods employed by the Guanches, as documented by the early chroniclers, only the following were identified as procedures whose effects might be recognizable in the bodies at this time:

1. Evisceration (some authors, as Chil y Naranjo, 1877, have observed signs of evisceration in the chest and abdominal walls of Guanche mummies). Some of the first references to evisceration were made by Abreu y Galindo (1632) and Viera y Clavijo (1776).
2. Foreign material in the body cavities of the thorax, abdomen or mouth and esophagus (this was first described by Espinosa, 1594).
3. Craniotomy (nobody observed this kind of operation in the Guanche mummies, it being only a literary reference of the sixteenth and seventeenth centuries).
4. Sand in subcutaneous tissues (identified by Brothwell, Sandison and Gray, 1969, in a Guanche mummy curated at the Museum of Archaeology and Ethnology of the University of Cambridge, England).

In our research on the mummy collection of the Archaeological Museum of Tenerife, the only mummification effort identifiable was the presence of foreign material in a significant quantity in the thoracic and abdominal cavity areas of some mummies (as the historian Espinosa, 1594, stated).

Ortega and Sánchez-Pinto (1992) identified this material as follows:

– Mineral content:
  • Red lapilli: it constitutes more than 90 % of the total content of the sample, with pieces ranging between 1 and 30 mm in diameter.
  • Pumice stone.
  • Soil.
– Vegetal remains:
  • Acicules (needles from *Pinus canariensis*).
  • Fragments of gramineous (grass) stems.
  • Seeds (in scarce number) from *Visnea mocanera*.
  • Remains of pollen from *Erica arborea* (heath).
  • Seeds from an unidentified crucifera.
  • Little pieces of vegetal carbon.
  • Fragments of *Bromus sp.*
  • Vestiges of dragon-tree blood (*Dracaena draco*).
– Animal remains:
  • Goat skin and hair (most probably coming from the hide wrappings).
  • Tendons of animal origin.
  • Solidified fat.

On the other hand, small lizard bones, mice droppings and thousands of fly pupae and larvae of *Coleoptera* and *Diptera* also were found.

In the opinion of Ortega and Sánchez-Pinto (1992), probably *Diptera* became attached when the body was still fresh. Later, in the cave, the mice produced their droppings inside the body. The Coleoptera are typical from stored collections. They probably attacked the mummies later.

Although evisceration (described by early Canarian historians as Abreu y Galindo, 1632) has been reported by several authors, Chil y Naranjo (1877) among others, the retention of viscera in some mummies of the museum excluded this technique in these specimens. Aufderheide et al. (1992a) found no evidence of abdominal incisions for purposes of evisceration in the other mummies of the collection, but none of the examined mummies had sufficient retention of enough anterior abdominal wall to exclude the possibility.

No evidence of sand stuffing, as Brothwell, Sandison, and Gray (1969) have reported in the Guanche mummy of the Cambridge Museum, was detected in the mummies of the Archaeological Museum of Tenerife. On the other hand, we could find no evidence of craniotomy for the purpose of extracting the brain.

With the method of mummification observed in the mummies of the Archaeological Museum of Tenerife the probable role of desiccation enhancement by sun exposure, however, should not be minimized. Certainly this island's climate, especially that of its southern slopes, would have provided the appropriate conditions frequently, and it is probable these were exploited whenever possible, as the historian Torriani (1592) affirmed. Unfortunately, no anatomic "markers" of such corporeal processing exist to permit recognition of its application (Aufderheide et al., 1992b).

It is clear that the majority of the embalmed corpses were wrapped in goat and possibly other animal skins. Historical sources state that the most important personalities of the island were wrapped in 10–15 animal skins and the number of layers decreased if the people embalmed belonged to a lower social stratus (but always belonging to the noble class of Tenerife).

Conservation and curation of Guanche mummies

When we discuss conservation and curation of mummies in a museum we must take into account several factors: the mummy itself, which needs its own conservation treatment; the room and/or storage area where the mummy will be located needing a special equipment to control the environment; the scientific studies we are going to make on the mummy; and the possible transportation of the specimen to other places in the future.

Conservation of the mummy

Each mummy is a different individual, a different "patient", who needs his own treatment. However, we can outline here a standard treatment. Due to the characteristics and conditions of Guanche mummies we avoid, when possible, any kind of wet treatment. The steps of the treatment are as follows:

1. Removing of the skin wrappings in order to let the mummy be exposed completely naked. Each layer or fragment of animal skin is labeled according to the original location on the mummy.
2. The wrappings are subjected to "dry-cleaning" with soft brushes and aspiration.
3. The same treatment of "dry-cleaning" is applied to the whole mummy.
4. If there are several separated fragments of the mummy, they are replaced in their original anatomical position with cotton strings. The same process is employed to reenforce the mummy, especially the legs and the arms. Those parts of the string that can be seen externally are coloured with acrilic (Maimeri) and natural pigments. It is also possible to fix some hard fragments, as bones, with water soluble adhesive (Mowilitch DCM 2).
5. Rewrapping of the mummy.

Sometimes it is necessary to use a wet treatment for the conservation of the mummies, mainly to treat parasites' attack. A wet sterilization treatment can be employed which uses 1 %–5 % solution of Pentachlorophenol (sodium salt) in pure alcohol. According to David and David (1989), this combination provides a double action: it acts as a good fungicide and a contact herbicide and, being dissolved in alcohol, its penetration ability is increased, as well as alcohol being a dehydrating agent. However, it is necessary to say that it is a very dangerous chemical combination, and the operator must protect himself with appropriate mask, clothes and gloves. The operation can be safely carried out in a large fume cupboard or using an airless, low-pressure spray or by simply brushing the solution on to the body surface.

In the opinion of David and David (1989) the wet sterilization process could also improve the skin tissue brittleness. Notwithstanding, for the associated leather artifacts, it is more appropriate to rehydrate the specimens using water fungicide and Polyethylene glycol. Freeze drying is absolutely vital to the success of this treatment.

Mummy room

This room is separated from the rest of the museum by a double door with an intermediate space in order to isolate the mummy environment (in the entrance and in the exit). The number of people who can visit the mummy room at the same time is limited to 15–20, being controlled by a guard. This prevent great fluctuations in the environmental conditions, and at the same time is a good security measure.

The show-cases where the mummies are exhibited are hermetically closed. We have rejected the use of silica gel due to the huge amount we need that can affect the design of the show case.

The mummy room, and also the museum store, have their own air-conditioned and lighting equipment. On the other hand, equipment for controlling the measurements of temperature and humidity (thermohygrographs) are required. Each show-case has its own thermohygrograph and two other machines are located in the room and in the door (in the intermediate space). The standard conditions of environment for the Guanche mummies are not very different from mummies of other places: temperature fluctuating between 18 and 20 °C, and humidity about 40 to 50 percent.

With these conditions the organic materials are neither affected by an excess of dryness nor microorganisms attacks.

With regard to the lighting equipment, in our museum is employed a system of high frequency fluorescence with control of the light intensity, located under the mummy base. In order to increase the lighting of the room we use incandescent lights located in the ceiling.

Regularly (every six months) the mummies and mummified remains of our museum are analysed in order to detect parasite attacks and to control the physical conditions of the specimens.

Scientific studies

In order to make future scientifc studies on the mummies we avoid the use of materials that can affect the chemical composition of the wrappings, soft and hard tissues. This is very important in the case of diet reconstruction by mean of trace elements and stable isotopes analyses, DNA investigation, paleoserological studies, and taphonomy.

Transportation of the mummies

When we must transport the mummies, special boxes are employed. Actually, these are two boxes: one inside the other and are made of pine wood. The space between both boxes is refilled with high density foam in order to absorb vibrations and blows. The walls, base and top of the lesser box are covered with polystyrene sheets of 10 cm width. Once the mummy is placed inside this special box and covered with silver tissue paper, all the rest of the space inside the box is filled with high density foam.

Scientific studies in Guanche mummies

During the development of *CRONOS Project. The Bioanthropology of Guanche Mummies*, the most advanced biomedical and biological techniques were employed to carry out the bioanthropological studies: chemical reconstruction of diet through trace elements and stable isotopes analyses, molecular genetics, radiology, macroscopic pathology, histopathology, etc. were successfully employed to try to reconstruct the Guanche life from their human remains, and dozens of scientists took part in the project over three years.

Dietary studies

Chemical reconstruction of Guanche diet was carried out through the analysis of trace elements (team of Dr. Aufderheide) and stable isotopes (team of Dr. Tieszen).

• Trace elements: 171 Guanche specimens from seven of the island's archaeological sites were sampled and analyzed for their content of strontium, zinc, and calcium. As Aufderheide et al. (1992b) point out, results indicate that the Guanche diet was rich in meat and dairy products. Vegetal items formed a lesser part of their diet while marine resources were only selectively harvested. It is important to note that the island's mummified subgroup (the elite of the Guanche society) consumed the most meat and dairy products.

Geographically there are differences in the diet of the two sides of the island, north and south: vegetal fractions among populations occupying Tenerife's better-watered northern parts were higher than those of the southern and more arid regions, probably reflecting climatically-influenced soil productivity differences. On the other hand, and coinciding with pathological features, demonstrated dietary differences between sites on the same side of the island probably reflect the degree of social isolation of the respective populations (Aufderheide et al., 1992b).

• Stable isotopes (13C and 15N): Samples of terrestrial and marine resources were analyzed for $\delta 13C$ and $\delta 15N$ values. Bones from pig, goat/sheep and dogs provided good pseudomorphs and collagen which differed slightly but significantly in $\delta 13C$ (–18 ‰ to –21 ‰). Marine animals ranged from –7.35 ‰ to –15.2 ‰ for carbon and 8.0 ‰ to 11.4 ‰ for nitrogen. Plankton feeders were the most negative for carbon and the least positive for nitrogen. Most modern marine fish were more negative for carbon than terrestrial sources and as positive as 13 ‰ for nitrogen. Human specimens provided excellent pseudomorphs and high quality collagen. There was very little variation in carbon isotopoic values among these samples with means ranging from –19.3 ‰ in Hoya Fría to –19.8 ‰ in Los Guanches (Tegueste). There were no differences related to site, location, or altitude. As Tieszen et al. (1992) point out, specimens from the wetter north side had a mean value of 8.80 ‰ in $\delta 15N$ compared to 9.77 ‰ for the south side. The bioapatite $\delta 13C$ values showed a similar homogeneity. The spacing between bioapatite and collagen ranged between 4.4 ‰ to 5.6 ‰ and did not vary by site nor by location. The Guanche, in the opinion of the previous authors, possessed very similar isotopic compositions regardless of site, location, or altitude suggesting similar

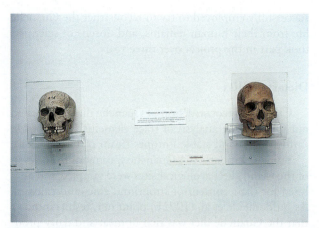

Fig. 2. Two examples of ancient skulls from Tenerife

Fig. 3. Reconstruction of a mummy inside a burial cave showing the „chajasco" or wood located between the corpse and the floor (Archaelogical Museum of Tenerife)

Fig. 4 (top). Infant mummy from "punta volcánica" (El Sauzal, northern Tenerife)

Fig. 5 (right). Partially mummified skull (adult male) from "Añaza" (Santa Cruz de Tenerife, north-eastern Tenerife) (frontal view)

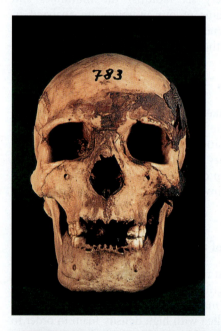

Fig. 6. Partially mummified skull (adolt male) from "Añaza" (Santa Cruz de Tenerife, north-eastern Tenerife) (lateral view)

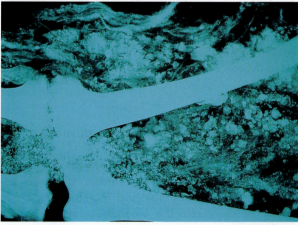

Fig. 7. Radiograph showing foreign material for the preservation of the corpse at the level of the lower limgs in a Guanche mummy from "Guia de Isora" (western Tenerife)

dietary dependencies. The negative carbon isotope values suggest a strong dependence on terrestrial resources. The consistently small collagen-bioapatite spasing suggests that a major portion of the assimilated carbon was derived from terrestrial animals, like pig and sheep/goats.

Molecular genetics

According to Salo et al. (1992) and Rogan et al. (1993), the main goal of the DNA studies in the Guanche population is to extract this genetic material and to probe the DNA as a means of determining how subgroups of these people are related. Such information could help determine their origin.

For most analyses of ancient DNA, it is necessary to make an amplification because the residual DNA from ancient remains is usually present in low yield, fragmented and otherwise modified, the target segment of DNA must be short and specific. However, as Salo et al. (1992) affirm, given sufficient sequence information (from work on contemporary humans) appropriate targets can be chosen and amplified by the polymerase chain reaction (PCR). The authors tried to analyze the polymorphic second exon region of the HLA-DPB gene which were amplified in two segments of 138 and 186 bp (including primers). Subsequently the amplified segments were reverse probed with nine DPB probes.

The goals of this work were:
1. To demonstrate the extraction of DNA from the soft and hard tissue remains.
2. To amplify a segment of mitochondrial DNA by PCR to establish that amplifiable DNA was extracted.
3. To amplify the HLA-DPB target segments by PCR.
4. To probe the amplified DPB segments.
5. To analyze the data obtained.

Radiology

It was only after World War II that radiology really gained recognition as a major tool in mummy studies, and in the 1960s a paleoradiologic boom started with Gray's work in England and the Egyptian expedition sponsored by the University of Michigan School of Dentistry (Rodríguez-Martín, 1992c). During the 1970s, several non-invasive techniques were introduced, the first being xeroradiography. Computed tomography was introduced in the field of mummy research in 1978. The first three dimensional CT reconstructions were made also in 1978 and 1988. Also in the 1980s, the first Magnetic Resonance Imaging was applied to the study of mummies.

These new techniques permit the study of mummies and mummified remains without damaging the display

Fig. 8. Mummified adult right foot (unknown origin) (Tenerife)

value of the specimens (with the exception of MRI, which needs rehydration of the sample). It is difficult, however, to assess many pathological conditions and morphological characters for the diagnosis of sex and age with only radiologic techniques: we also need macroscopic and microscopic pathology, immunology, biochemistry, and so on.

According to Brothwell et al. (1980), radiological studies on mummies have different goals:

1. To determine the presence or absence of human bones in the sarcophagus or bundle.
2. Sex and age determination (when possible).
3. Correlation of radiological findings with known embalming techniques and some pathological conditions.

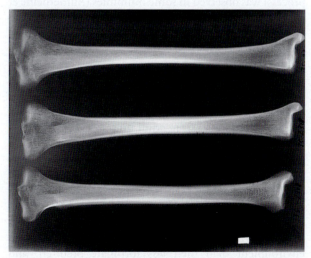

Fig. 9

4. Demonstration of objects and amulets inside the wrappings.
5. Demonstration of pathological findings (when possible).

During the development of the *CRONOS Project* the Guanche mummies of the Archaeological Museum of Tenerife were transported to the University Hospital of the Canaries where different radiological techniques were employed: conventional radiography with a Siemens 1000 Tridoros 150 machine; low kilovoltage technique; mammography (it is a high definition technique that is used to observe some details that are not clearly seen in the normal plain radiographs of soft tissues); and CT with a scanner Tomoscan 60/TX of Philips Medical System.

Following the indications of Notman (1992), flat radiographs were always taken before CT study. Sometimes, a low kilovoltage technique was employed to improve the images of the standard radiography.

The films were Kodak X-Omatic and Konica Medica (30×40 cm) and for CT, Agfa-Curix RP2.

• Technical parameters for adult mummies:
20–25 mAs.
60–65 Kv.
400 MA.
0.05–0.06 secs.
100 cm from the source of X-Rays to the film
11 cm from the object to the film.

For infant mummies low kilovoltage technique was used (35–40 Kv) in order to attain a better observation of metaphysis, epiphysis, and ossification centers.

• Radiological findings in Guanche mummies: As Notman (1992) affirms, the Guanche mummies demonstrated relatively poor soft tissue preservation but were otherwise reasonably intact.

Strange material (probably used in the embalming process) was observed in some mummies (Aufderheide et al., 1992a; Notman, 1992). Some layered debris and gravel were noted in the skull. The angle of the fluid level indicates that the body was buried in the supine position (Notman, 1992).

Few abnormalities were detected by the X-Ray study: only a fracture in a scapula and other fracture in a first metatarsal, and some low degrees of degenerative joint disease and spondylosis in adult mummies.

Few Harris lines (growth arrest lines) were present in the mummies while in the non-mummified individuals (and always depending on the site of the island they lived) appeared a relatively high frequency of those lines. For Notman (1992), "this intriguing finding could be confirmation that the mummies may have been privileged individuals who grew up in the healthier, more protected environment than their less fortunate brethren, and whose social status granted them a more formal burial".

Pathology

• Skeletal pathology: For this kind of study we employed traditional methods of observation in skeletal paleopathology (macroscopic examination, dissecting microscopy (10, 30, and 40 X) and, of course, radiology (mainly flat radiographs but also mammography) when it was necessary. The results we got were as follows (Rodríguez-Martín, 1992b):

1. Physical effort must have been variable among the Guanches, especially in males. There is a clear topographic difference in the distribution of degenerative joint disease which may indicate that there was differentiated physical activity depending on the geographic region of the island: in the south pastoralism was the main economical activity while in the north, besides pastoralism, agriculture and/or gathering played their economic role.

2. Although postcranial trauma lesions were not so frequent as those of the skull, the data we got confirm those of point one. The scarcity of postcranial trauma indicates a good adaptation to the terrain and an almost perfect knowledge of the geography of the island, although accidents must have been present before the conquest coinciding with the years of greater physical activity (18–35).

We can affirm that cranial trauma is one of the most fascinating fields of the Guanche paleopathology. Rodríguez-Martín et al. (1990), concluded after reviewing the series of skulls of the Museo Arqueológico de Tenerife:

a. The frequency of cranial injuries among the Guanches is high (with an average percentage of near 10 percent), especially in the south of the island.

b. Fractures appear only in adults with the male-female ratio showing that they are nearly twice as common in males.

c. The main type is the depressed fracture, irregular in shape, which clearly has a violent cause.

d. The anterior and lateral parts of the skull are much more frequently affected than the posterior part and the facial skeleton. This supports the hypothesis that the fractures were caused by violence (missil-stone shots and face-to-face fightings).

e. Healing occurs in most cases, and only 17 percent occurred at the perimortem period.

f. War must have been important in the prehistory of Tenerife.

3. The high frequency of osteochondritis and osteochondroses, that are caused by microtrauma on the joint cartilage, appearing at an early age of life (10–12 years), indicate that the young had also a vigorous life.

4. The very high frequency of congenital malformations, mainly spina bifida occulta (with a genetic component in its origin), points out to a high degree of inbreeding, more important in some parts of the north-

ern slope. This may indicate a certain degree of biological isolation.

5. The scarcity of infections of any kind, mainly specific infections; metabolic disturbances (with the exception of senile osteoporosis) and hematologic disorders (demonstrated by the low frequency of cribra orbitalia and porotic hyperostosis), together with a well marked sexual dimorphism in the great majority of the analyzed series, a relatively high stature, and the scarcity of other stress markers, indicate that most of the Guanche population of Tenerife was well adapted to the island's environment.

• Dental pathology: As Langsjoen (1992) affirms, native Guanches lived with extensive oral infection. 61 percent of adult molars experienced significant resorption of alveolar bone. Dental attrition became pathologic by early adulthood in the form of pulp exposures and proximal contact breakdown. The pulp necrosis and alveolar abscess formation caused by these two conditions generated a 5.4 percent frequency of alveolar fistulae. The abrasive quality of plant foods together with abrasive elements, such as mill stone grit, incorporated into it during preparation are implicated in the pathologic severity of attrition. Molars and premolars which had lost proximal contact due to attrition were particularly vulnerable to loss of periodontal attachment and subsequent cemento-enamel junction caries.

There was no crown caries. Adolescents were caries-free, caries was a predominantly adult phenomenon. The inhabitants of the northern slope experienced significantly more dental caries than the people living in the more arid southern slope. The data strongly suggest that proteolytic and aciduric components of the caries process can exist independently as well as silmutaneously in the oral environment (Langsjoen, 1992).

• Histopathology: As Aufderheide et al. (1992a) point out, soft tissue was not preserved well in the bodies as a whole, although that of isolated extremities and heads often demonstrated excellent preservation. Few viscera remained for study. The lungs of two mummies were sampled and revealed substantial anthracosis with both diffuse and localized fibrosis of undetermined etiology. As the previous authors state, together with the observations of Brothwell et al. (1969), this suggests Guanche cave inhabitants were indeed commonly smoke-polluted.

Summary

Guanche mummies have yielded new light to the Tenerife's and Canarian prehistory. Applying new techniques and methodologies in biomedical, bioanthropological, and curatorial techniques these unvaluable specimens can tell us new things about the daily life of our ancestors, their diseases, foodways, familial relationships, and adaptation to the environment of the archipelago. We cannot forget that mummies are not single objects in the stores or rooms of the museums: in fact, they were human beings long time ago.

Zusammenfassung

Die Guantschenmumien haben ein neues Licht auf die Urgeschichte Teneriffas und der Kanarischen Inseln geworfen. Durch die Anwendung neuer Techniken und Methoden in der Biomedizin, der Bioanthropologie und der Konservierung geben uns diese wertvollen Funde neue Informationen über das tägliche Leben unserer Vorfahren, ihre Krankheiten, ihre Ernährungsgewohnheiten, ihre familiären Beziehungen und ihre Anpassung an die Umwelt der Inselgruppe. Im Rahmen der hier präsentierten Ergebnisse sollte nicht vergessen werden, daß die Mumien nicht als „Objekte" behandelt werden, sondern sie stellen die sterbliche Überreste von Menschen dar, die wir als unsere unmittelbaren Vorfahren ansprechen sollten.

Résumé

Des momies de la culture des Guanche révèlent des nouveaux aspects sur la préhistoire de Tenerife et des Canaries. En applicant des nouvelles techniques et méthodologies en biomédecine, bioanthropologie et conservation, ces spécimens précieux peuvent nous donner des nouvelles informations sur la vie quotidienne de nos ancêtres, leurs maladies, leur nutrition, leurs relations familiales et leur adaptation à l'environnement de l'archipel. Il ne faut pas oublier que les momies ne sont pas des objets singuliers dans les caves et salles des musées: en fait, elles étaient des êtres humains dans une époque lointaine.

Riassunto

L'amalgama fra biologia umana e cultura, riflessa nella mummia, costituisce un importante campo d'investigazione. La lunga tradizione di studi sulle mummie in diverse discipline scientifiche si deve alla conservazione dei tessuti molli e a la magnifica persistenza della forma esterna del cadavere.

Guanche é il nome tradizionale degli antichi abitanti di Tenerife. Questa gente arrivó sull'isola originari a dalle tribu Berbere del Nord-Est dell'Africa e arrivo all'Arcipelago Canario si ebbe intorno al primo secolo.

I Guanche erano bianchi, alti, robusti e ben adattati all'ambiente di Tenerife.

Le mummie Guanche di Tenerife sono state famose da quando i conquistatori spagnoli le scoprirono nelle grotte sepolcrali dell'isola nel 1496.

Dei vari metodi di mumificazione impiegati dai Guanche, documentati nell'antiche cronache, i procedimenti seguenti furono identificati sui corpi: eviscerazione; presenza di materiale estraneo nella cavitá corporale, e di sabbia nel tessuto sottocutaneo.

Dopo un trattamento completo di conservazione arrivando a capo grazie a specialisti di Spagna e Gran Bretagna, le mummie Guanche furono studiate durante gli ultimi cinque anni in un gran progetto denominato "Progetto CRONOS, Bioantropologia delle Mummie Guanche" e i risultati furono presentati dai membri di un gruppo multi disciplinare e internazionale di investigatori nel Primo Congreso Mondiale di Studio Sulle Mummie (Puerto de la Cruz, Tenerife) in febbraio 1992. Le principali conclusioni furono:

1. Esistevano differenze dietetiche fra diversi luoghi della isola. Le frazioni vegetali fra le popolazioni delle regioni umide del Nord erano maggiori che quelle piú aride del Sud.
2. Fu estratto il DNA dalle mummie Guanche per studiare le relazioni dei sotto gruppi di popolazioni e per dimostrare la loro origine.
3. Lo sforzo fisico dovrebbe essere variabile fra i Guanche. Esiste una chiara differenza topografica nella distribuzione della malattia articolare degenerativa, microtrauma e trauma: nel sud il pascolo fu la principale attivitá economica, importante ruolo economico.
4. La guerra doveva essere stata importante nella preistoria di Tenerife perché la frequenza dei traumi cranici causati dalla violenza é molto elevata.
5. La scarsitá di infezioni, disturbi metabolici e alterazioni ematologiche insieme alla scarsitá di indicatori di stress metabolico indicano che la maggior parte della popolazione Guanche si era ben adattata all'ambiente dell'isola.

In conclusione, non possiamo dimenticare che le mummie non sono semplici oggetti nei magazzini o nelle sale dei musei: furono esseri viventi tanto tempo fa e da questo che possano continuare a raccontarci storie sulle loro vite e la vita della loro popolazione.

References

Abreu y Galindo, Fr. J,. de (1977 [1632]): *Historia de la Conquista de las Siete Islas de Canaria.* Santa Cruz de Tenerife: Goya Ediciones.

Arco-Aguilar, M., del (1976): El Enterramiento Canario Prehispánico. *Anuario de Estudios Atlánticos,* 22: 13–112.

Aufderheide, A. C., Rodríguez-Martín, C., Torbenson, M. (1992a): Anatomic Findings in Studies of Guanche Mummified Human Remains from Tenerife, Canary Islands. In: *Proceedings of the I World Congress on Mummy Studies.* Santa Cruz de Tenerife: Museo Arqueológico y Etnográfico de Tenerife-Organismo Autónomo de Museos y Centros, pp. 113–124.

Aufderheide, A. C., Rodríguez-Martín, C., Estevez-Gonzalez, F., Torbenson, M. (1992b): Chemical Dietary Reconstruction of Tenerife's Guanche Diet Using Skeletal Trace Element Content. In: *Proceedings of the I World Congress on Mummy Studies.* Santa Cruz de Tenerife: Museo Arqueológico y Etnográfico de Tenerife-Organismo Autónomo de Museos y Centros, pp. 33–40.

Berthelot, S. (1879): *Antiquités Canariennes.* Paris: Pion.

Brothwell, D., Molleson, T., Gray, P. H. K., Harcourt, R. (1980): La Aplicación de los Rayos X al Estudio de Materiales Arqueológicos. In: Brothwell, D. & Higgs, E. (eds): *Ciencia en Arqueología.* Madrid: Fondo de Cultura Económica, pp. 533–545.

Brothwell, D., Sandison, A. T., Gray, P. H. K. (1969): Human Biological Observations on a Guanche Mummy with Anthracosis. *American Journal of Physical Anthropology,* 30: 333–347.

Chil y Naranjo, G. (1877): *Estudios Históricos, Climatológicos y Patológicos de las Islas Canarias.* Vol. 1. Las Palmas de Gran Canaria: Miranda.

Criado-Hernandez, C., Clavijo-Redondo, M. (1992): Características Geográficas de los Enterramientos con Momias de la Isla de Tenerife. In: *Proceedings of the I World Congress on Mummy Studies.* Santa Cruz de Tenerife: Museo Arqueológico y Etnográfico de Tenerife-Organismo Autónomo de Museos y Centros, pp. 209–212.

David, A., David, R. (1989): *Preliminary Report on the Human Mummified Remains in the Museum of Tenerife: Notes on Present and Future Preservation.* Manchester (manuscript).

Diego-Cuscoy, L. (1968): *Los Guanches. Vida y Cultura del Primitivo Habitante de Tenerife.* Santa Cruz de Tenerife: Publ. Museo Arqueológico de Tenerife.

Espinosa, Fr. A., de (1968 [1594]): *Historia de Nuestra Señora de Candelaria.* Santa Cruz de Tenerife: Goya Ediciones.

Garcia-Talavera Casañas, F. (1992): La Estatura de los Guanches. In: *Proceedings of the I World Congress on Mummy Studies.* Santa Cruz de Tenerife: Museo Arqueológico y Etnográfico de Tenerife-Organismo Autónomo de Museos y Centros, pp. 177–186.

Hooton, E. A. (1925): *The Ancient Inhabitants of the Canary Islands.* Cambridge (MA): Harvard African Studies, vol. VII.

Langsjoen, O. M. (1992): Dental Pathology Among the Prehistoric Guanches of the Island of Tenerife. In: *Proceedings of the I World Congress on Mummy Studies.* Santa Cruz de Tenerife: Museo Arqueológico y Etnográfico de Tenerife-Organismo Autónomo de Museos y Centros, pp. 79–92.

Notman, D. (1992): Paleoradiology of the Guanches of the Canary Islands. In: *Proceedings of the I World Congress on Mummy Studies.* Santa Cruz de Tenerife: Museo Arqueológico y Etnográfico de Tenerife-Organismo Autónomo de Museos y Centros, pp. 99–104.

Ortega, G., Sanchez-Pinto, L. (1992): Análisis de los Materiales de Relleno de las Momias Guanches. In: *Proceedings of the I World Congress on Mummy Studies.* Santa Cruz de Tenerife: Museo Arqueológico y Etnográfico de Tenerife-Organismo Autónomo de Museos y Centros, pp. 145–150.

Rodríguez-Martín, C. (1992a): Una Historia de las Momias Guanches. In: *Proceedings of the I World Congress on Mummy Studies.* Santa Cruz de Tenerife: Museo Arqueológico y Etnográfico de Tenerife-Organismo Autónomo de Museos y Centros, pp. 151–162.

Rodríguez-Martín, C. (1992b): Osteopatología del Habitante Prehispánico de Tenerife, Islas Canarias. In: *Proceedings of the I World Congress on Mummy Studies.* Santa Cruz de Tenerife: Museo Arqueológico y Etnográfico de Tenerife-Organismo Autónomo de Museos y Centros, pp. 65–78.

Rodríguez-Martín, C. (1992c): Radiological Approach to Mummies and Mummified Remains. The Case of Guanche Mummies. *IXth European Meeting of the Paleopathology Association.* Barcelona (in press).

Rodríguez-Martín, C., Gonzalez-Anton, R., Estevez-Gonzalez, F. (1990): Cranial Injuries in the Guanche Population of Tenerife (Canary Islands). A Biocultural Interpretation. In: Davies WV and Walker R (Eds.): *Colloquium. Biological Anthropology and the Study of Ancient Egypt.* London: British Museum Press, pp. 130–135.

Rogan, P. K., Lentz, S. R., Rodríguez-Martín, C., Estevez, F., Gonzalez-Anton, R. (1993): Identification of Ancient Nucleic Acids in Preserved Soft Tissue from Aboriginal People of the Canary Islands (Guanches). Hershey (PA): *Symposium. New Frontiers in Biomedical Research* (poster).

Ruiz-Gomez de Fez, M., Rosario-Adrian, M. C., Arco-Aguilar, M., del (1992): Estudio de los Ajuares Funerarios de Tenerife. In: *Proceedings of the I World Congress on Mummy Studies.* Santa Cruz de Tenerife: Museo Arqueológico y Etnográfico de Tenerife-Organismo Autónomo de Museos y Centros, pp. 167–171.

Salo, W., Foo, I., Aufderheide, A. C. (1992): Determining Relatedness Among the Aboriginal People of the Canary Islands by Analysis of Their DNA. In: *Proceedings of the I World Congress on Mummy Studies*. Santa Cruz de Tenerife: Museo Arqueológico y Etnográfico de Tenerife-Organismo Autónomo de Museos y Centros, pp. 105–112.

Schwidetzky, I. (1963): *La Población Prehispánica de las Islas Canarias*. Santa Cruz de Tenerife: Publ. Museo Arqueológico de Tenerife.

Tieszen, L., Matzner, S., Buseman, S. K. (1992): Dietary Reconstruction Based on Stable Isotopes (13C and 15N) of the Guanches of Prehispanic Tenerife (Canary Islands). In: *Proceedings of the I World Congress on Mummy Studies*. Santa Cruz de Tenerife: Museo Arqueológico y Etnográfico de Tenerife-Organismo Autónomo de Museos y Centros, pp. 41–57.

Torriani, L. (1978 [1592]): *Descripción e Historia del Reino de las Islas Canarias*. Santa Cruz de Tenerife: Goya Ediciones.

Viera y Clavijo, J. de (1967 [1776]: *Noticias de la Historia General de las Islas Canarias*. Vol. I. 6th ed. Santa Cruz de Tenerife: Goya Ediciones.

Correspondence: Dr. Conrado Rodríguez-Martín, Instituto Canario de Paleopatologia y Bioantropologia, O. A. M. C., Cabildo de Tenerife, Apdo. 853, 38080 Santa Cruz de Tenerife, Canary Islands, Spain.

Natural and artificial 13th–19th century mummies in Italy

G. Fornaciari[1] **and L. Capasso**[2]

[1] Institute of Pathology, University of Pisa Medical School, Pisa, Italy
[2] Laboratory of Anthropology, National Archaeological Museum, Chieti, Italy

Introduction

Contrary to current opinion, the series of mummies in Italy, and of single mummies in particular, is relatively numerous (Di Colo 1910; Terribile and Corrain 1986; Fulcheri 1991). These mummies are distributed over the entire Italian territory, from Friuli, in northern Italy (Aufderheide and Aufderheide 1991), to Sicily (Fornaciari and Gamba 1993), and generally prevail in southern Italy, where the most remarkable collections are found (Fig. 1). Burials date from the medieval period, through the Renaissance, and up to modern times, with a higher incidence between the 17th and the 19th centuries (Table 1), all representing precious paleopathological material.

Table 1. Collections of mummies in Italy

Site	No.	Type	Century
VENZONE (Northern Italy)	15	NATURAL	14th–18th
URBANIA (Central Italy)	18	NATURAL	17th–19th
MEDICI MAUSOLEUM (S. Lorenzo, Florence)	39	NATURAL & ARTIFICIAL	16th–18th
FERENTILLO (Central Italy)	16	NATURAL	18th–19th
NAVELLI (Central Italy)	hundreds	NATURAL	13th–19th
ARAGONESE MAUSOLEUM (S. Domenico, Naples)	31	NATURAL & ARTIFICIAL	15th–19th
ALTAVILLA IRPINA (Southern Italy)	hundreds	NATURAL	18th–19th
SAVOCA (Sicily)	tens	NATURAL	17th–18th
COMISO (Sicily)	50	NATURAL	18th–19th
CAPUCHINS' CATACOMBS (Palermo)	thousands	NATURAL	16th–20th

Samples of each type vary from a few dozens to several thousands of individuals (Table 1), as is the case of the Catacombs of the Capuchins of Palermo (Di Colo 1910) and most probably the Church of San Bartolomeo di Navelli in the province of l'Aquila, central Italy, which has only recently been discovered (Capasso and Di Tota 1991).

Generally speaking, the collections can be grouped under two different typological categories: natural and artificial mummies. The natural mummies, as for example the mummy of Saint Zita in Lucca (Tuscany) (Fornaciari et al. 1989e), were preserved for climatic and environmental reasons and without man's direct intervention, and represent the majority of the Italian collections, while the artificial mummies (such as the royal mummies of the Aragonese Mausoleum of Saint Domenico Maggiore in Naples) (Fornaciari 1986) were an exclusive prerogative of some groups of individuals of a high social class or of personages considered as important by the community, as for example kings and saints.

However, in some cases, there is documentary evidence of the particular treatment applied on the bodies in order to facilitate their drying process and preservation.

Special "dripping" methods of the cadaveric sewage were frequently used in southern Italy. Until the last century, when a person of social importance died, his body was placed in the vault of a church and left there for several months. The body was probably fixed in a seated position, and large pottery vases named "cantarelle" collected the cadaveric fluids (Fig. 2). Even today, in Naples, the saying "drain off" is an omen of death by which people are sometimes addressed (Fornaciari 1984a).

After a few months, the body, still flexible but no longer draining, was laid horizontally in special "tubs" covered with soil which, being volcanic in origin and thus rich in minerals, completed the processes of dehydration and mummification. The completely desiccated and mummified body was then dressed and placed in its coffin (Fornaciari and Gamba 1993).

Fig. 1. Distribution of ancient mummies in Italy

We shall now illustrate the two series which have been more exhaustively studied, those of the Church of Saint Maria della Grazia in Comiso (Sicily) and those of the Abbey of Saint Domenico Maggiore in Naples.

The series of Santa Maria della Grazia in Comiso (Sicily): an example of "middle-class" mummies

A mortuary chapel called the "Chapel of the Dead", annexed at the beginning of the 18th century to the church of Santa Maria della Grazia in Comiso (Sicily), contains 50 mummified bodies lying in niches opened in the right and left lateral walls, at the sides, and above the entrance (Fornaciari and Gamba 1993) (Fig. 3).

The bodies lie in a slanting position, with their faces always turned towards the inside of the building. The individuals are almost all males of different age, varying from young to adult to old. There are Capuchin friars and laymen belonging to the Third Order of the Capuchins. Twenty mummies are labelled with the individual's name and date of death, dated between 1742 and 1838. Many of those without a label may be more ancient, dating back to the 17th century when the church was erected. All the mummies, except one, were wearing monastic clothes. This one exception, whose body had been placed in a vertical position in the middle of the left wall, wore civilian clothes characteristic of the 18th century.

Examination of the bodies allowed us to state that the mummies were natural, that is the individuals had not undergone any kind of treatment, either evisceration or craniotomy. Their natural mummification was probably due to the hot dry climate of Comiso, which is located at the same latitude as Tunis.

Evidence that the bodies in the chapel were dressed only after being mummified was due to the discovery that all the monastic clothes presented a posterior longitudinal cut running from the bottom to the hood, and this cut must have been made to facilitate the dressing of a dry and therefore very stiff human body.

As concerns the strictly paleopathological situation, we must point out the individual No. 3, the provincial Bernardino del Comiso who died in 1742 at about 50 years of age, affected by severe arteriosclerosis with calcifications of the lumbar aorta and iliac arteries; individual No. 5, aged 35, shows an enormous enlargement of the thyroid gland, histologically a colloid goiter; individual No. 13, who died at the age of about 45, was affected by massive splenomegaly with infarctual areas; individual No. 21, aged 30–35, who shows lung fibrosis with multiple, apical calcifications, was probably affected by tuberculosis; individual No. 31, who died at about 60

Fig. 2. Church of Annunziata (Comiso, Sicily): crypt with the typical "cantarelle" for dripping of the bodies (18th century)

Fig. 3. Church of S. Maria della Grazia in Comiso (Sicily): posterior wall with the mummy niches (18th–19th centuries)

years of age, shows diffuse acariasis with hyperkeratosis and abundant eggs, nymphs and mites in different stages of development: the "clinical" picture of the disease, which was treated "in vita" by sulphur ointments which englobed parasites and eggs, is not clear (Gutierrez 1990); we also have a case of colon diverticulosis, two cases of inguino-scrotal hernia and a case of varicose veins, with ulcers of the lower limbs (Fornaciari and Gamba 1993).

The mummies of the abbey of San Domenico Maggiore in Naples: an example of "noble" mummies

The Abbey of San Domenico Maggiore dates back to the beginning of the 14th century and is one of the largest and most important churches in Naples. Saint Thomas Aquinas used to teach in the annexed convent of the Dominicans; the humanist Giovanni Pontano and the philosophers Tommaso Campanella and Giordano Bruno studied in this Abbey.

In a suspended gateway close to the vault, about 5 metres high, the monumental Sacristy of San Domenico Maggiore holds 38 wooden coffins containing the bodies of 10 Aragonese princes and other Neapolitan nobles, who died in the 15th and in the 16th centuries. At first, these coffins were scattered everywhere in the church, but were gathered in the Sacristy in 1594 on the orders of Philip II, King of Spain (Miele 1977).

The sarcophagi, richly dressed in silk, brocade or other precious fabrics, are disposed in two rows, one above the other. The lower row is made up of smaller coffins, mostly of unknown people, while the upper row includes larger coffins, some still bearing the coats-of-arms and the names of the personages buried in them, and so can be easily identified. They include the Aragonese kings Alfonso I (died 1458), Ferrante I (died 1494), Ferrante II (died 1496), Queen Giovanna IV (died 1518) and the Marquis of Pescara Francesco Ferdinando of Avalos (see cover image of this volume here), who won the famous battle of Pavia against the French king François I in 1525 and died the same year (Miele 1977).

A first survey showed that the coffins contained mostly mummified and very well preserved bodies.

A typical burial of S. Domenico Maggiore is that of Pietro of Aragon, III Duke of Montalto, who died in 1552 at the age of 12 (Fornaciari 1984 b); it consists in a large external sarcophagus (or "ark"), containing a second coarse wooden coffin with the richly dressed body, in 16th century style; there are plants, such as laurel, rosemary and box placed on the body, which lies on a layer of small lime fragments, for the drainage of the cadaveric fluids. The mummy, with its hands crossed on the pubis, appears in an excellent state of preservation. The colour of the skin is generally light brown. The head is hairless and the lips, which are drawn back showing the anterior teeth, are shrunk.

This series of mummies is unique in Italy not only for the antiquity and state of preservation of the bodies, but also for the historical importance of the personages, whose lives and causes of death are well known (for example malaria for king Ferrante II, pulmonary tuberculosis for the marquis of Pescara). It was therefore possible and extremely interesting to compare the paleopathological findings with the historical data. Up to now, the only mummies of this type known in Europe were those of the Hapsburg princes in Vienna (Kleiss 1977).

From 1984 to 1987 all the sarcophagi were carefully examined by a team from the Institute of Pathology of Pisa University (Fornaciari 1984 a).

The clothes, sometimes extremely precious, and the jewels of the buried individuals have been carefully recovered and will soon be restored and displayed by the Superintendence of Naples at the Museum of Capodimonte.

The mummies were first radiographed and then submitted to anthropological and autoptic examination on site. The laboratory studies were performed in Pisa.

We shall now briefly summarize the results so far achieved: 38 sarcophagi were explored, of which 8 were found to be empty, while one contained a double deposition. All depositions resulted to be more or less disturbed. There were 27 primary depositions and 4 secondary depositions, or redepositions (Table 2).

Table 2. Basilica of S. Domenico Maggiore (Naples): general situation of the depositions

TOTAL No. INDIVIDUALS	31		
PRIMARY DEPOSITIONS	27 (87.1 %)		
SECONDARY DEPOSITIONS	4 (12.9 %)		
NATURAL MUMMIES	12 (38.7 %)	SKELETONIZED	5 (41.7 %)
ARTIFICIAL MUMMIES	15 (48.4 %)	SKELETONIZED	1 (6.7 %)

Of all the individuals examined, 15, equal to 48.4 % of the totality, had been submitted to embalming, while 12 equal to 38.7 % had not been treated (Table 2). Therefore, there was a prevalence of embalmed individuals.

This fact is certainly not surprising, considering the high social class of the individuals buried in San Domenico. We know from the physician Ulisse Aldrovandi that, during the Renaissance, "the European kings and great personages used to entrust embalming of their bodies to their doctors and surgeons" (Aldrovandi 1602). As regards San Domenico, there is a document informing us about the embalming of the body of Antonio of Aragon, IV Duke of Montalto (died 1584): "Lo corpo e in sacrestia imbalsamato" (his body embalmed in the sacristy) (Vultaggio 1984).

The skeletonized individuals were 6, corresponding to 22.2 % of the totality; however, among the embalmed individuals there was only one case of skeletonization, corresponding to 6.7 % (Table 2). Therefore, the importance of the embalming process for the preservation, at least partial, of the bodies, is evident.

As for the embalming method, 9 individuals out of 14, among whom 5 children and 4 adults, result to have been eviscerated by a long anterior incision running from the neck basis to the pubic symphisis (Table 3). In order to penetrate the thorax, the sternum was cut or sawn or even removed; otherwise, the ribs or costal cartilages were cut sideways with the aid of shears; in one case both operations were performed. The sternum was generally cut in newborn babies and sucklings, and less often in adults where, as in modern autopsies, the costal cartilages on each side of the sternum were cut and the anterior thoracic wall removed. The other 5 individuals, including 1 child and 4 adults, show only an abdominal incision, running from the xiphoid process of the sternum to the pubis (Table 3). In this case the thorax was necessarily eviscerated through the diaphragm.

Table 3. Basilica of S. Domenico Maggiore (Naples): types of evisceration in artificial mummies

JUGULO-PUBIC INCISION	9
XIPHO-PUBIC INCISION	5
UMBILICAL-TRANSVERSE (associated with the above)	4

In 13 out of 15 cases of embalmed individuals, the brain was removed by craniotomy, which was horizontal and circular in 7 cases, or posterior, and often circular, in 6 cases (Table 4).

Table 4. Basilica of S. Domenico Maggiore (Naples): types of craniotomy in artificial mummies

ARTIFICIAL MUMMIES	15
HORIZONTAL CRANIOTOMY	7
POSTERIOR CRANIOTOMY	6
WITHOUT CRANIOTOMY	2

In 4 cases, including 2 newborn babies and 2 adults, extended unfleshing was observed of the muscular masses, at the level of the dorsum, the glutaei, and the limbs.

The material used for embalming (Table 5) mainly consisted in resinous substances, present in 10 cases, wool or similar material, and clay or earthy substances, present in 6 cases; lime was found in 4 cases, leaves or twigs in 3 cases, while tow, sponges and mercury were only used twice; finally, in 4 cases, 2 new-born babies and 2 adults, the body was wrapped in bandages soaked in resinous substances.

Wood for the inner coffins was determined in 16 cases of primary depositions (Table 6); there was a high incidence of pine-wood, used in 9 depositions, followed by poplar (4 depositions) and chestnut wood (3 depositions). Even in the case of the coffins, preference was shown for resinous substances, which were thought to favour the preservation of the bodies.

Table 6. Basilica of S. Domenico Maggiore (Naples): types of wood used for inner coffins in primary depositions

	NATURAL MUMMIES	ARTIFICIAL MUMMIES	TOTAL
FIR	5	4	9 (56.2 %)
POPLAR	2	2	4 (25.0 %)
CHESTNUT	0	3	3 (18.7 %)

We are in the presence of very complex evisceration and embalming methods, showing long-practised and diffused customs. As already said, some well preserved individuals show no apparent signs of embalming. The natural mummification of the bodies is probably due to the warm and dry climate of Naples during summer. The disposition of the coffins, placed at about 5 m of height close to the windows of the sacristy, and the particular microclimatic conditions of S. Domenico Maggiore, may have contributed to the preservation of the bodies. Furthermore, a recent exploration of the crypts of S. Domenico has led to the discovery of two large rooms which were certainly devoted to the dehydration of the bodies. They were in fact provided with numbered spaces for the coffins and with thick sand beds, for the gathering of the cadaveric fluids, and with wide ventilation shafts. It is very likely that the mummies of San Domenico Maggiore underwent the type of natural mummification treatment practised in Naples until the last century, which we have already described.

As concerns the strictly paleopathological situation, we must point out 2 cases of infectious disease and 2 of neoplastic pathology.

The mummy of an anonymous 2-year-old boy, who died around the middle of the 16th century (radiocarbon dating is 1569±60), revealed a widespread vesiculopustular exanthema type eruption (Fig. 4). The macroscopic aspects and the regional distribution suggested smallpox. Light microscopy and indirect immunofluorescence

Table 5. Basilica of S. Domenico Maggiore (Naples): frequency of different embalming materials in artificial mummies

RESINS	WOOL	SOIL OR CLAY	LIME OR ASH	LEAVES OR TWIGS	TOW	SPONGES	MERCURY	COTTON
10	6	6	4	3	2	2	2	1
66.7 %	40.0 %	40.0 %	26.7 %	20.0 %	13.3 %	13.3 %	13.3 %	6.7 %

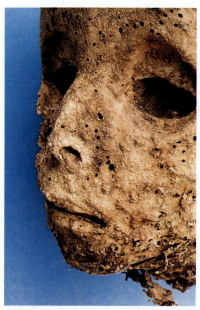

Fig. 4. Detail of face with vesiculo-pustular exanthema typical of smallpox

with anti-vaccinia-virus antibody confirmed this possibility. Electron microscopy revealed, among the residual bands of collagen fibres, pyknotic nuclei, and membrane remains with rare desmosomes, many egg-shaped, dense virus-like particles (250×150 nm), composed of a central dense region (or core) surrounded by an area of lower density (Fig. 5). After incubation with human anti-vaccinia-virus antiserum, followed by protein-A/gold complex immunostaining, the particles were completely covered by protein-A/gold (Fig. 6).

This showed that the antigenic structure of the viral particles was well preserved and that this Neapolitan child died of a severe form of smallpox some four centuries ago (Fornaciari and Marchetti 1986a, 1986b, 1986c, 1986d, 1986e; Fornaciari et al. 1989f).

Particularly interesting was the study of a case of treponematosis in the mummy of Mary of Aragon (1503–1568), marquess of Vasto in southern Italy. Famed for her beauty, this noblewoman of the Italian Renaissance belonged to the intellectual and religious circles of Ischia;

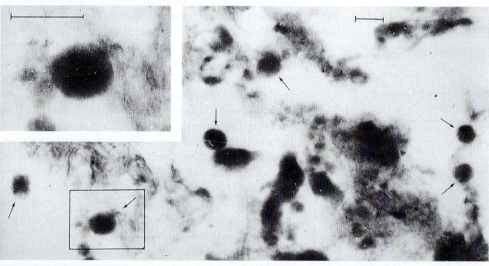

Fig. 5. Electron micrograph of a skin lesion showing many egg-shaped, dense structures (arrows) among the cellular debris. The inset (corresponding to framed area) provides better details of a particle of smallpox virus with its central dense core surrounded by halo of lower density (bars: 300 nm)

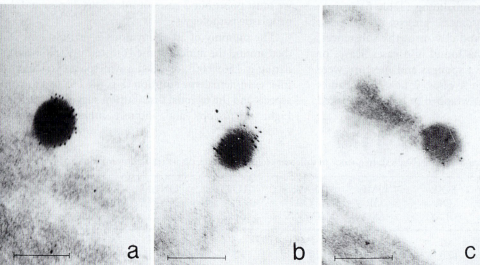

Fig. 6. Ultrathin section of skin incubated with anti-vaccinia virus antiserum followed by protein-A/gold. (a) Dense particle of smallpox virus completely covered by protein-A/gold; (b, c) partially labelled particles. Only a very few gold grains are scattered on the skin tissue (bars: 300 nm)

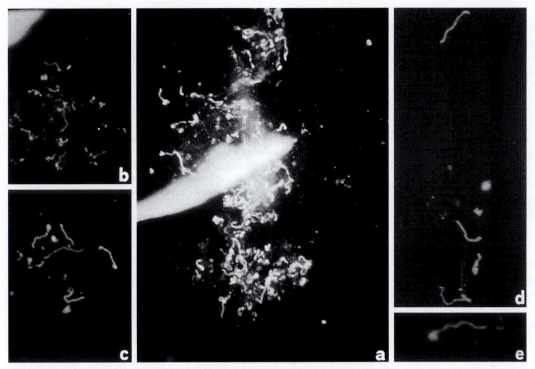

Fig. 7. Section of the cutaneous ulcer. Intense positive indirect immunofluorescence reaction with human anti-Treponema pallidum antibody: heaped (a, b) or isolated (d) treponemes of various size (c, e). Original magnification: (a, b) 250×; (c, d) 400×; (e) 1,000×

these included the poetess Vittoria Colonna, Michelangelo's friend.

The left arm of the mummy presented an oval 15×10 mm ulcer covered by a linen dressing with ivy leaves. Indirect immunofluorescence with human anti-treponema pallidum antibody identified a large number of filaments provided with strong yellow-green fluorescence and the morphological features of fluorescent treponemes (Fig. 7). An ultrastructural study evidenced morphologies typical of the spirochetes, as for example the axial fibril.

Immunohistochemical and ultrastructural findings thus demonstrated treponemal infection; the cutaneous ulcer is typical of third-stage luetic gumma. Venereal syphilis is the most probable diagnosis (Fornaciari et al. 1989b, 1989c, 1989d).

This discovery is important as it dates back to the 16th century and it can help clarify the biology of treponema in the epidemic phase of the disease.

Ferdinando Orsini, Duke of Gravina in Apulia, who died in 1549, shows a naso-orbital malignant tumour. We observed the complete destruction of the right nasal bone, with wide erosion of the upper orbital margin and the glabella. The retro-orbital bones were also destroyed. The histology showed solid neoplasia, with cords of spindle-shaped cells destroying compact and spongy bone and forming osseous lacunae, without bone reaction.

Fig. 8. Pseudo-glandular lumina in fibrous stroma. H & E, original magnification: 200×

The most probable diagnosis is a widely destroying skin epithelioma (Fornaciari et al. 1989a).

The autopsy performed on the artificial mummy of Ferrante I of Aragon, King of Naples, who died in 1494 at the age of 63, evidenced in the small pelvis a fragment of fibrous tissue which reached the dimensions of 6×4×1 cm after rehydration. Histologically, neoplastic epithelial cells disposed in cords, nests and glands (Fig. 8) were disseminated in a fibrous stroma containing scattered striated muscular fibers. The cells were tall, crowded, with abundant cytoplasm and pseudo-stratified pleomorphic hyperchromatic nuclei. The mucus was scarce and limited to pseudo-glandular formations, as shown by the specific staining (Alcian-blue). The application of a monoclonal antibody versus pancytokeratin showed strong intracytoplasmic immunoreactivity of the tumoral cells. The ultrastructural study evidenced well preserved pleomorphic nuclei with indented membranes.

These results clearly point out a mucinous adenocarcinoma infiltrating the muscular-fibrous layers of the small pelvis. It is impossible to establish, with the available data, the site of the primary neoplasm: the histological appearance suggests prostatic adenocarcinoma or an adenocarcinoma of the digestive tract (Fornaciari et al. 1993).

In conclusion, the paleopathological study of the mummies of eleven adult individuals from the Abbey of S. Domenico Maggiore, with good or excellent preservation, made it possible to diagnose two cases of cancer. Despite the very limited number of available specimens, we were in presence of an incidence of neoplastic pathology (18.8 %) similar to the one we find nowadays.

Although only a preliminary account, the present work shows the importance of the series of the Italian mummies, from a paleopathological point of view. The majority of these mummies have not yet been examined, but we think that further research in this field will provide extremely interesting results.

Summary

Numerous collections of mummies in Italy are unexpectedly distributed over the entire and in particular in the southern part of the country. Burials date from medieval to modern times, with a higher incidence between the 17th and 19th centuries. The mummies are both *artificial,* those submitted to the process of embalming, and *natural,* which represent the majority of the collections. Natural mummification was made easier by particular drainage methods of the cadaveric sewage frequently used in the crypts of the churches, especially in southern Italy.

The study of the natural mummies of the church of Santa Maria della Grazia in Comiso (Sicily, 18th–19th centuries) made possible the diagnosis of atherosclerosis, splenomegaly, tuberculosis, acariasis, colon diverticulosis, inguinal hernia and varicose ulcer.

Examination of the natural and artificial mummies of the Abbey of S. Domenico Maggiore in Naples (15th–17th centuries) clarified the embalming methods of the Renaissance, allowing the diagnosis of two cases of infectious disease (smallpox and venereal syphilis) and two of neoplastic pathology (epithelioma and adenocarcinoma).

The Italian mummies submitted to paleopathological studies have shown that modern biomedical techniques, such as immunohistochemistry and electronmicroscopy, can be applied to this type of material, providing extremely interesting results for the history of diseases.

Zusammenfassung

Im Gegensatz zur weit verbreiteten Ansicht, sind Mumien in Italien relativ zahlreich. Ihre Verteilung erstreckt sich auf das gesamte italienische Gebiet, mit einer gewissen Häufung in Süditalien. Die Datierungen erstrecken sich im allgemeinen auf einen Zeitraum, vom Mittelalter bis zur Neuzeit. Vom 17. bis zum 19. Jahrhundert findet man jedoch eine größere Inzidenz. Es handelt sich um künstliche Mumien, die also einem Einbalsamierungsprozeß unterzogen wurden, und natürliche Mumien, die die Mehrzahl der Exemplare darstellen.

Die natürliche Mumifizierung wird manchmal durch besondere Prozeduren der Dränage der Kadaverflüssigkeiten vereinfacht, die in den Kellern der Kirchen, besonders Süditaliens, praktiziert wurden. Die Untersuchungen an den natürlichen Mumien der Kirche St. Maria della Grazia in Comiso, Sizilien (18. bis 19. Jahrhundert) erlaubte folgende Diagnosen: Arteriosklerose, ein Schilddrüsenstruma, Splenomegalie, Lungentuberkulose, Akariasis, Kolondivertikel, Inguinalhernie und venöse Ulzera.

Das Studium der natürlichen und künstlichen Mumien der Basilika von St. Domenico Maggiore in Neapel (15. bis 17. Jahrhundert) hat uns Informationen über die Einbalsamierungstechniken der Renaissance erbracht und erlaubte uns die Diagnose von zwei Infektionskrankheiten (Variola und Syphilis) sowie von zwei Fällen von Tumoren (Epiteliom und Adenokarzinom).

Die paläopathologischen Studien der italienischen Mumien haben so gezeigt, das es bei so einer Art von Materialien möglich ist einige moderne biomedizinische Technologien anzuwenden, wie die Immunohistochemie und das Elektronenmikroskop mit Ergebnissen von höchstem Interesse für die Geschichte der Krankheiten.

Résumé

Contrairement à ce que l'on pense, les exemplaires de momies en Italie sont assez nombreux. Elles sont distribuées sur tout le territoire italien, avec une certaine prédominance dans le sud du pays. Elles sont datées généralement entre le moyen âge et l'ère moderne, avec une incidence majeure entre le XVIIe et le XIXe siècle. Ce sont des momies artificielles, soumises à un procès d'embaumement, et naturelles, celle ci etant le pourcentage plus élevé. Quelquefois la momification naturelle était facilitée par des techniques particulières de drainage des fluides cadavériques, effectuées dans les souterrains des eglises, surtout dans l'Italie du sud.

L'examen des momies naturelles de l'eglise de S. Maria della Grazia à Comiso, Sicile (XVIIIe–XIXe siècle) a mis en évidence les diagnostics suivants: artériosclerose, goître thyroidien, splénomégalie, tuberculose pulmonaire, acariose, diverticulose du colon, hernie inguinale et ulcères variqueux.

L'étude des momies, naturelles et artificielles, de la Basilique de S. Domenico Maggiore à Naples (XVe–XVIIe siècle) nous a donné des informations sur les techniques d'embaumement de la Renaissance et a permis le diagnostic de deux cas de maladie infectieuse (variole et syphilis vénérienne) et de deux cas de pathologie néoplastique (épithéliome et adénocarcinome).

Les études paléopathologiques effectuées sur les momies italiennes ont donc démontré qu'il est possible d'appliquer à ce type de matériel certaines modernes techniques biomédicales, comme l'immunohistochimie et la microscopie électronique, avec des résultats d'un grand intérêt pour l'histoire des maladies.

Riassunto

Contrariamente a quanto si potrebbe pensare, gli esemplari di mummie in Italia sono relativamente numerosi. La loro distribuzione comprende l'intero territorio italiano, con una certa prevalenza nell'Italia meridionale. La datazione è in genere compresa fra il Medioevo e l'Età Moderna, con una maggiore incidenza fra il XVII e il XIX secolo. Sono rappresentate mummie artificiali, sottoposte cioè ad un processo di imbalsamazione, e naturali, le quali costituiscono la maggioranza degli esemplari. La mummificazione naturale risulta talora facilitata da particolari procedimenti di drenaggio dei liquami cadaverici che venivano praticati nei sotterranei delle chiese, soprattutto dell'Italia meridionale.

L'esame delle mummie naturali della Chiesa di S. Maria della Grazia in Comiso, Sicilia, (XVIII–XIX secolo) ha permesso la diagnosi di: arteriosclerosi, gozzo tiroideo, splenomegalia, tubercolosi polmonare, acariasi, diverticolosi del colon, ernia inguinale e ulcere varicose.

Lo studio delle mummie, naturali ed artificiali, della Basilica di S. Domenico Maggiore a Napoli (XV–XVII secolo) ci ha fornito informazioni sulle tecniche di imbalsamazione di Età Rinascimentale e ha permesso la diagnosi di due casi di malattia infettiva (vaiolo e sifilide venerea) e di due casi di patologia neoplastica (epitelioma ed adenocarcinoma).

Gli studi paleopatologici effettuati sulle mummie italiane hanno così dimostrato che è possibile applicare a questo tipo di materiali alcune moderne tecnologie biomediche, come l'immunoistochimica e la microscopia elettronica, con risultati di altissimo interesse per la storia delle malattie.

References

Aldrovandi U., 1602 – cited in Gannal J. N., 1841, Histoire des embaumements. Paris, Desloges.

Aufderheide A. C., Aufderheide M. L., 1991 – Taphonomy of spontaneous ("natural") mummification with applications to the mummies of Venzone, Italy. In Human Paleopathology, Current Syntheses and Future Options. Washington, Smithsonian Institution Press, 79–86.

Capasso L., Di Tota G., 1991 – The human mummies of Navelli: natural mummification at new site in central Italy. Paleopath. Newslett., 75, 7–8.

Di Colo F., 1910 – L'imbalsamazione umana: manuale teorico-pratico. Milano, Hoepli.

Fornaciari G., 1984a – The mummies of the Abbey of Saint Domenico Maggiore in Naples: a plan of research. Paleopath. Newslett., 45, 9–10.

Fornaciari G., 1984b – The mummies of the Abbey of Saint Domenico Maggiore in Naples. Paleopath. Newslett., 47, 10–14.

Fornaciari G., 1986 – The mummies of the Abbey of Saint Domenico Maggiore in Naples: a preliminary survey. In V European Meeting of the Paleopathology Association, Siena, September 1984, 97–104.

Fornaciari G., Marchetti A., 1986a – Virus del vaiolo ancora integro in una mummia del XVI secolo: identificazione immunologica al microscopio elettronico (nota preliminare). Arch. Antrop. Etnol., CXVI, 221–225.

Fornaciari G., Marchetti A., 1986b – Intact Smallpox Virus Particles in an Italian Mummy of Sixteenth Century. Lancet, 8507, 625.

Fornaciari G., Marchetti A., 1986c – Des particles intactes du virus de la variole dans une momie italienne du XVI siècle. Journ. Intern. Méd., 79, 53.

Fornaciari G., Marchetti A., 1986d – Intact smallpox virus particles in an Italian mummy of the XVI century: an immuno-electron microscopic study. Paleopath. Newslett., 56, 7–12.

Fornaciari G., Marchetti A., 1986e – Italian Smallpox of the Sixteenth Century. Lancet, 8521/22, 1469–1470.

Fornaciari G., Bruno J., Corcione N., Tornaboni D., Castagna M., 1989a – Un cas de tumeur maligne primitive de la région naso-orbitaire dans une momie de la basilique de S. Domenico Maggiore à Naples (XVIe siècle). In "Advances in Paleopathology", Proceedings of the VII European Meeting of the Paleopathology Association (Lyon, September 1988), 65–69.

Fornaciari G., Castagna M., Tognetti A., Tornaboni D., Bruno J., 1989b – Syphilis in a Renaissance Italian Mummy. Lancet, 8663, 614.

Fornaciari G., Castagna M., Tognetti A., Tornaboni D., Bruno J., 1989c – Treponematosis (venereal syphilis ?) in an Italian mummy of the XVI century. Riv. Antrop., LXVII, 97–104.

Fornaciari G., Castagna M., Tognetti A., Tornaboni D., Bruno J., 1989d – Syphilis tertiaire dans une momie du XVIe siècle de la basilique de S. Domenico Maggiore à Naples: étude immuno-histochimique et ultrastructurelle. In "Advances in Paleopathology", Proceedings of the VII European Meeting of the Paleopathology Association (Lyon, September 1988), 75–80.

Fornaciari G., Spremolla G., Vergamini P. and Benedetti E., 1989e – Analysis of pulmonary tissue from a natural mummy of the XIII century (Saint Zita, Lucca, Tuscany, Italy) by FT-IR microspectroscopy. Paleopath. Newslett., 68, 5–8.

Fornaciari G., Tornaboni D., Castagna M., Bevilacqua G., Tognetti A., 1989f – Variole dans une momie du XVIe siècle de la basilique de S. Domenico Maggiore à Naples: étude immuno-histochimique, ultrastructurelle et biologie moléculaire. In "Advances in Paleopathology", Proceedings of the VII European Meeting of the Paleopathology Association (Lyon, September 1988), 97–100.

Fornaciari G., Gamba S., 1993 – The mummies of the church of S. Maria della Grazia in Comiso, Sicily (18th–19th century). Paleopath. Newslett., 81, 7–10.

Fulcheri E., 1991 – Il patologo di fronte al problema della perizia in corso di ricognizione sulle reliquie dei Santi. Pathologica, LXXXIII, 373–397.

Gutierrez Y., 1990 – Diagnostic pathology of parasitic infections with clinical correlations. Philadelphia, Lea & Febinger.

Kleiss E., 1977 – Some examples of natural mummies. Paleopath. Newslett., 20, 5–6.

Miele M., 1977 – La Basilica di S. Domenico Maggiore in Napoli. Napoli, Laurenziana.

Terribile V., Corrain C., 1986 – Pratiche imbalsamatorie in Europa. Pathologica, LXXVIII, 107–118.

Vultaggio C., 1984 – Personal communication.

Correspondence: Dr. G. Fornaciari, Institute of Pathology, University of Pisa Medical School, Via Roma 57, I-56126 Pisa, Italy.

The roman mummy of Grottarossa

A. Ascenzi[1], P. Bianco[1], R. Nicoletti[2], G. Ceccarini[2], M. Fornaseri[3], G. Graziani[3], M. R. Giuliani[4], R. Rosicarello[4], L. Ciuffarella[4], and H. Granger-Taylor[5]

[1] Department of Human Biopathology, "La Sapienza" University, I-00161, Rome, Italy
[2] Department of Chemistry, "La Sapienza" University, I-00185, Rome, Italy
[3] Department of Earth Sciences, "La Sapienza" University, I-00185, Rome, Italy
[4] Department of Vegetal Biology, "La Sapienza" University, I-00185, Rome, Italy
[5] British Museum, London, U.K.

Historical introduction

The Grottarossa mummy is only the second ever to have been discovered in Rome. According to Stefano Infessura (1723), the first Roman mummy – of a 12- or 13-year old girl – was found in 1485 in a grave near the via Appia about five miles from the city. After its removal, it was exhibited in the Palazzo dei Conservatori near the Capitol in Rome, but Pope Innocent VIII, was afraid of an outburst of popular fanaticism, and ordered it to be reburied at a secret site outside the Porta Pinciana, so that all knowledge of its precise whereabouts was lost.

Mummification has never been a Roman custom, and at present the Grottarossa mummy must be considered a unique specimen.

Grottarossa is a district on the outskirts of Rome, nine miles north of the Capitol along the via Cassia, and the mummy was discovered there on 5 February 1964. During building work, the mummy and its sarcophagus were inadvertently loaded on to a truck together with rubble to be dumped. When the driver realised he had been transporting a body, he immediately informed the police, who thought it might be an archaeological find and notified the "Sovrintendenza alle Antichità" (Monuments and Fine Arts Service, the government agency responsible for all such finds in Italy), and had the mummy sent to the Institute of Forensic Medicine at the "La Sapienza" University, Rome.

The location of the sarcophagus had been a space below a block of building materials measuring 3×2.6×2.6 m, which could have been the foundations of a funerary monument (Fig. 1). This explains why the mummy and its sarcophagus had escaped vandalism over a period of many centuries.

On February 12, encouraged by public interest and the press, Prof. C. Gerin, then Director of the Institute of Forensic Medicine, convened a press conference to present the mummy and give some preliminary comments on it based on external inspection, microscopic examination and a conventional x-ray study. He stated his opposition to an autopsy, which would have destroyed an exceptional finding in the history of the Roman world. At the same press conference archaeologists provided information about the sarcophagus itself, and the wrappings, jewellery and a doll that were found in it.

The sarcophagus was made of white marble, with a double-weathered lid opening at the front. It is rectangular, and has masks on its corners. Both the sarcophagus and its lid are decorated with ornamental carvings. There is a deer-hunting scene on the long, front side, which continues as a boar-hunting scene on the short, right side. According to E. Paribeni (1964), the scene is inspired by the Aeneas and Dido episode in Book IV of the *Aeneid*. A lion-hunting scene is shown on the lid opening. Africa, Venus and a fluvial divinity – the River Bagradas near Carthage, according to Andreae (1969) – are displayed in symbolic form on the short, left side; this could indicate the setting of the hunting scenes.

The mummy exhibited jewels having features in accordance with its youthful age: a pair of gold earings, a gold necklace with sapphires and a gold ring. Between the funerary items it is worth remembering an articulated ivory doll, having – as common at that time – features of an adult woman; a small amber shell-shaped box; a small amber pot; a little box with handle; a little amber die.

Considering the special features of the sarcophagus and funerary items, the Grottarossa girl must have lived in the second century A.D. either at the end of the reign of Hadrian, and/or the beginning of the Antonine era, as asserted by Andreae (1969), or during the reign of Hadrian, as argued by Sapelli (1979), or under the Anto-

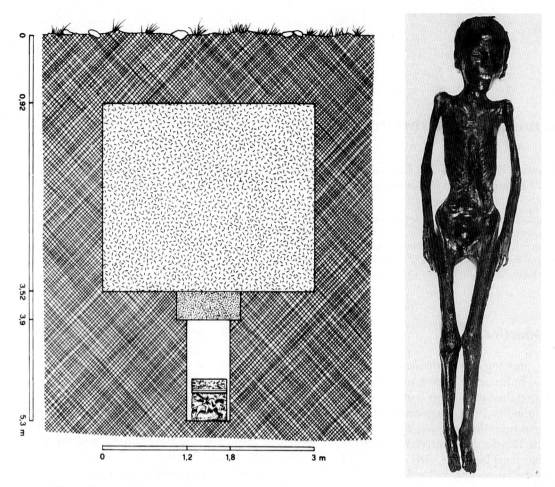

Fig. 1. The underground location of the mummy sarcophagus in a space below a block of building materials (left). The Grottarossa mummy (right)

nine emperors (160–180 A.D.), according to Bordenache Battaglia (1983).

The medical experts involved in the earlier examination of the mummy did not publish any account of their observations. From an archaeological viewpoint, a paper by Scamuzzi (1964) is of doubtful value since it contains a great many unsupported assertions (Bordenache Battaglia, 1983), as is an article by Castellani (1964). Fully analytical observations on the sarcophagus were published both by Andreae (1969) and Sapelli (1979). Chapter XII of Bordenache Battaglia's book is fundamental in providing information on the jewels, funerary items and sarcophagus. A note on the same subject is supplied by Pfeiler (1970).

At the end of February the mummy was taken to the Museo delle Terme in Rome. A few years later, when the museum was about to be moved, the mummy was returned to the Institute of Forensic Medicine, where it has remained till the present day. Its final destination will be the new Museum of the Roman World in Rome, which will be opening soon.

The unavailability of any interdisciplinary study on the Grottarossa mummy has been a disappointment to the scientific community. We therefore decided to reconsider the whole subject and all the available materials, using the following approaches: a) an anthropological and paleopathological study; b) x-ray analysis of the teeth; c) computerized tomography (CT) or CT scan, with sampling of superficial and deep tissues using CT-guided needle biopsies; d) light and electron microscopic examination of tissues; e) embalming chemistry; f) studies on the wrappings and other textiles; g) pollen analysis; h) a gemmological inquiry into the jewels and funerary items.

The results achieved by our specially created team appear for the first time in complete form in the present paper.

Materials and methods

The Grottarossa mummy was examined and the methods and devices of classical anthropometry were applied. A new conventional x-ray investigation and a CT scan were carried out. The panoramic x-ray technique was used to examine the state of the teeth. To do this, the mummy was held in a vertical position by being fixed to a wooden board from which the head and neck protruded.

As permission to carry out an autopsy had been refused, samples for the microscopic examination of some organs were obtained under CT control using a Bordier's needle. For histology, specimens were hydrated and fixed according to Sandison (1955) and embedded in paraffin or in glycol-methacrylate. 4–8 μm thick sections from paraffin and 1–2 μm thick sections from glycol-methacrylate were stained using the following methods: Hematoxylin-eosin, van Gieson for collagen, Weigert for elastic fibers, phosphotungstic acid-hematoxylin for fibrin, Ziel-Neelsen for acid-fast bacteria. For electron microscopy, hydrated and fixed specimens were post-fixed in OsO_4 and embedded in Araldite. Semithin sections for light microscopy survey were stained with methylene blue-azure II. Thin sections were contrasted with U/Pb.

Photographs taken during the first press conference in 1964 show that the upper limbs of the mummy were wrapped but by the time the mummy was delivered to us, no wrappings were left. Textile material was, however, found lying on each side of the body. It was impregnated to varying degrees by a brownish, aromatic substance and showed small clots of resin of the same colour. The textiles[1] turned out to comprise seven units. Four of these (Nos. 2, 3, 4, 5) corresponded to rolled up wrappings and two (Nos. 1, 6) had an incomplete tubular shape. A seventh (No. 7) was a shapeless cloth fragment folded many times over and compressed. It too was deeply impregnated by resin.

To study the weaving pattern and the fibers of the yarn making up the textile materials, the optical conventional microscope, the stereomicroscope and the scanning electron microscope were used (Gianolio, 1987). The fiber identification was also confirmed by microchemical tests using a solution of zinc-iodide reagent (methods ASTM D276–77).

Analytical chemical data about the mummifying procedure were obtained from (a) hair, (b) skin, (c) powder obtained by scratching the rectum, (d) textiles and the small clots sticking to them, (e) resinous material contained in a little box belonging to the funerary equipment.[2]

The sampling of material from the rectum was prompted by the remark of Pazzini (1964) that at the time of discovery aromatic liquid dripped from the anus.

Volatile aromatic substances for "head space analysis" were collected from the textiles.

All the above materials were processed as follows.

Solid samples

After extraction with exane and/or methanol, the solutions were submitted to gas-chromatography (GC) and gas-chromatography/mass-spectrometry (GC/MS). GC was performed using a Hewlett-Packard system, with helium as carrier gas and hydrogen flame detector. A silicone coated capillary column 20 m long and 0.32 mm in diameter was used. Starting at temperature of 50 °C, this was progressively raised to 250 °C at a rate of 3 °C/min.

GC/MS analysis was carried out using a similar capillary column operating the conditions reported above. Mass spectra were recorded both by electron impact at 70 eV and by the chemical ionization system.

To identify abietic acid derivatives, that is, the terpenoids considered as specific markers for the coniferous resin, standard procedures were used involving dissolution in 10 % methanol-ether, methylation with diazomethane before injecting in the gas-chromatograph, according to Schlenk and Gellerman (1960) and Mills and White (1989).

As suggested by the black colour of hairs and skin, the possibility that bitumen may have been used for mummification was considered; extraction of the solid material therefore was attempted using CS_2 and an analysis for demonstrating Mo, Ni and V (elements commonly present in bitumen) was carried out using inductively coupled plasma spectroscopy.

Volatile samples

After the textiles had been sealed in boxes, the volatile substances present were trapped in absorbing coal using a flux of nitrogen. After thermal de-absorbing the same substances were conveyed to the column of the gas-chromatograph where the initial temperature was 50 °C; it was then raised to 150 °C at a rate of 10 °C/min and finally to 250 °C at the rate of 3 °C/min.

Besides a small amber box containing resinous materials, the investigations on the funerary items and jewels were restricted to the golden necklace with sapphires and an amber die because the remaining items had already been studied at the archaeological laboratory at the "Sovrintendenza Roma I" (Andergassen, 1983).

In studying the necklace, consisting of thirteen blue sapphires joined with elongated golden pieces, care was taken to collect information about the possible origin of the sapphires. To achieve this a sapphire sample was subjected both to optical observations and standard gemmological tests. Its chemical composition was determined using an electron micro-probe (Jeol JSM-50A). Minor element concentrations were measured by carrying out the micro-analyses along alignments with scanning steps of 10 and 20 μm and a time count of 100 sec. Synthetic and natural standards were used, with online corrections performed according to a modified version of the MAGIC IV program (Colby, 1968).

The absorption spectrum in the visible region between 340 and 800 nm was recorded on a Varian Cary 219 double-beam spectrophotometer.

To determine the site of origin of the amber die, the surface was gently scratched with a needle to obtain small quantity of powder. Samples resulting of powder (3 mg) dispersed in KBr (150 mg) were tested in a Perkin Elmer 257 spectrophotometer under the following conditions: spectrum range 4,000 to 600 cm^{-1}; scanning time 6 min; slit 2 : 2. As standards, ambers of different origin were used.

A pollen analysis was carried out on the clots found sticking to the textiles, and the resinous material contained in a box that was part of the funerary equipment . The method applied was the conventional one: an alcohol wash to remove possible superficial contaminating pollen; dissolution in alcohol for 24 h; centrifugation; acetolysis according to the method of Erdtman (1969) modified by Moore and Webb (1978); conservation in glycerol.

Results

The Grottarossa mummy may be considered to be satisfactorily preserved (Fig. 1), although the right big toe, the distal phalanx of the left big toe, the distal phalanges of the 2nd, 3rd and 4th right toes, and the distal

---

[1] Extensive papers on textiles are in preparation.

[2] Here chemical methods and results are presented in a summarised form. An extensive paper on the subject is in preparation.

phalanx of the 2nd left toe are all missing. These *post mortem* amputations are most probably due to the traumatic action suffered by the body during the excavation of the grave and its removal to the dumping site.

All of the body is wrinkled by complete dehydration. As a result its weight is only 4,960 g in spite of an overall length of 120 cm. The muscles show conspicuous rigidity, so that the body can be lifted as if it was a board.

The tough, stiff skin reveals an intense brownish colour and its surface is free from any type of crystals. The few hairs are 3–5 cm in length. Their colour is black with a reddish tint. The ears are regularly developed. The eyes are wrinkled and collapsed, especially the right one, which is closed. The nose is deformed with the tip pushed backward and to the left. The wrinkled lips uncover the crown of incisors, which reveal a few defects in the enamel. In spite of its deformations the face shows a caucasian look.

The external genitalia are female, with very poorly preserved labia.

In the anterior aspect of the right thigh, a rectilinear incision 9 cm long sutured with sewing thread is to be attributed to the withdrawal of samples from the femur and the surrounding soft tissues soon after the discovery. There are no incisions over the remaining surface of the body. More specifically, this is true of the orbital, occipital, abdominal and perineal regions. Probably because of the wrappings, the chest volumes low and the sternum is displaced backward with respect to the anterior end of the ribs.

On the hands the whorls and ridges of the fingers are present, and the finger nails are well preserved.

The external examination of the mummy was completed by the collection of some anthropometric data, which are reported in Tables 1 and 2. It is worth pointing out that these measurements are somewhat atypical because they were not taken directly from the skeleton and at the same time the dehydration of the soft tissues does not allow any comparison with measurements obtained from a living body. The body height was measured going from the bregma to the middle of the inferior surface of the calcaneum. This last reference point was also chosen as representative of the ground surface for measuring the height of other anthropometric markers.

Table 1. Cranial and facial indexes with their corresponding conformational types

Indexes		Conformational types
Cephalix index	85.2	Brachycephaly
Height[a] to length index	71	Hypsicephaly
Height[a] to breadth index	83.3	Metriocephaly
Facial index	83	Mesoprosopic
Nasal index	64.9	Leptorrhine

[a] Auricular height

Table 2. Measurements in mm of body height, head, trunk, and limbs

Body height	Head	Trunk	Upper limb	Lower limb
1160	(1) 169	(4) 905	(8) 965	(13) 625
	(3) 144	(6) 560	(9) 750	(15) 295
	(6) 112	(27) 345	(10) 605	(16) 30
	(8) 80	(35) 200	(11) 485	(54) 595
	(13) 24	(40) 180	(45) 480	(55) 330
	(15) 120		(46) 360	(56) 265
	(18) 93		(47) 215	
	(21) 37		(48) 145	
			(49) 120	

The measurements are given using Martin's numbering placed here between brackets

Head: (1) maximum cranial length, (3) maximum cranial width, (6) maximum inter-zygomatic distance, (8) gonion-gonion distance, (13) inferior nasal width, (15) auricular height of the head, (18) nasion-gnathion distance, (21) nasal height.

Trunk: (4) height of the sternal jugular notch, (6) height of the symphysial crests, (27) anterior length of the trunk, (35) acromion-acromion distance, (40) distance between the iliac cristae.

Upper limb: (8) height of the acromion, (9) height of the humero-radial joint, (10) height of the radial styloid process, (11) height of the apex of the middle finger, (45) length of the upper limb, (46) length of the upper limb excluding hand, (47) length of the arm, (48) length of the forearm, (49) length of the hand.

Lower limb: (13) height of the anterior superior spine, (15) height of the knee joint, (16) height of the apex of the medial malleolus, (54) length of the lower limb excluding foot, (55) length of the thigh, (56) length of the leg.

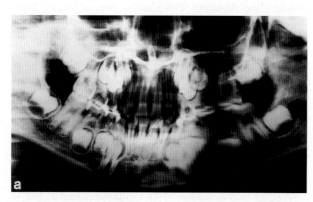

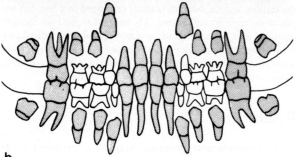

Fig. 2. Panoramic x-ray film (a) and sketch (b) of the teeth of the Grottarossa mummy

The conventional x-ray examination and the CT scan show a regularly developed skeleton. There is persistence of the epiphyseal cartilages in long bones. With the exception of the pisiphorm all the remaining primary ossification centers of the carpus are present, indicating that the age of the girl may well have been 8.

The panoramic x-ray film of the teeth (Fig. 2) shows that the upper and lower first permanent molars are fully erupted. On the other hand, the upper and lower deciduous canines and molars are persistent, while the permanent canines, premolars and second molars are unerupted. The third permanent are not appreciable. This stage of dental development corresponds to that of a subject about 8 years old.

In the skull the encephalon is persistent although deeply wrinkled and displaced into the occipital region. In a few areas a honeycomb appearance suggests a gaseous putrid state (Fig. 3). The histological examination of an encephalon sample taken through a little hole opened in the occipital bone showed only disorganised amorphous material without recognisable tissue structures.

At the x-ray examination thoracic and abdominal viscera are recognisable although frequently wrinkled and deformed. In the chest a central mass is assumed to represent a dessicated heart with partially collapsed lungs. This last finding is mainly due to a pneumothorax and to a minor extent to pleural effusions which appear at the CT scan as two opaque semilunar figures in the posterior and inferior regions, one on each side (Fig. 4). As the effusions are completely dehydrated, no change occurred in their seat and morphology when the body was moved from the supine to the prone position. In order to have information about the nature of the effusions, a bilateral puncture was carried out (Fig. 4). The cylindrical samples of the collected material, examined under the light and electron microscope, provided the following findings in going from the deepest to the superficial layers. (a) Well preserved lung tissue with bronchi, vessels and many small foci of anthracosis (Fig. 5a, c); (b) a pleural cavity fully filled with unorganised fibrils suggesting fibrin (Fig. 5b, d). The fibrils were mixed with myelin figures and an enormous number of varied bacterial species, probably not all pathogenic, but responsible for the putrid state; (c) structures lining the thoracic walls, such as muscles, fat-tissue and skin.

X-ray examination of the abdomen reveals opacities attributable to shrunk liver, spleen and kidneys, while

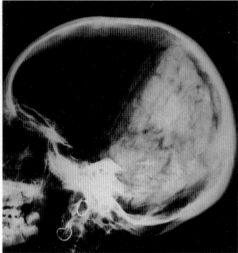
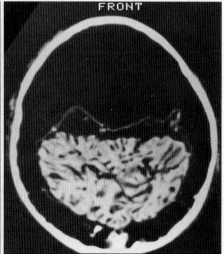

Fig. 3. X-ray lateral film of the skull of the Grottarossa mummy showing a shapeless opaque material in the occipital region (left). At CT scan the opaque material in the skull corresponds to the wrinkled and displaced encephalon showing gaseous putrid cavities (right)

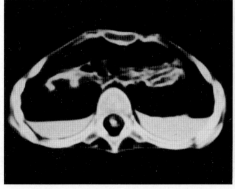
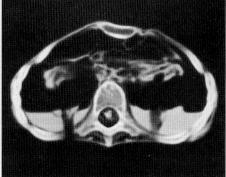

Fig. 4. CT scan of the mummy chest: Bilateral pleural effusion in the back side and collaps of the lung by pneumothorax (left). CT scan of the chest showing the two parietal discontinuities produced by the Bordier's needle. On the right side the discontinuity partially involves the collapsed lung (right)

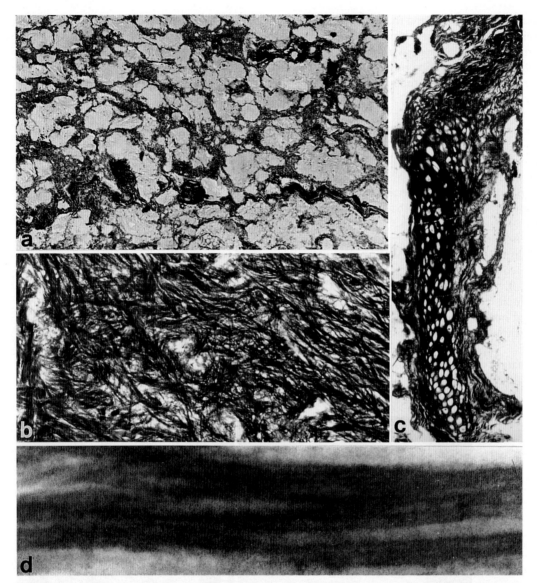

Fig. 5. (a) Microscopic section of the lung: alveolar cavities of irregular width with few foci of anthracotic pigment. ×110. (b) Thick interlacing of fibrin from pleural effusion. ×120. (c) Cartilaginous structure of a bronchial wall. ×110. (d) Fibrils of fibrin with residual crossbanding. TEM, ×130,000

the pancreas was not identified. Especially at the CT scan there was evidence of the stomach and intestine outlines, but it would be unsafe to state that the intestine persists in its totality or has been partially destroyed by putrid processes or the possible inoculation of aromatic fluid in the rectum.

Radiology has also revealed pathologic changes in the skeleton, which is uniformly porotic. The femurs in particular show obvious Harris' lines suggesting episodes of infection or malnutrition during life.

Under the optical microscope, textile material samples from units Nos. 1, 2, 3, 4, 5 and 6 showed features of linen fibers. They were 12–25 μm in diameter and showed a narrow central cavity, thickened walls and the so-called "beat marks" (Fig. 6b, c). The violet col-

our caused by staining reaction of the zinc-chloro-iodide solution confirmed the vegetable nature of this fibers (cellulosic material). Textile fibers from unit No. 7, made of protein material, assumed instead, a yellow colour caused by the reaction of the solution mentioned before. In fact these last fibers showed features of silk fibers (single filaments, smooth, nearly structureless) with an almost constant diameter averaging 20 μm (Fig. 6d).

Of the textile material found, the four rolled up wrappings (Nos. 2, 3, 4, 5) showed a circumference apparently corresponding to that of the upper limbs. They proved to be linen which has a rather coarse texture but retains an appreciable degree of elasticity. In each wrapping the selvedge is present only along one edge, while

Fig. 6. (a) Weaving pattern of the linen wrappings with Z-spun direction as seen under SEM. ×100. (b) Linen fiber with "beat marks". 220. (c) Linen fiber showing central cavity. ×350. (d) Silk fibers from the cloth in close contact with the skin of the mummy. ×100

the other edges are frayed. This finding raises the suspicion that the wrappings may have been recycled from worn-out tissue. Textile units Nos. 1 and 6 show fine linen without selvedge, possibly cut out of larger pieces of cloth. Small fragments of fine linen of the same type are found glued with resin to wrappings of coarse linen and proved that the wrappings were of various different qualities. All the linen textiles have a Z-spun direction for spinning (Fig. 6a).

Textile unit No. 7 turned out to consist of silk whose very bad state of preservation was due to deep imbibition with resin. The high resin content suggests that this textile was the remnant of a cloth in close contact with the skin of the mummy.

Lastly, no crystals to be attributed to natron were found in any of the textiles examined.

Going on to review the results of the chemical analyses whose aim was to yield data on the material used for mummification, one main finding was that the GC and GC/MS analyses both from the "head space" and from solutions of extracted solid material (Fig. 7) showed the presence of *p-cimene* ($C_{10}H_{14}$) and of sesquiterpenes (C_{15}) with a molecular weight ranging between 192 and 238, as revealed by chemical ionization. Electron impact spectra obtained were compared with those reported in the Wiley/NBS mass spectra cathalog (McLafferty and Stauffer, 1988). Only α-*cariophyllene* was identified, while the nature of the other sesquiterpenoids remained uncertain.

Abietic acid derivatives were identified in textiles, resin cloth, skin and rectum.

The existence of triterpenoids (C_{30}) can be ruled out.

The impossibility of obtaining a CS_2 soluble fraction from the solid material, the absence of characteristic straight-chain hydrocarbon pattern in the gas chromatograms and the presence of Mo, Ni and V in almost undetectable quantities (0.01, 0.09 and 0.08 ppm, respectively), all indicate that bitumen, if present, is a negligible component of the embalming material.

The resinous substance contained in the small box that was part of the funerary equipment showed no difference in composition with respect to that found in the other solid materials mentioned.

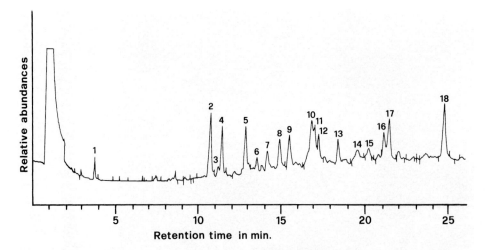

Fig. 7. Gas-chromatogram recorded from a metanol solution of the extracted material sticked to the wrappings. Peaks from 2 to 17 are representative of sesquiterpenes having molecular weights ranging between 192 to 238 as indicated by GC/MS analysis

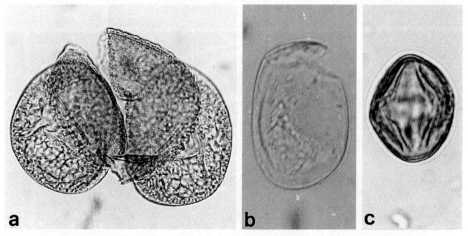

Fig. 8. (a) Pollen of *Pinus pinaster* type. ×670. (b) Pollen of *Juniperus* type. ×1,500. (c) Pollen of *Rosaceae* family. ×1,500

Pollen analysis gave inhomogeneous results. In some of the wrappings *Pinus pinaster* type (Fig. 8a) is dominant, with 35 pollen grains, followed by *Quercus* type with 20 grains; of the plants represented by one or two grains, the *Cupressaceae* (Fig. 8b) arouse some interest. In the remaining textiles *Pinus* and *Quercus* are sporadic: among other abundant pollen grains, myrrh (*Commiphora* sp.) deserves particular attention. Lastly, some pollen grains have been ascribed to the families of *Rutaceae*, *Chenopodiaceae* (both known as vermifuges) and *Cistaceae*. One grain of *Abies* and two grains of *Artemisia* were present too.

One finding to be stressed is the irregular crowding of the pollen grains of the same species at various points in the silk textile, suggesting possible human interference. The identification of such grains has been not easy. They can be probably ascribed to *Peganum* genus (*Zygophyllaceae*).

In the resinuous material contained in the small box, pollen was very scanty, although some grains of *Quercus*, *Carpinus*, *Pinus* and *Rosaceae* (Fig. 8c) were found.

As regards the necklace, a drilled sapphire was chosen as a sample. This is an asymmetrically cut cabochon with a pure hue ranging between middle and deep blue. The average refractive index is 1.76 and the specific gravity 4.03. At the hand spectroscopy the line of the iron at 450 nm is present. The optical microscope reveals a pseudo-hexagonal symmetry in conformity with the crystal growth steps, as proved by differences in the blue hue. Minute acicular stout needles of rutile are present too. They are associated in various ways, occasionally giving rise to sporadic milky stripes. There are sporadic 'feathers' of sealed fluid remnants. Opaque anhedral crystals (probably sulphides) and transparent dark crystals (probably pyroxenes) are also visible.

Since this set of findings was insufficient to allow identification of the original site of the sapphire, the method suggested by Poirot (1992) was applied. Using an electron probe the trace elements were identified and the absorption spectrum in the visible region was recorded, focusing in particular on the region between 460 and 360 μm. As these techniques showed the presence of gallium, chromium and titanium, and, besides the peak at 693 nm, a shift to 400 μm of the peak situated at 377 in Burmese sapphires, it was concluded that this sapphire probably came from Sri Lanka (Ceylon).

The amber die shows a brownish-red colour and under microscopic examination reveals swirl marks and globular cavities sometimes collected in clusters of crumps. At the range between 1,200 and 800 cm^{-1} the infrared spectrum provides the characteristic trend of the Baltic amber, although somewhat reduced in sharpness when compared with that of the yellow Baltic amber. This finding may support the view that the brownish-red colour of the die is a consequence of the antiquity of the amber.

Discussion

The investigations reported above refer to the following topics, which will now be discussed separately: (a) description of the Grottarossa mummy, (b) the cause of death, (c) the preservation technique, (d) the site where mummification was performed, (e) why the mummy should have been inhumed in Rome. On the first topic, there is little doubt that the mummy was an 8-year-old girl, as revealed by the dentition, body height (120 cm), and skeletal ossification (in particular there was no pisiform ossification center). Even allowing for some shrinkage, this last figure is in line with the modern standards laid down by the National U.S. Center for Health Statistics and commonly used in Italy, too, according to which 125+11 & −10 cm is the range of body height for a 8-year-old girl.

The morphological and anthropometric features all support the view that the girl was of caucasian race, even if her skin has been discoloured, taking on a dark brown tinge.

According to the eye-witness accounts of the few people who were able to examine the mummy immediately after its discovery, before it had been taken to the Institute of Forensic Medicine, the body was apparently well hydrated and its features resembled those of a white living subject. A drawing obtained from a photograph taken on this occasion and reproduced in the newspress is shown in Fig. 9. The same people who first examined the mummy stated that the body became wrinkled and discoloured while they were watching it. As a result, when the mummy was first presented at a press conference at the Institute of Forensic Medicine, its features were similar to those seen at the present time and described here. To explain this deterioration, it has been supposed that the sudden exposure to air of a mummy previously kept in an air-tight setting may have led to rapid dehydration, which could have activated the discolouring capability of the resin impregnating the skin and its wrappings.

In examining the problem of the origin of the Grottarossa mummy, it seems useful to consider the anthropometric data obtained from the cranium, especially its prominent brachicephally. The geographical distribution of the cephalic index yields evidence that brachicephal is a rare condition among the populations living around the southern Mediterranean coasts. This is especially true of Egypt, where dolicocephaly is prominent (Martin and Saller, 1957, 1959). This makes it unlikely that the girl was born of Egyptian parents; she probably had Italic parents who came from the center or possibly the north

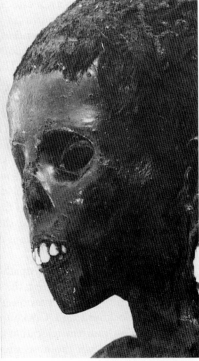

Fig. 9. Exact copy of a photograph of the Grottarossa mummy taken from an amateur few after the discovery and reproduced in some newspapers at that time (left). The present aspect of the mummy (right)

of the peninsula, where brachicephaly is common (Livi, 1896; Sergi G., 1905; Martin and Saller, 1959). This interpretation is strengthened by the appearance of the sarcophagus and funerary equipment which, according to Andreae (1969), Sapelli (1979) and Bordenache Battaglia (1983) are typically Roman.

Considering now the topic of the cause of death, there is no doubt that the main pathologic finding is a bilateral pleural effusion clearly showing the features of a pleuritis, possibly associated with a septic pneumothorax. One of two alternative pathogenetic mechanisms may account for this: (a) the pleuritis is secondary, that is, an infection originated in another organ, and then spread to the pleura; (b) the pleuritis is primitive, that is, the infection began in the pleura. The commonest cause of a secondary pleuritis spreading from an adjacent organ is pneumonia. In the samples obtained by puncturing the thoracic walls neither lung appeared to be affected. This finding does not rule out the possibility that the pleuritis originated from a pneumonia, because a single small sample is not representative of a whole lung. As regards the transmission of an infection from other sites, a tubercular pleuritis cannot be excluded, but cannot be demonstrated either. No tubercles were observed, and the high quantities of various types of microorganisms in the pleura do not at present constitute a clear demonstration of a tubercular bacillus.

Pleuritis is not the only disease demonstrable in the mummy; other pathologic conditions, such as Harris' lines and diffuse osteoporosis, have been detected. The general consensus of opinions is that Harris' lines suggest episodes of infection or malnutrition. In young subjects osteoporosis may be due to a variety of causes, most importantly, primary or secondary malnutrition. In the case of the Grottarossa girl, primary malnutrition is hardly credible; it seems much more likely that osteoporosis was the result of malnutrition induced by a disease. The possibility of a tubercular pleuritis discussed above could explain osteoporosis, especially if it made necessary a long stay in bed. However, yet another hypothesis should be considered. Osteoporosis possibly associated with Harris' lines may be the result of a juvenile diabetes of protracted course, as reported by Morrison and Bogan (1927) and later confirmed by other authors (Albright and Reifenstein, 1948); Hernberg, 1952; Berney, 1952; Menczel et al., 1972).

If the Grottarossa girl did suffer from diabetes, which reduces immunological resistance, she would have been particularly susceptible to infectious diseases (tubercular or not).

Two other results of the paleopathological studies deserve a word of comment. First, the x-ray image of the skull was identical to those previously observed in other mummies (Cockburn and Cockburn, 1980), and attributed to post-mortem relocation of embalming material instilled through the nose after extraction of the brain. The CT scan demonstrated in our case that the brain was in situ and shrunk, and it was indeed responsible for the crescent-like image in standard x-ray images.

Second, the finding of rather abundant anthracotic pigmentation in the lungs is of surprising in view of the age of the girl and of her environment.

As to the mummification technique used, it should be described sensu strictiori as "embalming" (i.e. "treatment with balms") as revealed by the absence of any trace of sectioning or residual thin crystals to be attributed to natron on the skin or in the wrappings, the persistence of all the internal organs and the aromatic odours exhaled from the body at the time of its discovery, with the presence of resin in the wrappings. This technique was commonly applied in Egypt during the last period, including the Roman era (Lucas, 1989; Cockburn and Cockburn, 1980; Proefke et al., 1992) when the Grottarossa girl was alive. On this point the statement of Sandison (1969) is clear: ". . . continued deterioration [in mummies] set in after the XXVIth Dynasty so that by Roman times preservation was usually mediocre and largely achieved by covering the body with hot resinous substances often described as bitumen" (p. 490). The chemical analyses confirm the view that this kind of treatment was used.

The presence of abietic acid derivatives provided evidence that coniferous resins were used in the embalming procedure. This conclusion is also supported by the presence of pollen grains of *Pinus pinaster* type found inside the resin clots. The finding of sesquiterpenoids cannot be univocally interpreted until they can be identified more precisely. Many compounds of this class are known to be contained in essential oils and some of them are typically found in *Cupressaceae*. The possible use of products from *Cupressaceae*, especially *Juniperus*, was therefore considered with special attention, all the more so since an oil from *Juniperus* is mentioned in Pliny the Elder as being used by the Romans for the preservation of corpses. There is already some evidence that this product was among those used to embalm the Grottarossa girl because α-cariophyllene was identified together with p-cimene in the "head space". Both substances are known to derive from *Juniperus*. On the other hand the presence of p-cimene could be explained on the basis of the greater chemical stability of this aromatic monoterpene, as compared with that of α-pinene, which was not found in the materials examined.

The finding of pollen grains from Myrrh (*Commiphora* sp.) raises the question whether it too was employed in this case. So far we have not found evidence of the characteristic terpenoids.

The resinous material found in the small box belonging to the funerary equipment did not differ from the materials mentioned above, but the presence of a pollen grain from *Rosaceae* might indicate the presence of a perfume.

The type of mummification shown by the Grottarossa girl closely resembles that used for the Egyptian boy who died between the ages of seven and nine, and has recently been studied by Proefke et al. (1992). In this last mummy bitumen was demonstrated to be present, and the ratios of metal characteristic of petroleum (vanadium, nickel and molybdenum) were similar to those reported for Dead Sea and Mesopotamian bitumens.

At this point it appears worth recalling that according to Lucas (1989):

"The word mummy [....] was applied at a late date to the embalmed bodies of the dead in Egypt, owing to the mistaken idea that because the body so preserved was black and looked as though it had been soaked in bitumen, therefore the preservative agent employed must always have been bitumen, which, however, was not so, though in one mummy of the Persian period bitumen has been found" (p. 271).

This is especially true of the Grottarossa mummy, which, in spite of the absence of bitumen, revealed an intense brownish colour attributable to the action of the resin. However it cannot be excluded that the discolouration produced by resins could have been by dehydration, if credence can be given to the eye-witness accounts that, when it was discovered, the mummy looked fresh and well hydrated, whereas shortly afterwards, when dehydration took place, the skin took on a deep brownish colour (Fig. 9).

The last topic to be discussed is the site of mummification. The lack of funerary inscriptions makes it difficult to be specific. According to the archaeologists (Paribeni, 1964; Andreae, 1969; Sapelli, 1979; Bordenache Battaglia, 1983) some of the scenes carved on the sarcophagus allude to Africa. They are: a symbol of that continent, the Aeneas and Dido episode, a lion hunting, and a fluvial divinity tentatively interpreted by Andreae (1969) as alluding to the Badagras river. But these scenes might do no more than allude to the circumstance that the girl was born or lived for a time in that continent. If so, mummification could have been carried out elsewhere, possibly in Rome. At that time the foreign faiths, especially the Egyptian ones, had made such inroads in Rome (Malaise 1912 a, b; Roulet, 1972) that an Egyptian-style mummification could have been performed there. At this point a question arises: does any specific finding indicate that the Grottarossa mummy was prepared in Rome? The lack of hard and fast clues to this problem makes it hard to give a positive answer.

The absence of bitumen on the body and wrappings does not justify the view that mummification was carried out in Rome, since Lucas (1989) argues that many Egyptian mummies were not treated with this substance.

Nor do the jewels or funerary equipment offer many clues to the identification of the site of mummification. According to the archaeologists (see Bordenache Battaglia, 1983) the jewels were typically Roman; on the other hand the sapphires in the necklace surely came from Sri Lanka (Ceylon). The same uncertainty applies to the funerary equipment, especially the die, despite the Baltic origin of the amber. Lucas (1989) states that sapphires were not used by the ancient Egyptians and: "That amber may have been used by the ancient Egyptians, especially at the late date, is not denied, but that all the objects termed amber are indeed amber has not been proven" (p. 387). But even if the jewels and the funerary equipment are to be considered of Roman workmanship, it is still possible that the girl died and was mummified in Egypt.

In contrast, when the textiles are considered, some well-grounded suspect arise that mummification has been carried out in Rome. This is the opinion of Granger-Taylor who on our invitation attentively examined all the textiles and provided us detailed informations about it. From her conclusive report we summarize here the following most important points.

The lump of silk textile, thoroughly impregnated with the embalming resin (No. 7), suggests it lay directly against the skin. We cannot know whether the girl was dressed in it or simply wrapped with it, but its integrally-woven contrasting band (or bands) makes it virtually certain to have been tunic. Roman tunics were made with the warp running horizontally in the made up garment and as a rule had two isolated contrasting bands, *clavi,* running from the shoulder to the hem (Granger-Taylor, 1982).

The method of grouping the warp threads for the band on silk textile is typical of textiles of the Roman period, more particularly of items woven on the two beams upright loom, and is found on recovered textiles of both wool and silk (Sheffer and Granger-Taylor, 1993). Contrasting bands with grouped warp threads occur on two other silk textiles (Granger-Taylor, 1987) and in a silk dalmatic preserved in the church of St. Ambrose in Milan which probably belonged to St. Ambrose himself who died 397 A.D. (Granger-Taylor, 1983).

As for the use of silk, we should not be surprised to find it in this context: although not acceptable for men until some time later, by the 2nd century A.D. Roman ladies had long been wearing silk clothing. The silk itself, called by the Roman *sericum* or "Chinese", was probably been woven in Rome as well as in the other large cities of the Empire.

As regards the linen textiles (No. 1–6), the crucial bit of evidence is the spin direction. They have Z-spun yarns in both warp and weft. In Roman Italy the normal direction for spinning wool and silk was Z while linen could be S- or Z-spun. In Egypt linen cloths were woven of yarns that were almost invariably S-spun. Elsewhere in the Near East wool was sometimes Z-spun but linen threads was again virtually always S-spun. The only Mediterranean country besides Italy where we have good evidence for the Z-spinning of linen in Classical Antiquity is Greece (Beckwith, 1954).

In conclusion, the textiles furnish data indicating that the site of the mummification of the Grottarossa girl may have been Rome. In this case, if the girl was Roman (or Italic) as suggested by her anthropological features, the most probable reason for her Egyptian-style mummification is that she or her parents were believers in the Egyptian religion, as a result of a period of residence in Africa, or because they had become familiar with it while living in Rome.

Acknowledgements

The authors are greatly indebted to the following colleagues: Prof. Gualdi G., Dr. Gualdi G.-F. and Prof. Leccisotti A. for the preparation of the conventional radiological and CT material; Prof. Cartoni G. for his supervision of gas-chromatographic and mass-spectrometric diagrams; Prof. Follieri M. for her supervision in the pollen analysis; Dr. Varoli R. for the coordination of the textile working group of the Italian Central Institute for Restoration, in collaboration with Dr. Anuradha Dej. A special thank is due to Dr. Balbi De Caro S. of the "Sovrintendenza alle Antichità Roma I" for the facilities she kindly offered in studying some funerary items. Finally the authors are obliged to Mr. Virgilii L. for his skilful help in preparing the photographic material. This study was supported in part by a research grant of the "La Sapienza" University, Rome.

Summary

During building work in 1964 the mummy of an 8-year-old girl was discovered at Grottarossa, on the northern outskirts of Rome. Since only one other mummy has ever been found in Rome, the virtually unique nature of this discovery suggested that a series of anthropological, archaeological, chemical and pathological investigations should be carried out to identify the reason for mummification, the procedure used, the cause of death, the medical conditions involved, and the girl's cultural background. The mummy can be dated between 150 and 200 A.D. The girl was a caucasian, probably of mid- or north Italic origin. Her body was embalmed and preserved using procedures characteristic of Egypt's Roman period; her brain and viscera were *in situ* and could be viewed easily by CT scan. She may have lived in Africa, but it cannot be concluded that she died there. She had suffered from several infectious or nutritional conditions, as shown by the series of Harris' lines in her long bones, associated with a certain degree of generalised osteopenia. The ultimate cause of death was a bilateral fibrinous pleuritis of uncertain nature. The site of and the reason for the mummification procedure are both discussed in detail and it appears likely that mummification was carried out in Rome or in Italy anyway.

Zusammenfassung

Im Jahre 1964 wurde bei der Aushebung eines Fundaments im Rahmen von Baumaßnahmen in der Ortschaft Grottarossa in der nördlichen Peripherie Roms die Mumie eines achtjährigen Mädchens entdeckt. Da es sich um den bisher einzigen archäologischen Fund dieser Art handelt, wurde eine Reihe von anthropologischen, archäologischen, chemischen und pathologischen Untersuchungen vorgenommen, um den Grund für die Mumifizierung, das angewandte Verfahren, die Todesursache und die damit verbundenen medizinischen Implikationen sowie schließlich das kulturelle Umfeld des Mädchens zu klären. Auf Grund der archäologischen Fundstücke wird die Mumie auf die zweite Hälfte des 2. Jhs. n. Chr. datiert. Das Mädchen, das morphologische Charakteristiken kaukasischen Typs aufweist, stammte wahrscheinlich aus Mittel- oder Norditalien. Der Körper war mittels Einbalsamierung konserviert worden, und zwar mit Hilfe eines während der Epoche der römischen Kolonisation Ägyptens vielfach angewandten Verfahrens, bei dem Gehirn und Eingeweide nicht entfernt wurden. Diese Organe konnten in der Tat mit Hilfe einer Computertomographie nachgewiesen werden. Es ist möglich, daß das Mädchen in Afrika gelebt hat, es muß aber nicht unbedingt dort gestorben sein. Harris-Linien auf den langen Knochen der Gliedmaßen deuten in Verbindung mit einer Osteopenie auf Infektionen und Mangelernährung hin. Die Todesursache ist in einer bilateralen fibrinosen Pleuritis unbestimmter Natur zu sehen. Der Ort, an dem die Mumifizierung durchgeführt wurde, sowie das Verfahren, das hierbei angewandt wurde, werden im Detail untersucht. Aus diesen Untersuchungen geht hervor, daß die Mumifizierung mit größter Wahrscheinlichkeit in Rom oder jedenfalls in Italien durchgeführt worden war.

Résumé

En 1964, au cours d'un creusement pour jeter les fondements d'une construction en localité Grottarossa, à la périférie septentrionale de Rome, la momie d'une fillette de 8 ans a été mise à jour. S'agissant d'une pièce archéologique actuellement unique dans son genre, une série de recherches anthropologiques, archéologiques, chimiques et pathologiques ont été mises au point dans le but d'établir la raison de la momification, le procédé appliqué, la cause de la mort ainsi que ses implications médicales et, enfin, le milieu culturel dans lequel la fillette avait vécu. D'après les données archéologiques la momie remonte à la deuxième moitiée du deuxième siècle apr. J. C. Ses traits morphologiques sont de type caucasien et il est possible qu'elle ait été originaire de l'Italie centrale ou septentrionale. Le corps avait été préservé par embaument, c'est à dire appliquant un procédé commun au cours de la période Romaine de l'Egypte. En effet le cerveau et les viscères n'avaient pas été enlevés, ci-bien qu'ils étaient appréciables à l'examen radiologique. L'enfant pouvait avoir vécu en Afrique, mais cela ne veut pas dire qu'elle y soit nécessairement décédée. Au niveau des os longs des membres, l'association de lignes de Harris et d'une condition d'ostéopénie font penser à des affections de type infectieux ou ayant rapport avec une mauvaise nutrition. La cause de la mort s'identifie en une pleurésie fibrineuse bilatérale d'origine incertaine. Le lieu où la momification a été effectuée et le procédé adopté sont discutés en détail, et il parait vraisemblable que la momification ait eu lieu à Rome ou, de toute façon, en Italie.

Riassunto

Nel 1964, durante lo scavo delle fondamenta per la costruzione di un edificio in località Grottarossa, alla periferia settentrionale di Roma, venne rinvenuta una mummia di bambina dell'età di 8 anni. Trattandosi di reperto archeologico attualmente unico nel suo genere, si è provveduto ad una serie di ricerche rispettivamente antropologiche, chimiche e patologiche nell'intento di stabilire il motivo della mummificazione, il procedimento usato, la causa della morte con relative implicazioni mediche e, da ultimo, l'ambiente culturale della bambina. Sulla base dei reperti archeologici la mummia viene fatta risalire alla seconda metà del secondo secolo d.C. La bambina che mostra caratteristiche morfologiche di tipo caucasico, era probabilmente originaria dell'Italia centrale o settentrionale. Il corpo era stato conservato per imbalsamazione, cioè facendo ricorso ad un procedimento comunemente applicato nel periodo Romano dell'Egitto. Infatti il cervello ed i visceri non erano stati asportati e si rendevano apprezzabili con la TAC. La bimba poteva essere vissuta in Africa, ma non si può ritenere che vi sia necessariamente deceduta. L'associazione, a livello delle ossa lunghe degli arti, di linee di Harris e di una condizione di osteopenia, orientano verso sofferenze di tipo infettivo e nutrizionale. La causa di morte va individuata in una pleurite fibrinosa bilaterale di natura indeterminata. La sede in cui la mummificazione è stata eseguita ed il procedimento adottato vengono discussi in dettaglio e da essi si evince che con ogni verosimiglianza la mummificazione abbia avuto luogo a Roma o, comunque, in Italia.

References

Albright F. and Reifenstein E. D., Jr (1948) Parathyroid glands and metabolic bone disease. Selected studies. Williams & Wilkins, Baltimore.
Andergassen W. (1983) Analisi delle ambre. G. Bordenache Battaglia: Corredi funerari di età imperiale e barbarica nel Museo Nazionale Romano. pp. 152–153. Edizioni Quasar, Roma.
Andreae B. (1969) Aeneas-Sarkophag. W. Helbig (ed.): Führer durch die Öffentlichen Sammlungen Klassischer Altertümer in Rom. 4th ed., vol. 3, pp. 66–69. Wasmuth, Tübingen.
Beckwith J. (1954) Textile fragments from classical antiquity. Illustrated London News Jan. 23: 114–115.
Berney P. W. (1952) Osteoporosis and diabetes mellitus. Report of case. J. Iova Med. Soc. 42: 10–12.
Bordenache Battaglia G. (1983) Corredi funerari di età imperiale e barbarica nel Museo Nazionale Romano. Edizioni Quasar, Roma.
Castellani O. (1964) La momie de Grottarossa (Rome). Rev. Archéol. du Centre 3: 132–142.
Cockburn A. and Cockburn E. (1980) Mummies, Disease, and Ancient Culture. Cambridge Univ. Press, Cambridge, London, New York, New Rochelle, Melbourne, Sydney.
Colby J. W. (1968): MAGIC IV. A computer program for quantitative electron microscope analysis. Bell Telephone Labs., Allenton, Pennsylvania.
Erdtman G. (1969) Handbook of Palynology. Munksgaard, Copenhagen.
Gianolio A. (1987) L'analisi delle fibre tessili. Zanichelli Editore, Bologna.
Granger-Taylor H. (1982) Weaving clothes in the Ancient World: The tunic and toga of the Arringatore. Textile History 13, 3–25.
Granger-Taylor H. (1983) Two Dalmatics of St. Ambrose? CIETA Bulletin 57–58, 127–163.
Granger-Taylor H. (1987) Two silk textiles from Rome and some thoughts on the Roman silk-weaving industry. CIETA Bulletin 65, 13–31.
Hernberg C. A. (1952) Skelettveränderungen bei Diabetes mellitus der Erwachsenen. Acta Med. Scandinav. 143: 1–14.
Infessura S. (1723) Diarium Urbis Romae. G. Eccardo (ed.): Corpus Historicum Medii Aevi. Vol. 2nd, pp. 1863/4–2015/6. Gleditschii, Lipsia.
Livi R. (1896) Antropometria militare. Parte I: Dati Antropologici ed Etnologici. Presso il Giornale Medico del Regio Esercito, Roma.
Lucas A. (1989) Ancient Egyptian Materials and Industries. 4th Ed. Histories and misteries of man, London.
Malaise M. (1972a) Inventaire préliminaire des documents égyptiens découverts en Italie. Brill, Leiden.
Malaise M. (1972b) Les conditions de pénétration et de diffusion des cultes égyptiens en Italie. Brill, Leiden.
McLafferty F. W. and Stauffer D. B. (1988) The Wiley/NBS Registry of Mass Spectral Data. John Wiley, New York.
Martin R. and Saller K. (1957, 1959) Lehrbuch der Anthropologie. Vol. 1st and 2nd. Fischer Verlag, Stuttgart.
Menczel J., Makin M., Robin G., Jaye I., and Naor E. (1972) Prevalence of diabetes mellitus in Jerusalem. Its association with presenile osteoporosis. Isr. J. Med. Sci. 8: 918–919.
Mills J. S., and White R. (1989) The identity of resins from the late bronze shipwreck at Uln Burum (KAS). Archaeometry 31: 37–44.
Moore P. D., and Webb J. A. (1978) An illustrated guide to pollen analysis. Hodder and Stoughton, London.
Morrison L. B., and Bogan I. K. (1927) Bone development in diabetic children. Roentgen study. Am. J. Med. Sc. 174: 313–319.
Paribeni E. (1964) Oral presentation during the press conference on the mummy of Grottarossa.
Pazzini A. (1964) A proposito della mummia di Grottarossa. Pagine d. Storia d. Med. 8: 3–24.
Pfeiler B. (1970) Römischer Goldschmuck. Footnote pp. 75–76. Von Zabern, Mainz.
Poirot J. P. (1992) Spectrométrie et fluorescence X, des aides pour la détermination de types de gisements de saphirs. Rev. de Gemmologie A. F. G. 110: 7–9.
Proefke M. L., Rinehart K. L., Raheel M., Ambrose S. H., and Wisseman S. U. (1992) Chemical Analysis of a Roman Period Egyptian Mummy. Analytical Chem. 64: 105–111.
Roullet A. (1972) The Egyptian and Egyptianizing Monuments of Imperial Rome. Brill, Leiden.
Sandison A. T. (1955) The histological examination of mummified materials. Stain Tech. 30: 277–283.
Sandison A. T. (1969) The study of mummified and dried human tissues. D. Brothwell and E. Higgs (eds.): Science in Archaeology. 2nd ed., pp. 490–502. Thames and Hudson, London.
Sapelli M. (1979) Sarcofago con caccia di Enea e Didone. A. Giuliani (ed.): Museo Nazionale Romano, Le sculture. Vol. I/1, pp. 318–324. De Luca, Roma.
Scamuzzi V. (1964) Studio sulla mummia di bambina, cosiddetta "mummia di Grottarossa", rinvenuta a Roma sulla via Cassia il 5/2/1964. Riv. Studi Classici 12: 264–280.
Schlenk H., and Gellerman J. L. (1960) Esterification of fatty acids on a small scale. Anal. Chem. 31: 37–44.
Sergi G. (1905) Die Variationen des menschlichen Schaedels und die Klassifikation der Rassen. Arch. Anthropologie 31: 111–121.
Sheffer A., and Granger-Taylor H. (1993) The Textiles: A Preliminary Selection. In J. Aviram, G. Foerster and E. Netzer (eds.): Masada V., Jerusalem: Israel Exploration Society, catalogue section G. (In press.)

Correspondence: Prof. Dr. Antonio Ascenzi, Università "La Sapienza", Sezione di Anatomia Patologica, Policlinico Umberto I°, Viale Regina Elena 324, I-00161, Rome, Italy.

Mummies of Saints: a particular category of Italian mummies

E. Fulcheri

To my father

Istituto Anatomia Patologica, Universita di Genova, Genova, Italy

Whenever anybody mentions mummies, the mind reverts straight away to Ancient Egypt and both experts and laymen start picturing the beautiful mummies of the Pharaohs still preserved in their splendid tombs. Everybody knows the most important and replete collections like those of the British Museum in London, the Archeological Museum in Cairo, and the Egyptian Museum in Turin. But it is not generally known that Italy has an outstanding collection of Egyptian mummies even in small town museums the length and breadth of the country. The results of an inventory we made fully support Italy's claim (1). There are 91 Egyptian Collections, 28 of which have complete or incomplete mummies, exhibited or in store, totalling 142 mummies and 214 fragments. In Turin alone there are 96 mummies and 85 fragments and as consequence many historical and anthropological studies especially those of the University Institute of Anthropology (2–3). However, apart from Egyptian mummies, we must also say that in Italy there is an unexpectedly large number of other mummies (4).

Between the Fifteenth and Seventeenth centuries, common people asked to be buried within monastic ground and when the environmental conditions were right, natural mummification took place. Thus, we find natural mummies allover Italy but especially in the South due to the particularly dry weather conditions. Yet the majority of mummies are to be found in the Catacombs of Cappuccini in Palermo, where about 1,850 bodies are collected, many of them mummified (5–7). Here, in addition to natural mummification, the Cappuccini used to mummify the bodies by putting them in ventilated chambers on special structures called "colatoi" (special strainers) thus giving origin to an effective yet somehow rudimentary art.

More mummies are present in Comiso, Naples, Ferentillo, Urbania, Navelli and Venzone, just to mention some of the most renowned groups. All these natural mummies, taken all together, make up an important biological file available for anthropological as well as paleopathological studies. Due to the high number of cases, it is possible to perform paleopathological studies on population samples rather than on single individuals.

A third category of mummies features those of Saints. Though insignificant as a population sample, they are very important as individual cases. Investigations on them are very important from a historical as well as paleopathological viewpoint, because of their correlations with the cultural and ethnic environment of their times. We shall now attempt at better understanding the size of this category, its time span and mummification methods.

But before going into detail, I would like to summarise some of the most important periods, as far as mummies are concerned, in Italian history. As is well known, the Romans did not embalm the dead although they built sumptuous tombs. Therefore, the anthropological material available includes skeletons or cremated bones. For some eras, the latter ones are the only remains available. The finding of mummified bodies was such an exceptional event that it was reported in the chronicles. A striking example is given by Tulliola, the daughter of Marcus Tullius Cicero who "haec enim post multos annos in sepulchro inventa, cum elevata esset, in cinerem cecidit" (Suessano 4 meteo com e vlt). But this cannot be considered a real mummification. It is only an example of exceptional preservation due to environmental factors and in this it is different from the mummy found in Grottarossa, near Rome, in 1964. As far as we know, this is the only example of a mummification during Roman times.

Many customs changed with the advent of Christianity while new funeral rites were introduced requiring body preservation. From the second and third century after Christ, the remains and relics of Christian Martyrs were collected and preserved thus constituting a remarkable heritage of human remains. Most of these Martyrs had been sacrificed under the Emperors Nero, Domitian, Antoninus Pius, Marcus Aurelius, Decius and Valerianus, Diocletian (from 41 to 313 after Christ) throughout the Empire and, of course, in Rome. Those of them who died in Rome were buried in the Roman catacombs of San Callisto, Priscilla, Sant' Agnese, Santa Domitilla and San Sebastiano. Later, many of the skeletal remains were transferred to major Basilicas in Rome, Italy and Europe. The cult of Saints and Martyrs was beginning and, subsequently, the cult of their bodies considered as a precious witness of faith.

From the first centuries A.D., many Bishops were also interred under the altars of the cathedral of their cities, where they had often brought the Faith and had protected it with their own martyrdom. Numerous bodies of Saints were collected for adoration in churches and chapels.

Whereas mainly skeletons only remain from the early centuries, from the Middle Ages onwards there are many mummies left. According to our brief review carried out in Italy, there are at least 315 eminent relics of Saints and Martyrs preserved, among which we know of 25 mummies (8–11). This estimate is undoubtedly below the real figure.

Mummification is known to take place either artificially, through embalming, or through natural causes under environmental conditions whereby autolysis and putrefaction are minimal and a fast dehydration process is ensured. Some special burial methods have sometimes clearly promoted natural mummification thus accounting for the many wonderfully preserved bodies of Saints.

We can take for example the case of Sant'Ubaldo da Gubbio (1084–1160) whose mummy is preserved in the Basilica bearing the same name on Mount Ingino (Fig. 1). When the old bishop died his body was in extremely favourable conditions for the onsetting of a natural mummification process. According to witnesses (12): "In eo siquidem ossa vix cum nervibus remanserant. Nam caro eius fere ex toto fuerat exausta, cute omnino detecta." Also, after death, his body remained for three days in an extremely favourable environment for natural mummification, since it was very hot and ventilated. "... calor immensus qui tunc ferventius solito mundum desuper exquoquebat ..." Under these conditions, the body of Sant' Ubaldo became mummified in such a perfect way that it is still possible to view it on the wonderful high altar of the Basilica.

Many bishops, priests and nuns were exhibited after their death for long periods of time to allow for special funeral honours and let people come and pay homage and show their devotion to them. An extreme example is the one of the Blessed Angelina da Spoleto, who died in 1450. According to the records (13) "after fifteen days since her death, blood was seen flowing out of her nostrils as if she were still alive". Undoubtedly, this phenomenon shows that putrefaction had clearly begun. But the interesting point here is that the bodies used to be exhibited for long periods, which, in the majority of cases, was not at all an advantage for their good preservation.

In some other cases, a very poor natural mummification would occur after an initial and partial decay, like in the case of Santa Savina Petrilli da Siena (1923) (14).

This is not the right place, nor would I be competent enough, to discuss the delicate and complex issue of natural mummification which, interpreted as a prodigious sign of incorruptibility, is often called for during Canonization trials. Here, it will be enough to note that there are many examples of naturally mummified bodies of Saints, from any age. Each of you will certainly add other cases to those already mentioned: Santa Lucia, (304), San Ciziaco (363), Sant' Anselmo da Lucca (1086), Sant'Ubaldo da Gubbio (1160), Blessed Beatrice d'Este (1226), Blessed Elena Enselmini (1231), Blessed Giordano Forzaté (1248), Santa Rosa da Viterbo (1252), Santa Zita (1278), Sant' Agnese da Montepulciano (1317), Sant'Odorico da Pordenone (1331), Santa Francesca Romana (1440), Sant Antonino da Firenze (1459), Blessed Margherita di Savoia (1464), Santa Caterina da Bologna (1465), Blessed Angela Merici (1540), San Gregorio Barbarigo (1627), Blessed Centurione Bracelli (1615), Savina Petrilli (1923) (15).

For some mummies it was possible to demonstrate that no embalming attempts had taken place: Sant Ubaldo da Gubbio (1160) (16), Santa Zita (1278) (17), Blessed Margherita di Savoia (1464) (18), Blessed Centurione Bracelli (1651) and Santa Savina Petrilli (1923) (14).

Sometimes, therefore, due to specific features of the tomb, corpses would mummify spontaneously. Not infrequently, however, the body was preserved artificially (19, 20). Especially in the Middle Ages, the bodies of Saints were sometimes mummified artificially to preserve them for veneration and cult which had often started when they were still alive and would go on for several years before their regular canonization by which they would be brought to the altar. In this way it was possible to exhibit the corpse for long periods after its death without damaging it. The case of San Bernardino da Siena, who was exhibited for 26 days, is exemplary (21). For all these reasons a certain number of artificially mummified subjects should be added to the number of naturally mummified Saints (Table 1).

Table 1. Natural mummies

Santa Lucia (304)	*
San Ciziaco (363)	*
Sant'Anselmo da Lucca (1086)	*
Sant'Ubaldo da Gubbio (1160)	*
Beata Beatrice d'Este (1226)	*
Beata Elena Enselmini (1231)	*
Beato Giordano Forzaté (1248)	*
Santa Rosa da Viterbo (1252)	*
Santa Zita da Lucca (1278)	*
Sant'Agnese da Montepulciano (1317)	*
Sant Odorico da Pordenone (1331)	*
Santa Francesca Romana (1440)	*
Sant Antonino da Firenze (1459)	*
Beata Margherita di Savoia (1464)	*
Santa Caterina da Bologna (1465)	*
Beata Angela Merici (1540)	*
San Gregorio Barbarigo (1627)	*
Beata Centurione Bracelli (1651)	*
Savina Petrilli (1923)	*
Total	19

Many popes (both Saints and not) were also embalmed and buried in the basilicas and in particular in the basilica par excellence, namely Saint Peter in Rome. This practice derives, in this case too, from the need to exhibit the Pope for many days after his death. During the Avignon period, the practice was very common and it is not rare even today, since Pope Pius XII was embalmed by Nuzzi as late as 1958 (20). Today, too, some renowned figures in the Church or whose beatification trials is under way have been embalmed. We shall just mention two of them whom we have personally followed, namely Don Orione (1940) (22–23) and Cardinal Slipjy (1984) (24).

The embalming practice was accepted on the basis of a very old Judaic-Christian custom; indeed we read in Saint John's Gospel: "Venit autem et Nicodemus, qui venerat ad Iesum nocte primum, ferens mixturam myrrhae et aloes, quasi libras centum. Acceperunt ergo corpus Iesu, et ligaverunt illud linteis cum aromatibus, sicut mos est iudaeis sepelire (Jo. 19, 39–40)." Further, Saint Mark's Gospel reads: "Et cum transisset sabbatum, Maria Magdalene, et Maria Iacobi, et Salome emerunt aromata ut venientes ungerent Iesum (16, 1)", while in Saint Luke's Gospel it is stated that "Et revertentes paraverunt aromata et unguenta: et sabbato quidem siluerunt secundum mandatum (23, 56)". In the Basilica of the Holy Sepulchre in Jerusalem the so-called "pietra dell'unzione" (anointing stone) is still preserved according to tradition.

The connection between these Jewish customs with Egyptian embalming traditions is evident. Jews are likely to have assimilated them during their long slavery in Egypt. Whatever its origin, however, the custom of anointing dead bodies with balsams became a usual practice among Jews.

Following such authoritative tradition, it is obvious that the bodies of Holy people or of those "in the odour of sanctity "were, out of devotion or respect, washed, anointed with spices and perfumed unguents, hence somehow "embalmed". We just refer to a few of the many cases reported by the chronicles (13): Sant'Emiliano Martyr of Trevi (302): "his body was buried by Christians amongst much tears with aromatic resins and precious perfumes and white linens"; San Ranaldo of Foligno, bishop of Nocera who died in 1222, "his holy body was buried in the Cathedral with aromatic resins and balsam by his canons and clerics who showed him extraordinary reverence"; Servant of God P. Bernardino of Colle Petrazzo in Todi, who died in 1594, "his body was ordered by Duke Cesio to be ointed with aromatic herbs and buried in his church on Monte Scopio".

No wonder, therefore, that many mummies of Saints, particularly well preserved, are actually only partially embalmed. As a matter of fact, in these cases, mummification took place naturally, favoured by the use of balsams and spices and by the environmental conditions inside the tomb.

An artificial, undoubtedly intentional mummification, by means of more complex embalming techniques, is documented only in rare cases. We know of the following ones: Santa Margherita da Cortona (1297), Santa Chiara da Montefalco (1308), Blessed Margherita Vergine da Città di Castello (1320), Santa Caterina da Siena (1380), San Bernardino da Siena (1444), Santa Rita da Cascia (1447) (11, 15, 19) (Table 2).

Table 2. Artificial mummies of Saints

Name	Sex	Year	Country
Santa Margherita da Cortona	F	1297	Toscana
Santa Chiara da Montefalco	F	1308	Umbria
Beata Margherita Vergine	F	1320	Umbria
Santa Caterina da Siena	F	1380	Toscana
San Bernardino da Siena	M	1444	Umbria
Santa Rita da Cascia	F	1447	Toscana

Some historical remarks are essential at this point before going on with the examination of the various cases.

From a first perusal of the above mentioned list, it is clear that artificially mummified Saints are more or less concentrated within a limited time range, from 1297 to 1447, namely within a span of two hundred years. Secondly, as to their geographic distribution, they are strictly limited to Umbria and Tuscany. Thirdly, almost all mummies of this type, with the only exception of San Bernardino da Siena, are women.

Thus it might be interesting to investigate whether there are any correlations accounting for this particular custom which appears to have established in a particular time span and in a very restricted environment, namely in monasteries and convents.

The group of mystics and, particularly, of mystic women, the majority of them nuns, which flourished particularly in Tuscany, Umbria and Lazio in the Middle Ages, though it later spread throughout Italy and Europe until the 19th century, deserves a special attention in the history of Italian Saints. Out of 135 mystic women Saints and Blessed women who lived in the Middle Ages, 56 of them lived in Tuscany (twenty-six) and Umbria (thirty). The mummified bodies of 7 of them are still preserved. For 5 of them an artificial mummification was documented to be conducted with special skill and care according to very similar rules and standards.

Thus, it could be surmised, partially with the support of existing records, that embalming practices were limited to a specific environment, namely convents, specially for women, and based on particular religious beliefs.

Now, going back to artificial mummification cases, we shall try and illustrate the current knowledge about

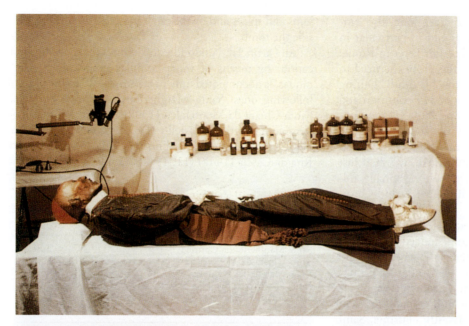

Fig. 1. Sant'Ubaldo da Gubbio (1084–1160): his mummy is preserved in the Basilica bearing the same name on Mount Ingino. The body during the last survey performed in 1977 which aimed at disinfecting and disinfesting the mummy

Fig. 2 (a, b). Santa Margherita da Cortona (1297). During the body examination we carried out in 1988, we could just observe the effect of balsams as well as direct embalming operations, like deep chest and abdomen incisions performed to eviscerate the body, as well as incisions along the large upper and lower limb muscles. All incisions had been sutured with whipstitches in a thick black thread

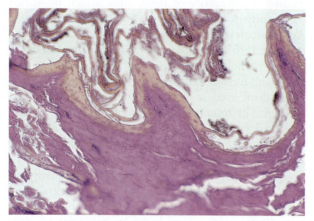

Fig. 3. Santa Margherita da Cortona (1297). Histological sample of skin well preserved with all its layers, epidermis, dermis and hypodermis. (Hematoxylin-Eosin, 250×)

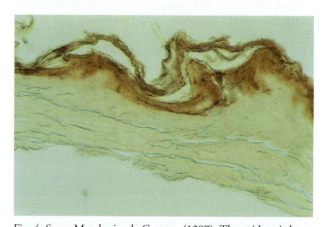

Fig. 4. Santa Margherita da Cortona (1297). The epidermis layers at the surface, intermediate and base were shown up immunohistochemically by anti-keratin antibodies with different molecular weight. (ABC Anti-Keratin, 250×)

them to better understand the problem and see whether there were other purposes, apart from body preservation, leading nuns to carry out real post mortem examinations with eviscerations on the bodies of their holy sisters.

The "Legend" by Brother Giunta Bevegnati (25), when speaking about the death of Santa Margherita da Cortona, in 1297, reads: "Then the people of Cortona, having received the news of her glorious departure, summoned the general Council to honour and glorify the Supreme Lord, then rushing with great emotion to the church of San Basilio, they embalmed her body, dressed it with a red tunic and buried it in a new sepulchre…" During the body examination we carried out in 1988 (26) we could just observe the effect of balsams as well as direct embalming operations, like deep chest and abdomen incisions performed to eviscerate the body, as well as incisions along the large upper and lower limb muscles. All incisions had been sutured with whipstitches in a thick black thread (Fig. 2a and 2b). The embalming of Margherita was never shrouded in mystery (27–30), on the contrary, it is often mentioned in survey reports since "it would show the high belief in her Holiness, in which she died, since the City wouldn't have taken the trouble to embalm a poor public sinner". As to embalming techniques, the survey report of 1719 generally mentions "unguents and spices" (… fuerit unguentis conditum, et aromatibus delibutum proxime post eius obitum, …). Thus, without more detailed records on her embalming, the ideas regarding the methods used to embalm Margherita were very confused and vague. From the previous remarks it can be inferred that the corpse was never thoroughly undressed to see the extensive incisions made during embalming. All surveyors kept on attributing the effects of embalming to unguents and spices alone, as was common in the past. Only with a proper scrupulously conducted inspection could more ligth have been thrown on this issue. Indeed, the 1719 report states: "et aperta parumper decentissime veste, qua dicta Beata Margarita operitur, viderunt crura pelle obducta (thus not skeletal, which was just enough evidence) ac pariter integra, et reliquo Corpori alligata, quod quidem Venerabile Corpus, totum, et integrum ad sensum visus, et tactus apparet." Therefore for the evidence of incorruptibility it was enough for the body to be "intact and joined together with all its limbs, tendons and features". Indeed, the corpse of Margherita, thanks to embalming but also for other reasons, met the above requirements in the best way.

Now, going back to well documented cases, we want to refer to Santa Rita da Cascia, who was embalmed since it was reported that the viscera were removed and the cavities were stuffed with rose perfumed cottonwool (11). The knife used to incise the body of San Bernardino is still preserved, thus documenting, in this case too, an intentional and complex preservation measure including the removing of the viscera (21).

The stripping of flesh and boiling of corpses was a frequent procedure in the Middle Ages, to postpone burial or allow for the transport of the body to far away places. This practice, however, became so common and applied also in cases where there was not such a need, that in 1300 Pope Boniface VIII issued the famous Bull "Detestandae feritatis abusum". By stripping the flesh off the corpse and with its subsequent boiling, it was often possible to have two graves: one in the place of death where the "soft" and most perishable parts were buried and a second one far away in place and also in time. Thus the partition of corpses of illustrious people became common to allow for the erection of authentic sepulchral munuments in different places. It also became customary (and it is still so in several parts of Europe) to remove the heart to be preserved in a separate urn thus erecting another shrine in memory of the deceased. In this light, no wonder that priests and nuns would not hesitate to partition the bodies of Saints to justify for a second burial and build mausoleums in different cities or create privileged devotional centres. Obviously, this "macabre" aspect as we would conceive it today, was by far overcome by deep religious beliefs. The relics were, just like today, distinguished into "reliquiae insignes", comprising the head, arms, legs or a vital organ, and the heart "par excellence"; "reliquiae notabiles" such as the hands and feet, and "reliquiae exiguae", when featuring only minor fragments. Therefore, since we are speaking about real eviscerations or dissections, we may better understand the reasons behind these customs, which stemmed out of the need to set up the so called "reliquiae insignes", namely the heart or other "noble" organs, or in which or for which a particular virtue used to shine while alive. During embalming with evisceration, it can be assumed that, apart from preservation, the embalmers also intended to remove the viscera to obtain relics or, as documented below, to achieve more subtle and nobler purposes.

A clear and exemplary case in the procurement of relics through eviscerations proper, is the one of the Blessed Cecilia Coppoli of Perugia, who died in 1500. The records state that (13) "after about seven years, new burials were made and her relics were disinterred. With great amazement, her body was found to be intact even with the brain as when she had passed away".

Going back to real embalming practices, we should mention other cases where evisceration proper was performed though the mummification of the holy corpse was not its only purpose.

Another particularly strange and interesting evisceration is the one of Santa Chiara da Montefalco (1308)(31). Sister Francesca, without any medical nor anatomic knowledge, while preparing the corpse for embalming, opened the body from the back with a razor and, going beyond the ribs, removed the heart and abdominal viscera. The heart appeared to be big, like a child's head: when opened, the figure of the cross was reported, the scourge and the spear and then three nails and, while the heart was shrinking and becoming dry, the column, the crown of thorns and the hyssop with the spunge. In the proceedings of the trial ordered by the Municipality of Montefalco in 1303 it is stated:

"Silicet cor ipsius beate Clare, in quo corde inventa fuit quedam crux de carne ad modum thau, in latere dextro dicti cordis in quodam loco depresso in ipsa carne ad modum dicte Crucis, nec infixa erat cum ipsa carne cordis, sed separata per se stabat, nisi quod in piede dicte Crucis erat quidam filus carnius satis exilis, qui ex ima parte conguntus erat cum pede dicte Crucis, et ex alia parte natus videbatur in ipso corde et ipsa cruce. Ex latere ipsius erat quodam foramen parvunculum ad modum percussionis lanciae. Ex parte vero sinistra prefati cordis erat quedam fusta de carne habens in summitate quinque nervunculos, que in nullo congruncta erat cum ipso corde. In ipso etiam corde ex interiori parte breviter continebatur totum misterium passionis, silicet lancia, et clavi, omnia de carne dicto cordi continebantur."

Further on, the document states that the dissection of the heart of their Abbess by the nuns was prompted by the very words of Chiara who would often say: "The cross of our Lord Jesus is deep in my heart." The nuns, a few days later, exhumed the viscera and in the gallbladder found three stones of equal size and weight which they immediately interpreted as the Holy Trinity. Frankly pathologic lesions which have modified the shape of papillary muscles, of tendons and valves in the heart can be postulated all too easily. This event is yet remarkable in the light of this typically medieval mysticism. Professor Baima Bollone of the University of Turin has recently conducted a thorough survey confirming these impressive lesions. A histological and ultrastructural examination was also performed on the myocardial tissue which resulted to be in excellent preservation conditions. Undoubtedly, though the attempts to preserve the corpse of Chiara who was deemed to be holy already at the time of her death, should not be surprising, the very fact that back in those times a real post mortem examination was performed is undoubtedly exceptional.

As already mentioned, in 1320, the body of the Blessed Margherita Vergine of Città di Castello was also embalmed (32). In the book "De imbalsamatione corporis huius virginis per Rectores Civitatis Castelli ordinata", we read:

"Et de mirabilibus quae occurrerunt cum dictum corpus imbalsamari deberet, et cum, scorsum positis interioribus, cor eius fuisset acceptum et a canna abscissum. Rectores Civitatis pro balsamo et aromatibus fratribus pecuniam contribuunt et interim multa miracula fiunt. Volentes igitur fratres aromatibus corpus condire, medicos cirurgos vocant…"

During embalming, the heart was examined and, "coram omnibus quaerens cor, coepit incidere cannam, a qua membrum praedictum dependet; et subito de eadem canna tres lapides mirabiles exierunt et apparuerunt". The cause of death can easily be inferred from the above description, yet, in this case too, the finding was interpreted, still in the light of medieval mysticism, as the symbol of the Holy Trinity, as was the case with Santa Chiara da Montefalco.

Still in Città di Castello, four hundred years later, in 1727, another post mortem was performed on the body of Veronica Giuliani and, again, in her heart, the symbols of the Passion of Christ were found. The heart indeed resulted to be "pierced from side to side".

It is often possible to find other autopsies ante litteram in chronicle reports which were performed in order to find prodigious signs (13). We shall now mention some significant examples of some subjects whose preservation conditions are currently unknown: in the case of Servant of God F. Raniere of Borgo di Sansepolcro, who died in 1589,

"the brothers wanted to prepare his body for embalming and the physician and the surgeon of the municipality of Todi in the presence of many people found three triangular stones in his gallbladder … in his heart a white scourge was found with five white lashes and a red noose at the top";

as to the Blessed Lucia of Norcia, who died in 1430 and was exhumed 169 years later, in 1599, "the coffin containing her holy body was opened and the body was found to be intact… in her heart there was a wound with a cross made of flesh".

In conclusion, we can state that embalming acquired particular features, since it aimed at finding evidence of holiness which would uphold the esteem and veneration of the subject among his/her followers.

Let us now consider the problem of paleopathological investigations on the bodies of Saints.

Examinations performed as an exception over the centuries are often simply inspective, sometimes recognitive, while only rarely are complete examinations, including paleopathological and anthropological aspects, allowed. Therefore, under these conditions, a real autopsy cannot be performed, while examination manœuvres or numerous samplings are not allowed since body integrity is paramount. However, during examinations, it is possible to take some targeted samples which may be of great interest (15).

Personally, I have documented the artificial mummification of Santa Margherita da Cortona both historically and histologically. During this examination, a well preserved skin still with all its layers, epidermis, dermis and hypodermis was histologically documented (Fig. 3). The epidermis layers at the surface, intermediate and base were shown up immunohistochemically by anti-keratin antibodies with different molecular weight (Fig. 4). Based upon these results and our previous experience regarding Egyptian mummies, we thought that it would be interesting to apply an immunohistochemical study of the skin with anti-keratin antibodies on natural mummies as well, and in investigations on mummies in general. Useful information on the condition of preservation of the mummy can be supplied by the histological examination of the skin (33–37). For example, bacterial or fungus growths can be detected together with the damage they have caused on the skin. Conversely, it is very difficult to learn about mummification processes which are more easily understood with immunohistochemical techniques. The reliability of the immunohistochemical approach on mummified tissues was investigated by some authors and also by us (38, 39).

Briefly speaking, the cytoskeleton components and the intermediate filaments, due to their resistance, could well be detected with immunohistochemical techniques employing Avidin-Biotin-Peroxidase Complex (40). Among cytoskeleton components, cytokeratins are the most resistant, thus permitting the study of all their subtypes (41). In well mummified natural and artificial bodies, the skin is always well preserved (with a strong positivity for Keratins) in not perfectly mummified bodies, the skin is severely damaged.

Epidermolysis, known to be one of the earliest modifications in post mortem changes is caused by autolytic processes occuring in damp environments. Therefore, epidermolysis, due to the specific environmental conditions of the tomb, constantly occurs even when mummification takes place at a later stage. Once epidermolysis is complete, only the dermis and hypodermis are left. The immunohistochemical identification of even thin basal layers of epithelial cells reveals incomplete epidermolysis and gives information about the mummification process and the degree of preservation by confirming that it is still relatively good. A complete absence of epidermis would mean inadequate mummification, unable to prevent autolysis or decay. In practice, the sampling of a very small skin area for histological and immunohistochemical investigations yields useful information and supplies a guideline for further research without damaging the body thus solving the theological and ethical problem of the preservation of Saints' bodies (42, 43).

In rare instances, during the examination, some pathological findings can occasionally emerge. We shall now report a few examples.

During the last survey performed in 1977 which aimed at disinfecting and disinfesting the mummy of Saint Ubaldo of Gubbio, Dr. V. Blasi performed also some X-ray examinations which showed the outcome of multiple fractures of the ribs and of the lower portion of the tibia-fibula in the right leg (44) (Fig. 5). This observation confirmed the historical record according to which: "armo semel, bis crure confracto" (12).

During the examination of the body of Santa Zita (died in 1278), Prof. Fornaciari from Pisa University

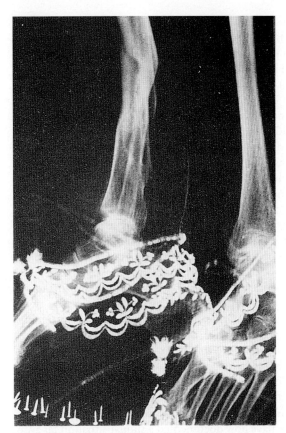

Fig. 5. Sant'Ubaldo da Gubbio (1084–1160): at X-ray examinations, multiple fractures of the lower portion of the tibia and fibula in the right leg were found

found the presence of a massive anthracosis in the pulmonary tissue, also documented at molecular level, with FT-IR microspectroscopy measurements (45).

When the body of Santa Margherita da Cortona, as I have already mentioned, was examined, we found a prominent fibrous thickening of the apical pleura which indicated a chronic pleuritis.

In the above cases the pathological findings substantiated the recorded lives of the Saints.

In rare instances, the examination calls for a true pathological investigation. Here I am pleased to mention the case of San Ciziaco, bishop martyred during the first centuries of the Church. His natural, partially skeletonized mummy, which was kept in the Cathedral of Ancona, was studied by Prof. Mariuzzi from the University of that city (46). Prof. Mariuzzi was able to demonstrate healed bone fractures and fractures without sign of healing on the right side of the cranium from the orbit to the maxilla up to the fossa cranica anterior. Furthermore, he found a great amount of lead along the trachea and the oesophagus, thus confirming the tradition according to which the Saint was martyred by forced ingestion of molten lead. "… et forcipe ferrea aperiri os eius, et infundit plumbum, ut interiora eius conflagrentur." These very interesting cases confirm that the mummies of the Saints can also be important for paleopathological studies.

In these cases, though extremely rare, paleopathology could be considered to be at the service of history and the search for truth.

At the end of our overview on the mummies of Saints we could also make some general remarks. First, we should point out that they are more frequent than expected, and most of them in excellent preservation conditions. The methods of embalming are numerous, since they include natural mummification as well as actual embalming with the use of balsams. In several cases they were completed with proper evisceration. Evisceration was often made to obtain eminent relics and to document prodigious events as part of medieval traditions which became rooted in a particular area and within peculiar socio-cultural environments.

To conclude this overview, we could also make another type of remark, perhaps too general to be rigorously and scientifically proved, yet anyway intriguing.

In the biological history of populations, the Egyptians are for sure a unique example since they have left a remarkable heritage of mummified bodies making up a wonderful biological archive. In addition to environmental factors, their religious customs and ceremonies have helped promote and favour this phenomenon.

We have now documented a similar situation far away in time and space from ancient Egypt, though much less conspicuous and extraordinary. As in the Egyptian world, in the Christian world too, religious habits and practices have contributed to the preservation of bodies through mummification. Cultural milieus which are so far apart have become similar, and though motivations are very different from a theological viewpoint, they are much less so if looking at the technical and practical aspects. Embalming has indeed rooted on a different ethnic and religious medium following a single thread of traditions mediated over the centuries by different populations – like the Jews for example – and, despite great transformations, showing no clearcut interruptions.

Acknowledgements

I would like to thank Ms. Gabriella Sonnewald for her accurate translation, Mr. Paolo Frega for the production of iconographic material, Ms. Anna Scandale for her technical collaboration and, last bust not least, Dr. Patrizia Baracchini for her cooperation in the investigations on mummified tissues.

Summary

In Italy there is a great number of mummies, most of them naturally mummified, due to climate conditions, but also artificial ones obtained with proper embalming processes. Mummies of Saints make up a particular category of mummies.

The cult of saints is still deeply rooted in the Italian tradition, and closely linked with the cult of relics and of Saints' bodies. Therefore, for religious and devotional reasons, many bodies of important people in the history of the Roman Church or people with outstanding virtues, who died in the 'odour of sanctity' used to be buried with particular care or even treated and preserved.

No wonder that the tradition to process the bodies of dead people to guarantee their better preservation is part of the Christian culture. This tradition has very old origins. Indeed we read in Saint John's Gospel : "Venit autem et Nicodemus, qui venerat ad Iesum nocte primum, ferens mixturam myrrhae et aloes, quasi libras centum. Acceperunt ergo corpus Iesu, et ligaverunt illud linteis cum aromatibus, sicut mos est iudaeis sepelire (Jo. 19, 39–40)." The practice of anointing corpses was common among Jews who are likely to have learnt and assimilated it during their long slavery in Egypt. Christians acquired this tradition and followed it even for a long time afterwards.

Some mummies of Saints are natural mummies. Mummification took place for natural reasons due to the particular conditions of the tomb, of the micro-climate and, more generally, of the local climate. In many cases, a perfect mummification was achieved, whereas in other instances a very poor natural mummification would occur after an initial and partial decay. An example of an excellent, well documented preservation, is the one of Sant'Ubaldo da Gubbio.

Conversely, other mummies derive from partial artificial mummification processes. The body was 'anointed' with unguents, perfumes and spices before burial, out of devotion but also to allow for a long public exposure before being buried.

A third category of mummies features proper artificial mummies and we were able to document the case of S. Margherita da Cortona. On her body, large incisions made during evisceration are still visible. In these cases a real dissection was made and viscera removed.

It should also be pointed out that eviscerations were not only made to better preserve the body but also to obtain eminent relics like the heart etc.

Special eviscerations were made for this particular purpose in the cases of S. Chiara da Montefalco and Santa Margherita da Città di Castello whose post mortem examination is a real forerunner in this practice. During the autopsy, as reported in detail by the chronicles and as can still be observed from the original findings, pathological lesions were found in the heart which were the cause of their long suffering and death. Yet, in the light of medieval mysticism, these lesions were interpreted as the signs of Christ Passion. Fractured papillary muscles can be observed in the shape of a cross, deformed heart valves in the form of the crown of thorns, fully ruptured and pierced hearts, the result of massive infarctions, resembling a spear wound.

These embalming practices can reliably be documented in six individuals. Based on some historical considerations, it would be suggestive to note that these practices have particularly spread and rooted in a well confined environment, namely convents, in a limited geographical area, Tuscany and Umbria.

To better understand the techniques employed in mummification and assess the preservation conditions of the findings, a histopathological and immunohistochemical examination of the skin was conducted in those cases where a survey was possible.

Following the conventional histopathological methods on mummified tissues, and by application of the Avidin-Biotin-Peroxidase Complex method with anti-keratin antibodies, several skin specimens were investigated. In artificial mummies, an excellent preservation of all the layers of the epidermis was proved thus confirming the excellent performance of mummification techniques employed.

Conversely, proper paleopathological examinations on the mummies of Saints are not possible in the majority of cases, due to extremely stringent rules and limitations allowing examinations for identification or preservation purposes only.

Four cases are reported referring to S. Ubaldo da Gubbio, S. Zita da Lucca, S. Margherita da Cortona and S. Ciziaco di Ancona. In these cases, peculiar lesions confirming the historical sources were observed through paleopathological examinations.

In conclusion, the chapter of mummies of Saints, though failing to supply Paleopathology with homogeneous material reflecting the population of a certain place at a certain time, is in any case important for historical and folklore investigations. Due to the high number of mummified subjects belonging to this category, however, a very peculiar habit and a line of tradition can be identified which, starting in ancient times and mediated by different populations, is an ideal bridge between Christian Medieval Italy and ancient Egypt.

Zusammenfassung

In Italien existiert eine Vielzahl von Mumien, die aufgrund der besonderen klimatischen Verhältnisse meistens auf natürlichem, anhand regelrechter Einbalsamierungsverfahren aber auch auf künstlichem Wege mumifiziert wurden. Die Mumien von Heiligen stellen eine besondere Mumienkategorie dar.

Der Heiligenkult war und ist noch heute tief in der italienischen Tradition verwurzelt und eng mit dem Kult ihrer Reliquien und Körper verbunden. Daher wurden aus Religions- und Devotionsgründen viele in der Kirchengeschichte bedeutende oder außergewöhnlich tugendhafte Personen, die „im Ruf der Heiligkeit" starben, nach einer besonderen Pflege oder Erhaltungsbehandlung beigesetzt.

Es darf nicht verwundern, daß in der christlichen Kultur die Tradition besteht, den Körper der Verstorbenen zu behandeln, um ihn vor der Verwesung zu schützen. Diese Tradition hat sehr antike Ursprünge; im Evangelium des Johannes lesen wir: „Venit autem et Nicodemus, qui venerat ad Iesum nocte primum, ferens mixturam myrrhae et aloes, quasi libras centum. Acceperunt ergo corpus Iesu et ligaverunt illud linteis cum aromatibus, sicut mos est iudaeis sepelire (Joh. 19, 39–40)." Wahrscheinlich war die Leichensalbung eine normale Gewohnheit bei den Hebräern, die sie während der langen Gefangenschaft in Ägypten gelernt und übernommen haben könnten. Die Christen machten sich diese Tradition zu eigen und erhielten sie jahrhundertelang.

Einige Heiligenmumien sind natürliche Mumien. Die Mumifizierung erfolgte auf natürlichem Wege aufgrund des besonderen Grabzustandes, des Mikroklimas und des örtlichen Klimas im allgemeinen. In vielen Fällen handelt es sich um eine vollkommene, in anderen hingegen um eine teilweise Mumifizierung, die nach einer anfänglichen unvollständigen Verwesung eingetreten ist. Ein Beispiel für eine ausgezeichnete, gut nachweisbare Erhaltung ist der heilige Ubaldo da Gubbio.

Andere Mumien weisen dagegen teilweise künstliche Mumifizierungen auf. Vor der Beerdigung wurde der Körper aus Devotion mit Balsam und Kräutern eingesalbt; überdies war es auf diese Weise möglich, ihn vor der Beisetzung lange öffentlich aufzubahren.

Eine dritte Gruppe besteht aus den rein künstlichen Mumien, zu denen der Fall der hl. Margherita da Cortona zu rechnen ist, den wir dokumentieren konnten. Auf ihrem Körper sind große Schnitte zur Entfernung der Eingeweide noch deutlich sichtbar. In diesen Fällen wurde eine regelrechte Sezierung zur Entfernung der Eingeweide durchgeführt.

Überdies wird darauf hingewiesen, daß die Entfernung von Hirn und Eingeweiden nicht nur zu einer besseren Mumifikation dienten, sondern auch wertvolle Reliquien lieferten wie das Herz usw.

Besondere Exenterationen in dieser Hinsicht wurden an der heiligen Chiara da Montefalco und der heiligen Margherita da Città di Castello durchgeführt, die einer regelrechten Autopsie ante litteram unterzogen wurden. Wie die Chroniken ausführlich berichten und wie man noch an den ursprünglichen Organen beobachten kann, wurde im Verlauf der Autopsie im Herzen der pathologische Befund festgestellt, der für das lange Leiden und den Exitus verantwortlich war. Dieser Befund wurde jedoch aus der Sicht der mittelalterlichen Mystik als Zeichen der Passion Christi interpretiert. Man fand in Kreuzform verletzte Papillarmuskeln, in Dornenkronenform verformte Herzklappen, Risse und Brüche, deren Form der Verletzung durch einen Lanzenstich ähnelt, die das gesamte Herz durchqueren und als Folge massiver Infarkte anzusehen sind.

Die Einbalsamierungsverfahren sind mit Sicherheit bei sechs Leichen nachweisbar, und aus einer Reihe historischer Beschreibungen geht eindrucksvoll hervor, daß sie vor allem in der begrenzten Welt des Klosters und in einem kleinen geographischen Gebiet, Toskana und Umbrien, verbreitet und verwurzelt waren.

Zum besseren Verständnis der Mumifizierungstechniken und des Erhaltungszustands der Gewebe wurde in den Fällen, in denen eine Untersuchung möglich war, vorgeschlagen, ein histopathologisches und immunhistochemisches Studium an der Haut durchzuführen. Unter Anwendung der klassischen histopathologischen Methoden an den mumifizierten Geweben und der Technik Avidin-Biotin-Peroxydase-Komplex mit Antikeratin-Antikörpern wurden zahlreiche Hautproben untersucht. Bei den künstlichen Mumien konnte eine hervorragende Erhaltung aller Hautschichten und folglich die Perfektion der verwendeten Mumifizierungstechniken nachgewiesen werden.

Reine paläopathologische Untersuchungen an den Heiligenmumien sind aufgrund der äußerst strengen Beschränkungen, die nur Forschungen zu Erkenntnis- oder Erhaltungszwecken zulassen, meist nicht möglich.

Wir haben vier Fälle beschrieben, die Heiligen Ubaldo da Gubbio, Zita da Lucca, Margherita da Cortona und Ciziaco di Ancona. In diesen Fällen konnte man anhand der paläopathologischen Forschungen besondere Verletzungen nachweisen und die geschichtlichen Quellen belegen.

Obwohl das Kapitel Heiligenmumien der Paläopathologie kein homogenes Material bieten kann, das Aussagen über die Bevölkerung zu einer bestimmten Zeit und an einem bestimmten Ort liefert, ist es für die Geschichts- und Sittenstudien durchaus wichtig. Die Zahl der Mumien, die dieser Kategorie angehören, ist erheblich und verweist auf einen besonderen Brauch und einen Traditionsreichtum, der in längst vergangenen Zeiten entstanden ist, von verschiedenen Völkern überliefert wurde und es gestattet, eine ideelle Brücke zwischen der Kultur des christlichen Italiens im Mittelalter und dem antiken Ägypten zu schlagen.

Résumé

En Italie, on peut trouver un grand nombre d'individus momifiés, la plupart naturellement vu les conditions climatiques particulières mais même artificiellement á la suite de véritables processus d'embaumement. Une catégorie particulière de momies est représentée par les momies de Saints.

Le culte des Saints a été et est encore profondément enraciné dans la tradition italienne et strictement lié au culte des reliques et de leurs corps. Ainsi, pour des raisons religieuses et de dévotion, de nombreux corps de personnes importantes pour l'histoire de l'Eglise ou des personnes excellentes par vertu, mortes "en odeur de sainteté" ont été ensevelies avec des soins particuliers et bien traitées et conservées.

Il ne faut pas s'étonner que, dans la culture chrétienne, il y ait la tradition de traiter le corps d'individus morts pour en garantir une meilleure conservation. Cette tradition a des origines très anciennes: d'après l'Evangile de Saint Jean, nous lisons: "Venit autem et Nicodemus, qui venerat ad Iesum nocte primum, ferens mixturam myrrhae et aloes, quasi libras centum. Acceperunt ergo corpus Iesu, et ligaverunt illud linteis cum aromatibus, sicut mos est iudaeis sepelire (Jean; 19, 39–40)." Probablement l'habitude d'oindre des cadavres était d'usage courant chez les juifs qui pouvaient l'avoir apprise et assimilée au cours de leur longue captivité en Egypte. Les chrétiens ont acquis cette tradition et l'ont gardée même après un certain laps de temps.

- Certaines momies des Saints sont des momies naturelles; la momification a eu lieu pour des causes naturelles vu les conditions particulières du sépulcre, du micro-climat et de façon plus générale, du climat du lieu. Dans de nombreux cas, il s'agit d'une momification parfaite alors que dans d'autres cas il s'agit d'une très mauvaise momification qui a eu lieu après une putréfaction initiale incomplète. Un exemple de conservation excellente, pouvant être documentée, est celle de Saint Ubaldo de Gubbio.
- D'autres momies représentent au contraire des momifications artificielles partielles. Le corps était en effet "oint" avec des baumes et des aromes avant d'être enseveli soit par dévotion soit pour permettre, avant la sépulture, une longue exposition au public.
- Un troisième groupe est représenté par les véritables momies artificielles et nous avons pu documenter le cas de Sainte Marguerite de Cortona. En effet, sur son corps, on peut encore bien voir des grandes incisions effectuées pendant l'éviscération. Dans ces cas, on effectuait une véritable dissection pour enlever les viscères.

Il reste d'ailleurs à mettre en évidence comment les éviscérations n'avaient pas le but de permettre une momification meilleure mais même de procurer des reliques insignes comme le coeur, etc.

Des éviscérations particulières, faites dans cette optique, sont celles de Sainte Claire de Montefalco et Sainte Marguerite de Città di Castello, soumises à une véritable autopsie ante litteram. Au cours de l'autopsie, comme le réfèrent de façon détaillée les chroniques et comme on peut encore l'observer sur les pièces originaires, on a trouvé dans le coeur les lésions pathologiques responsables des longues souffrances et de la mort. Cependant ces lésions, dans l'optique de la mystique médiévale, ont été interprétées comme les signes de la passion de Jésus Christ. En effet on a trouvé des muscles papillaires fracturés en forme de croix, des valvules cardiaques déformées en forme de couronne d'épines, blessures et ruptures de coeur sur toute l'épaisseur, résultats d'importants infarctus, ayant une forme semblable au signe qu'aurait pu laisser une blessure de lance.

Les pratiques d'embaumement peuvent sûrement être documentées sur six individus et, d'une série de considérations historiques, il semble suggestif de remarquer comment se sont diffusées et enracinées, de façon particulière, dans un milieu restreint, le monastique et, dans une zone géographique limitée, la zone de la Toscane et de l'Ombrie.

Pour mieux comprendre les techniques utilisées pour la momification et connaître l'état de conservation des pièces, on a proposé, pour les cas où cela est possible, d'effectuer une reconnaissance, une étude histopathologique et immuno-histo-chimique de la peau. En suivant les méthodes classiques d'histopathologie des tissus momifiés et en utilisant la technique d'Avidine Biotine Peroxydase Complex avec des anticorps anti kératine, on a étudié de nombreux échantillons de peau. Chez les momies artificielles, on a démontré une excellente conservation de l'épiderme dans toutes ses couches en démontrant la perfection des techniques utilisées pour la momification.

Au contraire, on ne peut pas faire de véritables enquêtes paléopathologiques sur les momies des Saints, le plus souvent, à cause des contraintes très rigides qui ne permettent que des enquêtes ayant un but récognitif ou de conservation.

On a reporté quatre cas, concernant Saint Ubaldo de Gubbio, Sainte Zita de Lucques, Sainte Marguerite de Cortona et Saint Cyriaque d'Ancone. Dans ces cas, les enquêtes paléopathologiques ont pu mettre en évidence des lésions particulières et confirmer les sources historiques.

Pour conclure le chapitre des momies des Saints, même s'il n'offre pas, pour la paléopathologie, un matériel homogène et réfléchissant les populations d'un temps et un lieu déterminé, est toutefois important pour les études historiques et de coutumes. Ce qui est important toutefois, c'est le nombre d'individus momifiés appartenant à cette catégorie qui permet d'identifier une coutume très particulière et une traditions qui, ayant vu le jour dans des temps très éloignés et ayant été composés par des populations différentes, permet de relancer un pont idéal entre la civilisation de l'Italie chrétienne du Moyen-Ége et l'ancien Egypte.

Riassunto

In Italia si può trovare un gran numero di soggetti mummificati, per lo più naturalmente, date le particolari condizioni climatiche, ma anche artificialmente, a seguito di veri e proprii processi di imbalsamazione. Una particolare categoria di mummie è rappresentata dalle mummie dei Santi.

Il culto dei Santi è stato ed è tuttora profondamente radicato nella tradizione italiana; collegato strettamente al culto delle reliquie e dei loro corpi. Così, per motivi religiosi e devozionali, molti corpi di persone importanti nella storia della Chiesa o persone eccellenti per virtù, morte "in odore di santità", vennero sepolte con particolari cure o trattati e conservati.

Non deve stupire che nella cultura cristiana vi sia la tradizione di trattare il corpo di soggetti defunti al fine di garantirne una migliore conservazione. Questa tradizione ha origini molto antiche: dal Vangelo di San Giovanni leggiamo: "Venit autem et Nicodemus, qui venerat ad Iesum nocte primum, ferens mixturam myrrhae et aloes, quasi libras centum. Acceperunt ergo corpus Iesu, et ligaverunt illud linteis cum aromatibus, sicut mos est iudaeis sepelire (Gio. 19, 39–40)." Probabilmente la pratica dell'unzione dei cadaveri era d'uso corrente tra gli ebrei che potevano averla imparata ed assimilata durante la lunga prigionia in Egitto. I cristiani fecero opria questa tradizione e la mantennero anche a distanza di tempo.

Alcune mummie dei Santi sono mummie naturali; la mummificazione avvenne per cause naturali date le particolari condizioni del sepolcro, del micro clima e più genericamente, del clima del luogo. In molti casi si tratta di una perfetta mummificazione mentre in altri casi si tratta di una pessima mummificazione avvenuta dopo iniziale putrefazione incompleta. Un esempio di eccellente conservazione, ben documentabile, è quello di Sant' Ubaldo da Gubbio.

Altre mummie rappresentano invece parziali mummificazioni artificiali. Il corpo veniva infatti "unto" con balsami ed aromi prima di essere sepolto, sia per devozione che per permettere, prima della sepoltura, un lunga esposizione al pubblico.

Un terzo gruppo è rappresentato dalle vere mummie artificiali e noi abbiamo potuto documentare il caso di S. Margherita da Cortona. Sul corpo infatti sono ancora ben visibili le ampie incisioni effettuate durante l'eviscerazione. In questi casi veniva effettuata una vera e propria dissezione con rimozione dei visceri.

Resta per altro da evidenziare come le eviscerazioni non avessero il solo scopo di consentire una migliore mummificazione ma anche quello di procurare reliquie insigni come il cuore etc.

Particolari eviscerazioni effettuate in tale ottica sono quelle di S. Chiara da Montefalco e Santa Margherita da Città di Castello, sottoposte ad una vera e propria autopsia ante litteram. Nel corso dell'autopsia, come dettagliatamente riferiscono le cronache e come è ancora possibile osservare sui reperti originari, vennero trovate nel cuore le lesioni patologiche responsabili delle lunghe sofferenze e dell'exitus. Tali lesioni però, nell'ottica della mistica medioevale, vennero interpretate come i segni della passione di Cristo. Si trovarono infatti muscoli papillari fratturati a forma di croce, valvole cardiache deformate a forma di corona di spine, trafitture e rotture di cuore a tutto spessore, esiti di massivi infarti, di forma simile al segno che avrebbe potuto lasciare una ferita di lancia.

Tali pratiche imbalsamatorie sono documentabili con sicurezza in sei soggetti e, da una serie di considerazioni storiche, sembra suggestivo notare come si siano diffuse e radicate in modo particolare in un ambiente limitato, quello monastico, ed in una ristretta area geografica, quella Tosco-Umbra.

Al fine di meglio comprendere le tecniche impiegate per la mummificazione e conoscere lo stato di conservazione dei reperti, è stato proposto, per i casi ove sia stato possibile effettuare una ricognizione, uno studio istopatologico ed immunoistochimico sulla cute.

Seguendo le classiche metodiche di istopatologia dei tessuti mummificati, e mediante l'impiego della tecnica Avidina Biotina Perossidasi Complex con anticorpi anti cheratina, sono stati studiati numerosi campioni di cute. Nelle mummie artificiali è stata dimostrata una eccellente conservazione dell'epidermide in tutti i suoi strati dimostrando la perfezione delle tecniche di mummificazione impiegate.

Vere e proprie indagini paleopatologiche sulle mummie dei Santi non sono invece per lo più possibili a causa dei vincoli estremamente rigidi che consentono solamente indagini a scopo ricognitivo o conservativo.

Sono riportati quattro casi, relativi a S. Ubaldo da Gubbio, S. Zita da Lucca, S. Margherita da Cortona e S. Ciziaco di Ancona. In tali casi le indagini paleopatologiche hanno potuto evidenziare lesioni peculiari e confermare le fonti storiche.

In conclusione il capitolo delle mummie dei Santi, anche se non offre per la Paleopatologia un materiale omogeneo e rispecchiante le popolazioni di un determinato tempo e luogo, è tuttavia importante per gli studi storici e di costume. Resta rilevante il numero di soggetti mummificati che appartengono a questa categoria, tale da identificare un particolarissimo costume ed un filone di tradizioni che, originatosi in tempi molto lontani e mediato da popolazioni differenti, consente di riallacciare un ponte ideale tra la civiltà dell'Italia cristiana del medioevo e l'antico Egitto.

References

1. Fulcheri E., Baracchini P., Doro Garetto T., Pastorino A. M., Rabino Massa E.: Le mummie dell'antico egitto custodite nei musei italiani. Stima preliminare dell'entità del patrimonio museologico e considerazioni sul problema della conservazione di esse. Mus. Scient. Museol. sa. XI, 1–11, 1994.
2. Delorenzi E., Grilletto R.: Le mummie del Museo Egizio. Istituto Editoriale Cisalpino-La Goliardica, Milano 1989.
3. Grilletto R.: Catalogo Generale del Museo Egizio di Torino. Serie seconda – Collezioni. Supplemento al Volume VI. Torino 1991.
4. Fulcheri E.: Italian mummies over the centuries. XVIII International Congress of the International Academy of Pathology, Buenos Aires 1990.
5. Farella F.: Cenni Storici della Chiesa e delle Catacombe dei Cappuccini di Palermo. Ed. Fiamma Serafica, Palermo 1982.
6. Amato G. M.: De principe templo panormitano. Panor. Bapt. Aiccardo, 1728.
7. Antonio da Castellammare: Le catacombe, ossia la grande sepoltura dei Cappuccini in Palermo. Fiamma Serafica, Palermo, 1938.
8. Bibliotecha Sanctorum: Ed. Giovanni XXIII, Pontificia Università Lateranense, Roma 1968.
9. Bellotta I.: I Santi Patroni d'Italia. Newton Compton Ed., Roma 1988.
10. Dizionario dei Santi: Dizionari T E A, UTET, Torino 1989.
11. Cattabiani A.: Santi d'Italia. Rizzoli, Milano 1993.
12. Giordano: Vita beati Ubaldi Eugubini episcopi. Traduzione di Don A. M. Fanucci, "Famiglia dei Santantoniari", Gubbio 1992.
13. Jacobilli L.: Vite de Santi e Beati dell'Umbria. Forni, Bologna 1971.
14. Nolli G., Gabrielli H., Venturini M., Benedettucci M., Fulcheri E.: Relazione della ricognizione e del trattamento conservativo effettuato sui resti di S. Savina Petrilli. Roma 1988.
15. Fulcheri E.: Il patologo di fronte al problema della perizia in corso di ricognizione sulle reliquie dei Santi. Pathologica 83, 373–397, 1991.
16. Nolli G., Gabrielli H., Venturini M., Dati F.: Relazione del trattamento effettuato sul corpo di S. Ubaldo. Roma 1977. In Braccini U. F. La mano di S. Ubaldo, Santuario di S. Ubaldo, Gubbio 1993.
17. Fornaciari G., Spremola G., Vergamini P., Benedetti E.: Analisi di tessuto polmonare di un corpo mummificato del XIII secolo (Santa Zita, Lucca, Tuscany, Italy) mediante microspettroscopia FT-IR. IV Meeting of the Adriatic Society of Pathology. Ravenna 1989.
18. Fornaca L., Vanzetti F., Zucchi A.: Perizia scientifico-medica effettuata sulla salma della Beata Margherita di Savoia. Alba 1938.
19. Terribile W. M. V., Corrain C.: Pratiche imbalsamatorie in Europa. Pathologica, 78, 107–118, 1986.
20. Grilletto R.: La splendida vita delle mummie. Sugarco Milano, 1987.
21. Origo L.: San Bernardino da Siena ed il suo tempo. Rusconi, Milano 1982.
22. Nolli G.: Intervento conservativo sul corpo del Beato Luigi Orione. Roma.
23. Alciati G., Rippa Bonati M., Fulcheri E.: Ricognizione effettuata sul corpo mummificato del Beato Don Orione. Tortona 1990.

24. Holli G., Gabrielli H., Venturini M., Fulcheri E.: Cardinale Josyf Slipyi. Relazioni sul trattamento conservativo eseguito sul suo corpo. Roma, Elettrongraf, 1987.
25. Fra Giunta Bevegnati: La leggenda della vita e dei miracoli di Santa Margherita da Cortona. Nuova traduzione dal latino con prefazione e note di P. Eliodoro Mariani O. F. M. LIEF, Vicenza 1978.
26. Fulcheri E., Baracchini P.: Relazione della ricognizione effettuata sul corpo di Santa Margherita da Cortona. Cortona 1989.
27. Dicta testium Auctoritate Apostolica examinatorum anno 1719: Super primo miraculo. Perennis incorruptionis, et integritatis corporis eiusdem Beatae.
28. Visitatio et Recognitio Corporis Beatae Margaritae, Auctoritate Apostolica peracta anno 1719 a Iudicibus remissorialibus super primo miraculo perennis Incorruptionis et integritatis corporis eiusdem Beatae.
29. Visitatio et Recognitio Corporis Beatae Margaritae Auctoritate Apostolica peracta Anno 1634 a iudicibus Remissorialibus Super primo Miraculo Perennis Incorruptionis et integritatis Corporis eiusdem Beatae.
30. Sacra Rituum Congregatione Eminentissimo, et Reverendissimo D. Corradino Cortonen. Canonizationis B. Margaritae de Cortona. Positio super dubio. Romae, Typis Reu.Cam. Apostolicae MDCCXXIII.
31. Nessi S.: Chiara da Montefalco Badessa del Monastero di S. Croce. Panetto e Petrelli Spoleto, Montefalco 1981.
32. Analecta Bollandiana, 19. Vita Beatae Margaritae Virginis de Civitate Castelli. Roma, 1900.
33. Sandison A. T.: The histological examination of mummilied matherial. Stain Technol. 30, 277–283, 1955.
34. Fulcheri E., Rabino Massa E., Fenoglio C.: Improvement in the histological technique for mummified tissue. Verh.Dtsch. Ges.Path. 69, 471, 1985.
35. Terribile W. M. V.: Metodo ed apparechiatura per lo studio istologico dei tessuti mummificati. Pathologica, 79, 781–787, 1987.
36. Tapp E.: Histology and histopathology of the Manchester mummies. In Proceedings of the Science in Egyptology Symposia. R. David ed. Manchester University Press, Manchester 1986.
37. Tapp E.: Disease and the Manchester Mummies. The pathologist's role. In David A. R., Tapp E. Evidence embalmed. Modern medicine and the mummies of ancient Egypt. Manchester University Press, Manchester, Chapt. 4, 78–95, 1984.
38. Krypczyk A., Tapp E.: Immunohistochemistry and electron microscopy of Egyptian mummies. In Proceedings of the Science in Egyptology Symposia. R. Davided. Manchester University Press, Manchester 1986.
39. Fulcheri E., Rabino Massa E.: Immunohistochemistry in mummified tissues; a problem in paleopathology. Proceedings VI European Meeting of the Paleopathology Association, Madrid 1986.
40. Hsu S. M., Raine L., Fanger H.: Use of avidin-biotin-peroxidase complex (ABC) inimmuno peroxidase techniques: a comparison between ABC and unlabele dantibody (PAP) procedures. J Histochem Cytochem, 29, 577–580, 1981.
41. Battifora H.: The biology of the keratins and their diagnostic application. Capt. 8 in DeLellis R. Advances in Immunohistochemistry. Raven Press, New York, 1988.
42. Fulcheri E., Baracchini P., Rabino Massa E.: Use de l'immunoistochimique avec anticorps anti-keratins dans l'etude de la peau mummifiee. XVIII Colloque des anthropologistes de langue francaise, L'Escala, 1987.
43. Baracchini P., Holli G., Gabrielli H., Venturini M., Benedettucci M., Fulcheri E.: Studio immunoistochimico delle cheratine nella cute di soggetti artificialmente o naturalmente mummificati. IV Meeting of the Adriatic Society of Pathology. Ravenna 1989.
44. Blasi V.: Referto radiologico su alcune radiografie del Venerato Corpo di S. Ubaldo. Gubbio 1975. In Braccini U. F.: La mano di S. Ubaldo, Santuario di S. Ubaldo, Gubbio 1993.
45. Fornaciari G., Spremola G., Vergamini P., Benedetti E.: Analysis of pulmonary tissue from a natural mummy of the XIII century (Sain Zita, Lucca, Tuscany, Italy) by FT-IR microspectroscopy. Paleopathology Newsletter, 68, 5–8, 1989.
46. Mariuzzi G.: Causa mortis of a Christian Martyr. Proceedings of XV International Congress Academy of Pathology. Miami Beach 1984.

Correspondence: Dr. E. Fulcheri, Istituto Anatomia Patologica, Universita di Genova, Via De Toni 14, I-16132 Genova, Italy.

Mummification in the Middle Ages

B. Kaufmann

Anthropologisches Forschungsinstitut, Aesch/BI, Switzerland

Introduction

It seems that the art or technique of artificial mummification has never been forgotten entirely in Western and Middle Europe since the Early Middle Ages, even if one can rarely find any physical evidence. It is rather a scientific gap than contemporary technical ignorance as this subject has rarely been investigated. Yet since the discovery of the Iceman from the Similaun-glacier fortunately also this subject has met more attention.

In the following, I will present some cases from the Middle Ages, specially from the 13th to the 16th century, that have been found in the German speaking part of Switzerland. That most of the cases come from the city of Basel is not astonishing as the population was generally well-off. Moreover it was only a two day's journey away from Strassburg which at that time was one of the strongholds for Late Medieval mummification, according to unpublished researches by Dr. Dorothee Rippmann.

Queen Anna of Habsburg

The earliest written evidence of a burial in mummified state can be found in the "Collectanea historica" by Christian Wurstisen, a history of the Basel Cathedral with its adjoining buildings, published in 1585. In his description of the grave of Anna of Habsburg (see Fig. 1), born countess of Hohenberg and Haigerloch in Württemberg, wife to King Rudolf of Habsburg, he reports that she died in Vienna in 1281 and had desired to be buried in the Cathedral of Basel. Wurstisen writes about the process of her mummification:

„Also entweidet man ihren leichnam, füllet ihr den bauch mit äschen aus, balsamiert ihr das angesicht und uebrigen glieder, verwickelt sie in ein gewächsen tuch, legt ihren kostlichen seidin gewand an, setzt ihr auf das verschleiert haupt ein vergülte cron, hengt ihr ein kleinot an den hals, legt sie also rügglings in ein buchbäuminen sarch, und füret sie mit 40 pferden aus Oesterreich gen Basel…"

("Thus one removes her internal organs, fills her body with ash, balms her face and limbs, wraps her with wax cloth, clothes her with a silken garment, puts on her veiled head a gilded crown, hangs a locket around her neck, beds her in a coffin made from beechwood, and conducts her with forty horses from Austria to Basel…")

Resuming the process we find the following operations:
– after death the internal organs are removed
– head and body are embalmed (most probably with soda-lye)
– the body is wrapped in cloth impregnated with wax
– over this bandage one puts the proper clothing
– the body is laid on its back into a coffin of beechwood

One cannot tell how long took the whole process of mummification. Queen Anna died on the 23rd of February 1281, the journey from Vienna to Basel usually took 40 days, and she was buried on the 20th March of the same year.

According to two sources cited by Wurstisen the mummification was very thorough and successful: At her burial in the cathedral the coffin was opened and her body exposed in the presence of the king, three bishops and all of the 1200 priests of the diocese. Then she was laid to rest in her elevated tomb in the choir, together with the exhumed remains of her son Carolus who had died in 1276 at the age of six months and was buried at the same time in the same cathedral.

In 1510 the grave was opened for the first time. Again Wurstisen reports:

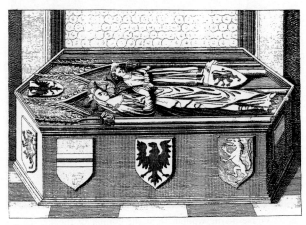

Fig. 1. Tomb of Queen Anna in the cathedral of Basel

„Im 1510. jar beisse die thumbherren der wunderfitz, das sie das königklich grab öfneten: funden darinn der königin cörper in guter ordnung, und neben ihren ein unordentlich häuflein gebein, von dem jungen herrlin Carolo."

("In 1510 the canons opened the royal grave out of curiosity: they found the queen's body in good condition and at her side the young lord Carolus in an unorderly heap of bones.")

Thus the mummification had kept for 230 years. The jewels of the queen were removed and confiscated for the treasure of the cathedral.

The tomb was opened again towards the end of the 18th century when the remains were transferred to the monastery of St. Blasien in the Black Forest which had been re-erected after a devastating fire. By the graves opening only few skeletal remains were preserved.

The burial in the church of the Franciscan friars at Basel

Site of the burial

During excavations in the church of the Franciscan Friars at Basel, rather at the beginning of the campaign in 1975, one found a mummified body in the grave number 15. We requested its deposition in the anthropological department of the Natural History Museum for examination and possible conservation (see Fig. 2).

The mummified body lay nude in a conical coffin. The body was embedded on wood chips; presumably they were once covered with cloth to form some kind of upholstery. In the coffin one found fragments of four different fabrics; one of them was some printed cloth, most probably from her original garment. All textile remains as well as the coffin were handed over to the Historical Museum in Basel.

Date of the burial

The make of the coffin dates the burial to around the middle of the 16th century, briefly after Reformation. The Franciscan Friars already had left the monastery, but the church still served as burial ground for well-off citizens. Until 1794 the nave was used for religious service as well as for burying people, whereas the choir had been turned into a storage place, partly from grain, and from 1799 to 1840 salt was stored in the whole church. Since 1894 it is the home of the Historical Museum.

Examination of the mummy

The anthropological department of the Natural History Museum is well equipped with nice storage room for the collection, but else there is only little space. Therefore the mummy was temporarily stored in an empty building near the museum. In the following winter the director of the museum gave us the permit to examine the mummy in one of the unheated rooms of that house.

State of preservation

The mummy was relatively undamaged and in good condition, except for sections of the head (see Fig. 3) and for the missing feet. On the frontal and temporal parts of the skull the scalp with the hair was missing, but it was intact at the rear side. The mummy gave the impression of being made of leather: skin and muscles were dehydrated and showed the typical middle to dark brown shades of leather. At first the pathologist, the late Professor S. Scheidegger († 1989), examined the surface and noticed small red particles on the thighs. The analysis at the Institute for Physical Chemistry of Basel University proved them to be red cinnabar.

For the dissection the body was turned onto its belly: as the mummy was intended to be used in exhibitions we wanted to avoid any obvious damages. The internal organs were removed through a cut in the lumbar region. The whole dissection, which took two days, was filmed

Fig. 2. Grave of the woman from the church of the Franciscan Friars in Basel

to keep records of the process. Afterwards the opening was stitched together.

Some of the removed organs were immersed in a solution of formalin, while the rest was taken, together with the mummy, to the laboratory of the Kantonsmuseum Baselland in Liestal where for a month they were dehydrated in a high-vacuum-tube and then treated with fungicides. The technical details will be explained later on. As there were neither interest nor money for any publication and as we did not dispose of a laboratory or any suitable rooms, we had to postpone further examinations. The mummy now rests in our institute in Aesch.

Anthropological examination

The numerous red particles on the skin, mainly on the thighs, lead to the conclusion that presumably the woman died of syphilis, or rather of the treatment with an overdose of paste of mercury. In this context it has to be reminded that the city physician of Basel, Paracelsus, was very much against the common practice to adminster high doses of mercury. Also the periostitis on the tibias and fibulas give a further indication for the diagnosis of syphilis: the bones are covered with vertical grooves and with vascular impressions. The internal organs have not yet been examined.

Technique of mummification

On the technique of mummification we cannot give precise details. At first we supposed it to be accidental: the extreme amount of mercury could have induced a primary conservation of the body which later, after oxidation of the mercury to cinnabar, would have been reinforced by the salt. Yet the humidity of the place speaks against this hypothesis: many of the floods from the nearby river Birs had reached the level of the church floor and had thus covered the burials. In the meantime we learned about several intended mummifications in Basel in the Late Middle Ages and Early Modern Age: during excavations one found three mummified bodies of adults in the Church of the Dominican Friars and five children in the church of Saint Leonhard. Further researches gave us written evidence of a deceased participant of the Concil of Basel (1431–1449) who was mummified and buried in the Charterhouse. The most spectacular case is the fate of the mummy of an adult man from the church of Saint Leonhard: it was the body of the anabaptist David Joris (Fig. 4). As a rich immigrant from the Netherlands he was welcome in Basel. He died as a highly respected protestant in 1556 and was buried in Saint Leonhard. When five years later his true religion became public, the courts ordered the exhumation of the body. It became obvious that the body had been mummified, and it was

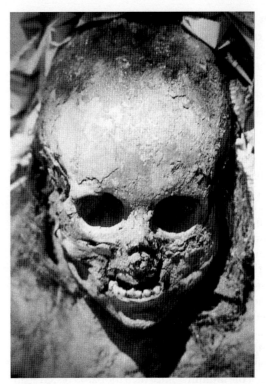

Fig. 3. Head of the mummy from the church of the Franciscan Friars

Fig. 4. Portrait of the anabaptist David Joris

still intact, but of a yellowish colour. As was the custom for heretics, the body was drawn and quartered and publicly burnt in front of a large crowd of people.

Johann Philipp of Hohensax, Baron of Sennwald (1550–1596)

The family of Hohensax was important for Swiss history, mainly in the Eastern part of the country. Two of his ancestors are documented in the "Manesse Liederhandschrift" as minnesingers, and his father was head of the Papal Guard and became honoured with the Banner of Julius.

When we transferred his mummified body from Sennwald to Basel in September 1979, the mortal remains of the baron (Fig. 5) had already its own history. The baron had died in a family feud, and was buried in the family vault in the church of Sennwald in the Upper Rhine Valley in the canton of St. Gallen. A fire in 1730 made it necessary to build an new church, and during construction some workmen found the intact body of the baron which was transferred to the new tower. But soon the news spread into the nearby catholic Vorarlberg: by night they stole the intact body and brought that presumably holy man to Dornbirn. Only after the Swiss authorities could prove that the man had died as a good protestant and thus could impossibly be holy, the catholics from Vorarlberg decided to give back the mummy. Since then the baron rested again in the tower of the church of Sennwald.

Examination of the deceased

The body was first externally examined in the anthropological department of the Natural History Museum in Basel by our pathologist, Professor Scheidegger, and myself.

State of preservation:
– mummified body of a leathery consistence
– no scalp nor hair
– the face is well preserved; only the sockets are sunken due to dehydration of the eye-balls
– circular impression on the neck
– the skin near the breastbone, on the left shoulder and the left side of the chest is freshly damaged
– recent rupture on the right elbow, skin defects on the right hand
– loss of the left hand; the forearm only consists of radius and ulna
– strong damages of the skin of the pelvic area and on the thighs; partial loss of the genitals
– loss of the mummified flesh below the knees; those parts are only represented by both tibias.

Dissection

This mummy was opened from the belly area by a horizontal cut beneath the bent forearms and two vertical incisions down to the hip bones (see Fig. 6). Thus resulted a big window that allowed to remove the internal organs without damaging much of the mummy. Also in this mummy all the internal organs were present. Most of them were treated as dried objects while few of them were conserved in formalin.

Conservation and partial reconstruction (Figs. 7–9)

Like the mummy from the church of the Franciscan Friars the body of the "Black Knight" was transferred to the labaratory of the Kantonsmuseum Baselland for conservation and partial reconstruction. The details of the conservation are recorded in the report of the head of the laboratory, Kurt Hunziker:

"29th of Nov. 1979 reception of the mummy …; brought by Dr. B. Kaufmann
13th of Dec. 1979 first treatment with a disinfectant (Schnelldesinfektionsspray W. I. Z., quaternäre Ammoniumbase Benzoesäureprophylester Isopropanol)

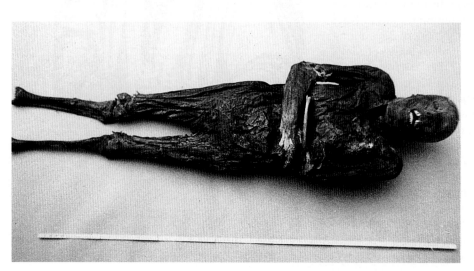

Fig. 5. Mummy of the Baron of Hohensax

14th of Dec. 1979	photographies mummy put into the high-vacuum-tube vacuum of 400 torr: checking for any fissures through the window
27th of Dec. 1979	lowering of the pressure to 300 Torr, control through window
18th of Dec. 1979	" 200 Torr "
19th of Dec. 1979	" 100 Torr "
20th of Dec. 1979	" 50 Torr "
21st of Dec. 1979	" 1 Torr "
26th of Dec. 1979	control of the vacuum rising pressure to 15 Torr control through window lowering again to 10–1 Torr
2nd of Jan. 1980	slow ventilation with nitrogen: removal of the mummy from the tube; control of the mummy; putting it back into the tube lowering to 100 Torr
3rd of Jan. 1980	" 50 Torr
4th of Jan. 1980	" 1 Torr
7th of Jan. 1980	rising to 7 Torr lowering to 10–1 Torr
8th of Jan. 1980	control of the vacuum
10th of Jan. 1980	slow ventilation with nitrogen; removal of the mummy treatment with Xylophen SC (chlorinate carbohydrate and organic metal binding, Lindan)
Febr. to May 1980	weekly control of the mummy
June 1980	discussion with Dr. Kaufmann for further procedure symposium of German anthropologists at the Natural History Museum in Basel: transport of the mummy to Basel where it stayed for about three months
October 1980	Dr. Kaufmann brings the mummy to the laboratory to discuss the reconstruction of the left arm and the right hand as well as missing parts on the neck and the back.

The mummy stays in storage room 15a of the Kantonsmuseum Baselland without any further treatment.

March 1981	fabrication of negative moulds for the arms (the model was Heinz Stebler, designer at the Kantonsmuseum Baselland); copies made from coloured Beracryl which are attached to the mummy. correction of the sockets and lips with Beracryl missing parts of the neck and the chest completed with Beracryl gluing together any loose skin parts missing parts of the back are stuffed fissures in the area of the lower belly are corrected

At the request of Dr. Kaufmann we x-rayed the skull for the possible cause of death. One can clearly recognize that with such a skull fracture our "patient" (the baron) couldn't write any letters, as was reported in the original sources."

Identity

We have great difficulties to identify this mummy with the baron Johann Philipp of Hohensax, even if there had never been any doubts before. We cannot match our pathological results with the official documents about his death.

The official version of the injuries of the Baron:

4th of May 1596	quarrel between Johann Philipp and his nephew Georg Ulrich of Hohensax in the pub at Salez: Georg Ulrich wounds his uncle with a hunting knife: several cuts on the head and on the chest

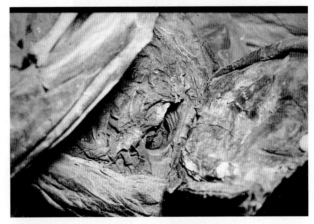

Fig. 6. Mummy of the Baron of Hohensax, view into the opened belly

Fig. 7. Fabrication of a mould for the arms

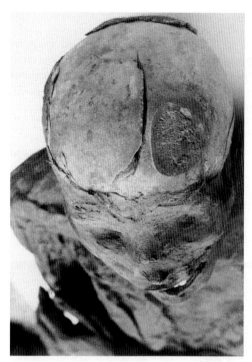

Fig. 8. Injuries at the head

	Johann Philipp is brought back to his castle Forstegg
7th of May 1596	Johann Philipp reports the course of events in a letter, written by himself, to the Assembly of the Swiss Confederation
12th of May 1596	Johann Philipp unexpectedly dies and is buried in the family vault at the church of Sennwald.

Palaeopathological examination

- There are two cuts on the skull that could not have been caused by a knife. Injuries of this kind lead to unconsciousness and marked perturbations; according to Professor Scheidegger it would have been impossible for the baron to write or even dictate a letter only three days after the incident.
- The stabs in the chest cannot be found, neither on the front nor on the rear side of the trunk.
- On the neck there are prominent bulges as they occur in death by hanging, or else through hanging up fresh corpses.

This technique of mummification consists on hanging a body in a dry, but windy place – usually a church-tower. The method has been explained to Professor Scheidegger by the poet Werner Bergengrün in a conversation: In the Baltic countries, the home of the poet, this method had been used up to the turn to the 20th century; the only disadvantage being a strong smell.

Our investigations about the identity of the Black Knife had to be stopped due to the wishes of the local authorities. Yet we do continue to collect further clues and will publish the results as soon as we have got enough material for one or the other hypothesis.

Neither in the case of the woman from the church of the Franciscan Friars nor in the one of the Black Knight of Sennwald had we been able to find the method of mummification. But at present we are convinced that both are cases of intended artificial mummification.

Summary

The art or trade of mummification in Middle and West Europe does not seem to have totally vanished at any time since the Early Middle Ages, although we only have a few references to it. The reason for this is the insufficient amount of research carried out.

In this essay a few mummified bodies, either artificially (or intentionally) from the area of German-speaking Switzerland are presented, with a concentration in the town of Basle. This is not particular astonishing, because Basle was a well-populated and wealthy town, and only a two-day march from Strassburg, which was probably a centre of embalming up to the modern times.

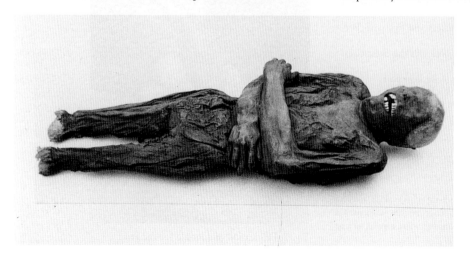

Fig. 9. Mummy of the Baron of Hohensax after restoration

The earliest example of mummification in Basle is the burial of Anna of Habsburg, Countess of Hohenberg and Haigerloch and wife of King Rudolf of Habsburg, who died in 1281 in Vienna, but she wanted to be buried in the cathedral of Basle. The mummification of the body, which took place in Vienna, was effective for over 200 years; at the first opening of her grave she was physically intact.

From the time of the Basle Council (1431–1449) the mummification of a cardinal is known; afterwards he was buried in the "Kartäuserkirche".

From bibliographical data we also know of the mummification of David Joris, an anabaptist, who lived in the 16th century and whose body – still intact – was exhumed 5 years after his death and – as a heretic – divided into four parts and cremated.

There is also the mummy of a grown up woman in the "Barfüsserkirche" from the time after the Reformation in Basle (1529). The body was preserved in a relatively good condition; her intestines were present. Probably the woman died of syphilis and before her death she received an overdose of mercury; we presume that this overdose of mercury influenced the conservation favourably.

At the present time I know of only two mummies outside of Basle. The body of a woman from Leuk/VS, whose body and clothing are in a bad state of preservation and are now being restored.

The most important and most thoroughly studied mummy, which has been examined best, may be the one of Johann Philipp of Hohensax, Baron of Sennwald (1550–1596), a very learned and accomplished commander-in-chief. The dead man was examined in 1979, and was partly restored and conserved. Since 1980 he rests in a little room of the cemetery chapel of Sennwald.

Zusammenfassung

Die Kunst oder das Handwerk des Mumifizierens scheint in Mittel- und Westeuropa seit dem Frühmittelalter nie ganz verloren gegangen zu sein, wenn wir auch nur wenige Hinweise darauf haben. Dies scheint eine Forschungslücke zu sein, wurde doch diesem Gebiet bisher kaum Aufmerksamkeit geschenkt.

In dieser Arbeit werden einige künstlich (absichtlich) mumifizierte Personen aus dem Gebiet der deutschsprachigen Schweiz vorgestellt, wobei sich ein Schwerpunkt im Bereich der Stadt Basel abzeichnet. Dies ist nicht weiter erstaunlich, da Basel eine bevölkerungsreiche und wohlhabende Stadt war und nur Zwei Tagesreisen von Straßburg entfernt war, das in der frühen Neuzeit vermutlich ein Zentrum des Einbalsamierens war.

Zeitlich finden wir in Basel als frühesten Beleg die Bestattung der Anna von Habsburg, geborene Gräfin von Hohenberg und Haigerloch. Die Gemahlin König Rudolfs von Habsburg starb 1281 in Wien, wollte aber im Basler Münster bestattet werden. Die in Wien vorgenommene Mumifizierung des Körpers hielt über 200 Jahre; bei der ersten Öffnung ihres Grabes war sie körperlich noch in unversehrtem Zustand erhalten.

Aus der Zeit des Basler Konzils (1431–1449) ist uns die Mumifizierung eines Kardinals überliefert; er wurde anschließend in der Kartäuserkirche bestattet. Ebenfalls nur aus der Literatur bekannt ist die Mumifizierung von David Joris, einem Wiedertäufer, der im 16. Jahrhundert lebte und dessen noch intakter Körper 5 Jahre nach seinem Tod ausgegraben und – als Häretiker – geviertteilt und verbrannt worden ist.

Ebenfalls aus der Zeit nach der Reformation in Basel (1529) ist die Mumie einer erwachsenen Frau aus der Barfüsserkirche. Der Leib war relativ gut und vollständig erhalten; auch lagen noch alle Eingeweide vor. Die Frau war vermutlich an Syphilis gestorben und hatte vorher eine starke Dosis Quecksilber bekommen; wir vermuten, daß dieser starke Quecksilbergehalt die Erhaltung günstig beeinflußt hat.

Außerhalb Basels sind mir zur Zeit nur 2 Mumien bekannt:

Der Leib einer Frau aus Leuk VS, deren Körper und Kleidung in einem schlechten Erhaltungszustand vorliegen und die zur Zeit restauriert wird.

Die wichtigste und bestuntersuchte Mumie dürfte aber die von Johann Philipp von Hohensax, Freiherr zu Sennwald (1550–1596) sein, einem sehr gelehrten und tüchtigen Heerführer. Der Tote wurde 1979 untersucht, teilweise ergänzt und konserviert. Seit 1980 liegt er in einem kleinen Raum in der Friedhofskapelle zu Sennwald.

Résumé

En Europe Centrale et Occidentale, l'art ou la profession d'embaumer les morts n'a pas complètement disparu; nous n'en avons que quelques preuves. Nous attirons votre attention sur le manque de recherches dans ce domaine scientifique jusqu'à ce jour.

Dans cette étude, on présente quelques corps momifiés, conservés artificiellement, de la région Suisse Allemande. Puisque Bâle, ..., a fournie la plupart des momies n'a rien d'étonnant car elle se trouvait à deux jours de voyage de Strassbourg qui, jadis, était vraisemblablement considérée un grand centre dans ce secteur professionnel.

Les premiers signes de cette pratique à Bâle remontent à l'enhumation d'Anne de Habsbourg, née comtesse de Hohenburg et de Haigerloch. L'épouse du roi Rodolphe de Habsbourg mourut à Vienne, en 1281, mais voulut se faire ensevelir dans la cathédrale de Bâle. L'embaumement de son corps, réalisé à Vienne, se conserva plus de deux cents ans et apparut en bon état, sans dommage, lors de la première ouverture du tombeau.

A l'époque du Concile de Bâle (1431–1449), selon la tradition on embauma le corps d'un cardinal qui fût ensuite enhumé dans l'église des Chartreux.

Grâce à la littérature, on connaît le cas de David Joris, qui fût momifié; cet anabaptiste vécut au 16e siècle et son corps se conserva encore intact 5 ans après sa mort jusqu'à son exhumation: En tant qu'hérétique, il fût écartelé, puis brulé.

A Bâle, on a découvert la momie d'une femme adulte, datant également du temps de la Réforme. Son corps était resté relativement bien conservé et entier avec les intestins intacts. Selon toute apparence, cette femme fût emportée par la syphilis, ayant absorbé auparavant une forte dose de mercure, laquelle favorisa probablement le bon état et la conservation de la momie.

De cette époque, en dehors de Bâle, je ne peux mentionner, à ma connaissance, que deux momies:

La momie d'une femme embaumée dans son costume, trouvés en très mauvaises conditions, à Leuk en Valais. On effectue le travail de restauration.

La momie la plus connue et la mieux analysée pourrait être celle de Jean Philippe de Hohensax, seigneur de Sennwald (1550–1596), savant et capitaine de haut mérite. 1979 on a analysé ses restes mortels; on les a partiellement reconstitués et conservés. Cette momie repose dans une petite chapelle du cimetière de Sennwald depuis 1980.

Riassunto

In Europa Centrale e Occidentale, l'arte o la professione d'imbalsamare i morti non è completamente sparita; noi ne abbiamo soltanto qualche prova. – Noi attiriamo la vostra attenzione sul assenza di ricerche in questo settore scientifico fino ad oggi.

In questo studio, si presenta qualche corpo mummificato, conservati artificialmente nella regione Svizzera tedesca, anzi tutto a Basilea, agglomerato molto popolare e prospero, questo non è sorprendente, poichè Basilea si trovava a due giorni di viaggio da Strasburgo che in quei tempi era verosimilmente considerata un grande centro specializzato in quel settore professionale. I primi segni di questa attività a Basilea rimontano al inumazione d'Anna di Asburgo, nata contessa di Hohenburg e Haigerloch. La sposa del re Rodolfo di Asburgo morta a Vienna nel 1281, volle farsi seppellire nella cattedrale di Basilea. L'imbalsamento del suo corpo, realizzato a Vienna, si conservò più di duecento anni e la mummia apparve in buon stato, senza danni, nel corso della prima apertura del sepolcro.

Nel periodo del Concilio di Basilea (1431–1449), secondo la tradizione si imbalsamò il corpo d'un Cardinale che fù in seguito inumato nella Chiesa dei Certosini.

Grazie alla letteratura si conosce il caso di David Joris, il mummificato, questo anabattista visse nel 16 secolo e il sui corpo si conservò ancora intatto 5 anni dopo la sua morte, fino alla sua esumazione. In quanto eretico, fù tagliato in quattro e poi bruciato.

A Basilea, si è scoperto la mummia d'una donna adulta, datata ugualmente del tempo della Riforma. Il suo corpo era relativamente ben conservato, intero con gl'intestini intatti. Secondo tutte le apparenze, questa donna morì a causa della sifilide; avendo assorbito prima una forte dose di mercurio, il quale probabilmente favorì il buon stato di conservazione della salma.

In quel epoca, fuori Basilea, io non posso che citare, a mia conoscenza, che due mummie:

Una donna imbalsamata nel suo costume, trovata in un cattivo stato, a Leuk (Vallese). Si effetua un lavoro di restauro.

La mummia più conosciuta e la meglio analizzata potrebbe essere quella di Jean Filippo di Hohensax, signore di Sennwald (1550–1596), erudito e capitano di altissimi meriti. 1979, si sono analizzati i resti mortali; sono stati parzialmente ricostituiti e conservati. Questa mummia riposa in una piccola sala, situata nella cappella del cimitero di Sennwald dal 1980.

References

Meier, E. A.: Basler Almanach. 2 Bde. Basel 1988.
Wurstisen, Chr.: Basler Chronik, Original Basel 1580, 665 S. (reprint Genf 1978).

Correspondence: Dr. Bruno Kaufmann, Anthropologisches Forschungsinstitut, St. Jakobstrasse 30, CH-4147 Aesch/BI, Switzerland.

The corpse from the Porchabella-glacier in the Grisons, Switzerland (community of Bergün)

B. Kaufmann

Anthropologisches Forschungsinstitut, Aesch/BL, Switzerland

Introduction

In the morning of the first of September 1988, at 8:30, Rudolf Käser, the keeper of the Kesch mountain hut (belonging to the Swiss Alpine Club) called the district police at Filisur: on the Porchabella-glacier some human remains and objects of mountaineering-equipment had been found (Figs. 1, 2). The skeletal remains together with the objects had been discovered and handed over to the hut-keeper on the twenty third of August at about 8 o'clock in the evening by Marco Sommerau from Latsch, a small village in the Grisons at the entrance to the Tuor Valley. As in the meantime it had started to snow, the tracks had been covered up until the end of August. On the same day the police officer J. Zippert set off for the mountains, accompanied by Peter Raffainer from Bergün, the chief of the rescue squad of the Swiss Alpine Club. At the Kesch-hut, the wife of the keeper led them to the site of discovery. The police officer wrote in his report from the 8th of October 1988:

"These remains are human without doubt, and they have been photographed in their original position; I refer to the adjoining sheet of pictures (Fig. 3). Afterwards they were collected and taken away. The other day they were given for further examination to the police officer Giger from the department of criminal investigation of the canton of Grisons. There were no objects that would have allowed a reliable determination of its age or even an identification.

But it can be assumed that these remains are very old and come from a male person.

It cannot be excluded that in due time the glacier will expose further remains of the body. For future comparison copies of the photographs have been left with the records at the police station."

According to the police report one had found:
– two large human bones, presumably tibia and fibula
– some small bones, possibly from a foot
– one hat, brown, without any special characteristics
– parts from a pair of leather shoes
– one wooden bowl, 16.5 cm in diameter, no special characteristics
– several remains of cloth.

The ensuing investigations in the civil records and the questioning of elderly people from Bergün and of the district officer from Samedan in the Engadin, the other side of the glacier valley, yielded no indications of any missing persons. The police report was sent on to Chur and the investigations were temporarily suspended. The remains stayed at the station of the district police at Chur; the archeology department of Grisons was not notified.

Nearly four years later, on the 18th of August 1992 at 9:20 a. m., the Archeological Department of the Grisons was informed, that a human body had been found on the Piz Kesch, and one had also informed the keeper of the Kesch-hut. In the meantime the public had become sensitive for such finds through the discovery of the Iceman. The local archeologist, Dr. Jürg Rageth, immediately contacted Mrs Käser, the wife of the keeper of the Kesch-hut, who in turn informed him of the findings in 1988. In the meantime several objects had been handed in at the hut:

– a wooden comb
– several wooden beads and further objects.

The following day, the 19th of August, Dr. Rageth visited the police station and received the stored objects as well as a copy of the police report. He also notified the anthropologist who announced his visit for the following day.

At the same time two employees of the Archeological Service, Mr Alois Defuns and Mr Sandro Lazzeri, climbed the rather steep path towards the Kesch-hut to contact Mr and Mrs Käser, the keepers of the hut. There they also met the former district forester, Fortunat Juvalt from Bergün: for years he had explored the Porchabella-glacier, and presumably he is the best expert of the area. Together they investigated the breaking end-piece of the glacier and after some looking around they found the site with the remains of the body. It was situated around 30 metres above the brim of the glacier and still contained parts of bones and rests of textiles. On an area of about twenty to three metres

Fig. 1. Topography of the site (LK 1: 25 000, Albulapass [1237], Bundesamt f. Landestopographie 1991)

one found the following items from below (North) to above (South): some finger or toe bones, in the middle of the site some ribs and vertebras, in the upper part the hip-bones, large pieces of cloth and a wooden comb, and after two metres the remains of the skull. Before collecting the remains the site was documented with drawings and photographs. The remains could be recovered rather easily with the exception of a femur: it had to be cut out from the ice within a block of about fifty to seventy to thirty centimetres. Finally one collected the various small cloth rests that were scattered around as well as some hair and parts of muscle and skin. Within the skull there were still some whitish parts of the brain.

The skeletal remains

On the 25th of August 1992 I went to Haldenstein where I saw the finds and learned about the circumstances. After initial discussion I received the human remains for further examination and for conservation:

- the objects from 1988: two teeth, one tibia and one fibula, four finger or toe bones
- two plastic containers with hair and flesh, cooled in ice water
- the skeletal remains that had been found the day before
- a tuft of hair, washed

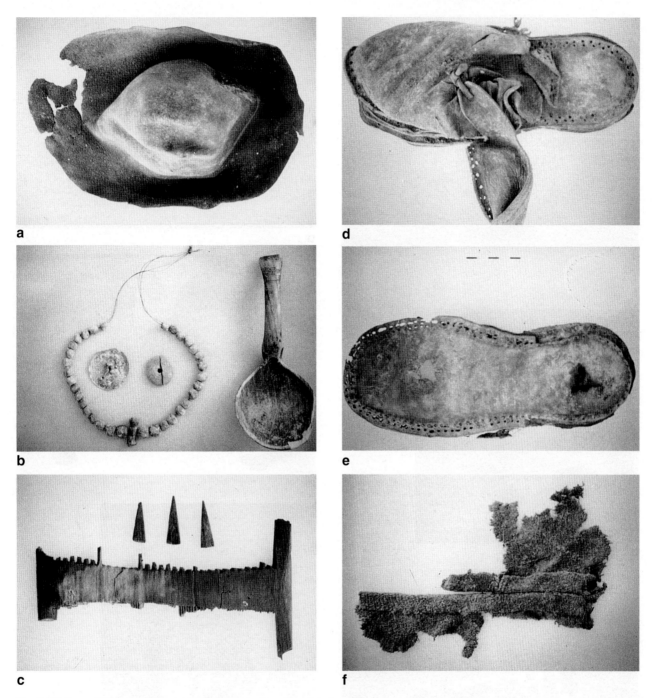

Fig. 2. Objects of the equipment. (a) Hat, (b) chain from the neck, buttons, wooden spoon, (c) comb, (d) shoe, (e) shoe, (f) remains of clothing

– a small plastic box with remains of the brain, still cooled in glacial ice.

The bones were so soft that they bent very easily. After an initial examination at the Institute for Anthropological Research in Aesch I put them into our storehouse at Eiken AG for drying in a relatively damp surrounding. The still frozen brain I put into the deep freezer (–18 °C) whereas I separated the soft body parts into hair, skin and muscles as well as textile remains.

Conservation

On the 29th of August I put the second group, the soft parts, into a solution with 10 % propanol which was exchanged daily:

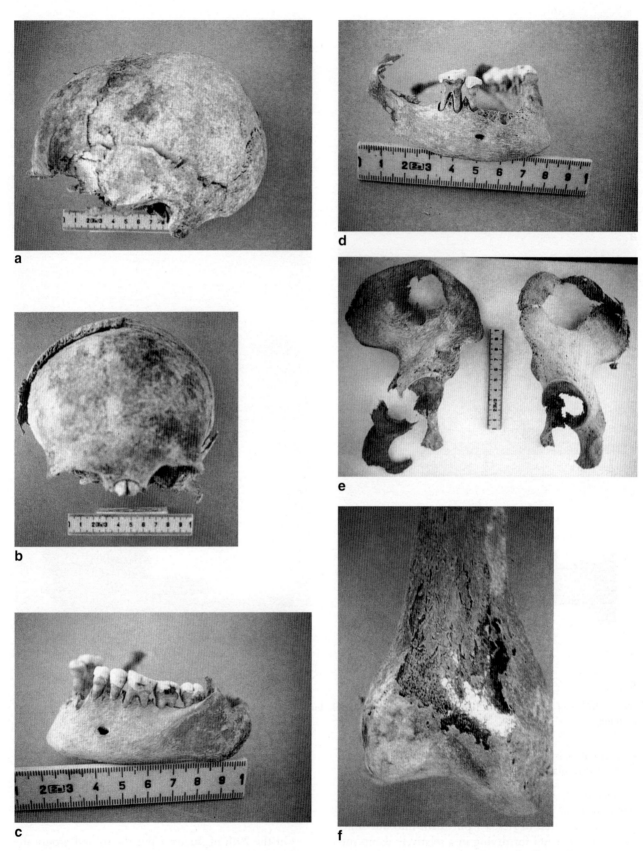

Fig. 3. Remains of the dead. (a) Skull, parietal view, (b) skull, frontal view, (c) mandible, (d) mandible, (e) hip bones, (f) femur with globules of bone-fat

- on the 7th of September they were put into a solution with 50 % propanol
- 8th of September: solution with 75 % propanol
- 10th of September: solution with 90 % propanol
- 11th of September: 100 % propanol
- 24th of September: 100 % propanol
- until the end of October the bath of 100 % propanol was changed weekly
  (Propanol: 2-propanol pro Analysi, Merck Nr. 9634.6010; CH3CH(OH)CH3)

The textile rests and the hair were removed from the propanol bath at the end of October and slowly dried. The fabric was examined for characteristics of material and fabrication, and given back to the Archeological Service.

Bones

As mentioned above I had deposited the bones in our storehouse to let them dry slowly; I had decided for this procedure, because the bones found in 1988 apparently had not been treated in any way: they were just left to dry naturally and were in good condition.

At my first control I had noticed a strong cadaverous smell, but otherwise no negative signs.

A week later (9th of September) the bones were completely covered by a layer of white wax; again the smell had considerably increased. Thus I was forced to react quickly and took the bones to the laboratory in Aesch. They were still rather soft and flexible: the consistency had rather lessened. Therefore I omitted the first step and immediately put them into a solution with 50 % propanol. I changed now the bath weekly; not daily as I had done before (see Table 1).

The exchange of propanol usually took place on Saturdays as I wanted to spare my assistants that foul smell: despite ample ventilation in a closed system it was still very strong.

On the 27th of October I took the bones from the propanol-bath and let them dry very slowly within a week to avoid fissures. In the meantime they had become firmer but they were still moderately flexible. On some of the bones I noticed the whitish and transparent periosteum which I removed to dry separately. At some protected areas while globules had appeared that I interpreted as a remainder of bone-fat; they could easily be removed with pincers.

Table 1. Conservation of bones

– 7th of September	50 % propanol	
– 17th of September	75 % propanol	bones soft, foul smell
– 24th of September	90 % propanol	bones soft, no smell
– 4th of October	100 % propanol	bones soft, the smell had returned
– 11th, 18th and 25th of October: weekly change of propanol		

Anthropological interpretation

After the drying process the bones could be catalogued.

Skull

- Calvaria with few defects at the base and in incomplete os temporale dexter
- a small fragment of the upper alveolar ridge
- 6 teeth from the upper jaw: one first incisor, one second incisor, two canines, one first premolar (with two root ends), isolated root of another premolar
- mandible with incomplete rami on both sides and a defect in the area of the frontal teeth
- 7 teeth of the lower jaw: one first premolar, but second premolars, two first molars, second and third molar of the left side

Trunk

- fragments from 6 cervical, 11 thoracic and 2 lumbar vertebras; remains of the first, second and fifth sacral vertebras
- 10 fragments of the rib cage
- the left clavicula is incomplete while the right one is missing
- both shoulder blades are incomplete, but the left one is in a better state
- the breastbone is missing
- large parts of the hip-bones

Arms

- left humerus (the right one is missing)
- the left radius is practically complete while the right one consists mainly of the shaft
- both ulnas have badly damaged or missing articular ends
- hands: 3 carpals, 7 metacarpals, 7 upper, 2 middle and 2 lower phalanges

Legs

- the left femur is practically complete while the right one is missing
- none of the patellas had been found
- the right tibia is complete, the left one is missing
- both fibulas are complete
- feet: 1 tarsal bone, 3 metatarsals and 2 upper phalanges

Soft parts

I assumed that the small whitish lump in the skull was a rest of the brain. It had been brought to our institute in its still frozen state and is still in the deep-freezer (–18 °C). These parts haven't yet been analysed.

Hair

The textile remains as well as the soft tissues were tightly entangled with the hair of the head; separation before conservation in alcohol was therefore not possible. The treatment did, however, discolour lightly the hair: at first it was darker, a kind of auburn, and now it had changed to a light brown. Besides the hair of the head there seem to have been preserved some rests of the hair in the armpits; pubic hair has not been found (yet?).

The hair has been examined microscopically, but a thorough analysis is still waiting. At magnification with the factors 16 and 40 one could distinguish rests of the scalp, some dandruff and rarely some nits of lice. There were also some small particles of cloth and of the felt hat.

Soft tissues

All the soft parts seem to come from the trunk and the upper arm. Some objects gave the impression of muscle-fibres, but partly also of larger vascular structures. The analysis will be carried out by an anatomist. In the meantime these remains will stay in a bath of 100 % di-propanol that will be changed every four months.

Postmortal deformations

In order to study the processes of decomposition the skull and the bones were x-rayed entirely.

The upper face is mostly missing while the skull and the mandible are in relatively good condition; on the latter only the rami are really damaged. Also the vertebras show differing grades of deformation: the best condition is found in some cervical vertebras, including the dens axis, and in the lower lumbar area; the worst condition show the thoracic vertebras. Some of the ribs are practically complete while the majority of them is missing altogether. Some ribs are considerably decalcified. The left side of the trunk with the clavicle and the shoulder blade is well preserved while the right side is in poor condition; the breast bone is missing.

Also the left arm is pretty much complete, and the right one, again, consists only of few remains.

The pelvic bones are quite well preserved expect for some dissolved sections on both iliums. The right femur shows a separation of the upper joint; the left femur and the patellas are missing. The left tibia and fibula (found in 1988) had been dried naturally, and there are still some muscle-fibres clinging to the shafts; yet the x-rays show no difference to the right fibula which has been conserved through alcohol. Also the small bones of the hands and feet do not display any peculiarities.

In general it can be noticed that:
– the left side of the skeleton is better preserved than the right side; this possibly indicates that the body had been resting sideways and not on the back
– the joints of the long bones are often separated in the area of the epiphyseal joints; it can thus be considered that the ossification has not yet been completely terminated.

Age of death

To determine the age of death we applied the methods common in physical anthropology. As the person is relatively young, one can regard the results as pretty reliable.

Skull

– all sutures are still open, except for the spheno-basilar-joint: 20 to 35 years
– the roots of the third molars of the mandible are still open: approximately 15 to 25 years
– the teeth show no abrasion except for the canines and the first and second molars: approximately 15 to 25 years

Postcranial bones

– all the epiphyses are closed, but the joints can still be observed; the separations in this area are secondary: approximately 18 to 25 years
– the upper ridge of the ilium is ossified onto the main bone: at least 21 years

Thus we have a span between 18 and 25 years; especially the spheno-basilar-joint and the ridge of the ilium point to an age above 20 years. The age can be determined as being approximately between 20 and 23 years.

Sexual dimorphism

For the determination of the gender we also took the methods that are used in Western Europe, according to the recommendations by Ferembach et al. (1979). All cranial characteristics were well within the female range. On the pelvis the sulcus praeauricularis was missing – this trait is usually only found in males. But the remaining traits were again unmistakably female; the pubic angle could not be observed, as the pubic bones were incomplete and quite deformed postmortally. The overall criteria and the sexual dimorphism (skull –1.3, n=12; pelvis –0.7, n=6) plainly indicate a woman; this is underlined by the slender bone formation.

Questions of dating

Only two references can be used to solve the question of the time of death: neither the situation in the glacier nor any objects like coins or parts of the equipment allow an exact dating.

Second dentition: the old sequence of tooth eruption

Since Roman times in Switzerland we can observe a change in sequence of the erupting teeth of the second dentition; the second molar is the decisive factor. In the old sequence the second molar follows the first one, while in the new one the second molar is always the penultimate tooth to come forth (the sequence of the incisors has only very recently been obeserved in a few cases). This change of sequences has been mainly researched by the late professor Roland Bay from the dentistry department of Basel University, and it shows typical patterns that even allow an approximate dating in larger burial sites (see Table 2).

These numbers are based on the examination of around 1,000 cases from Roman and post-Roman burials; I do not know whether these results could be applied to areas other than Switzerland. The data on the new incisor-sequence is based on private observations that have not yet been published (M. Dokladal, Brno and my own observations); it seems that the incidence of this latest variant is retarded for about 30 years in comparison to the development in Eastern Middle Europe (Czechia, Slovakia, Hungary).

As the body from Porchabella shows the old sequence we can approximately date into Late Middle Ages or Early Modern Period. I have to point to the fact that from around 300 observable individuals from the Early Modern Age only one still showed the old sequence. Yet this alone would not allow for any exact dating.

Table 2. Dentition sequences in Switzerland

period		old sequence	new sequence	I-sequence
Roman	– 5th century	100 %	rare	0 %
Early Middle Ages	– 8th century	75 %	25 %	0 %
High Middle Ages	– 12th century	50 %	50 %	0 %
Late Middle Ages	– 16th century	25 %	75 %	0 %
Modern Ages	from 16th cent.	0 %	100 %	0 %
Present Time	1950–	1–2 %	~ 99 %	0 %
Present Time	1980–	1–2 %	~ 98 %	1–2 %

Shoes

The second clue are the shoes of the dead woman. One of them was still relatively complete, even if the soles were quite worn, with a hole in it. The Service of Archeology had sent them before conservation to an expert for historical footwear at the Bally-shoe-museum at Schönenwerd AG.

According to his preliminary report they are Austrian shoes from the 16th century. He had evaluated the fabrication of the soles and of the thread, the seams and the embellishments. The final report still has to wait, as the shoes are now at the Schweizerische Landesmuseum in Zürich for conservation.

The location in the glacier

As everywhere in the Alps also the glaciers in the Grisons are receding since several decades. This process even has accelerated in the last six years, as the growth is being stopped due to minimal snowfall above 3,000 metres (information from Mr F. Juvalt). The actual recession at the tip of the Porchabella glacier is about 10 m per year. The site where the body was found in 1992 is now already beyond the reach of the glacier. In 1935 the thickness of the ice at this point was still over 30 metres.

Where did the body get into the glacier?

We do not know yet whether that young woman was on her way from Bergün to the Engadin – to Madulain or Zuoz – or whether she had chosen the path from the Engadin to Bergün. Yet it is certain that she met death on the side of Bergün. As she was found in the part of the glacier leading to Bergün (the nowadays isolated Northern part of the glacier is leading into the Engadin) she most probably chose, or had intended to choose, the passage near the Porta d'Es-Cha (3,008 metres). The exact circumstances cannot be reconstructed anymore.

On the assumption that the woman died near the passage, her body got transported about 1,800 metres; the difference in altitude being nearly 400 metres. At the present running speed this would mean a time span of 200 to 300 years at the most – one of two centuries less than the dating according to the sequence of dentition and the shoes. Thus the body had been held up at some point or the body is younger than we assumed.

It might be interesting to look into the name of the glacier. It is first mentioned with this name in 1540; the language is late Romansh and means "nice pig". In everyday use this designates a prostitute. This is surely an uncommon name for a mountain and a glacier.

Acknowledgements

I want to thank several people and institutions for various informations:

- Bündner Kantonspolizei Chur and Filisur
- Mr Käser and his assistants from Keschhütte of the SAC
- Mr Kropf from the Bally-Schuhmuseum Schönenwerd
- Mr Fortunat Juvalt, Bergün
- Prof. Dr. M. Dokladal, University of Brno, Czechia

Special thanks go to the "Archäologischer Dienst Graubünden" for their permit to investigate and publish the case, as well as to the "Institut für Alpine Vorzeit, Innsbruck" who gave us the possibility to present the find. My thanks also go to the staff of our institute who helped me in many ways.

Summary

In 1988 and 1992 the remains of a woman and her equipment were found in the ice of the "Porchabella-glacier" (community of Bergün, canton Grisons, Switzerland).

In contrast to the well-preserved objects of her equipment only single parts of the soft body parts (mainly shoulder and left arm) were preserved, whereas the skeletal remains were almost completely intact.

We concluded the age of the woman to be nearly 22 after the anthropological analysis. As far as her clothing and equipment are concerned an Austrian origins can be postulated.

She may have lived in the 16th or 17th century, because of the old sequence of tooth eruption (order of dentition: M1, M2, I1, I2), and this is also confirmed by her type of shoes.

From the movement of the glacier this dating seems to be reliable, even though we cannot exclude the possibility that the body is little more than 200 years old.

We will expect more detailed information after the final examinations have been completed in the course of 1994.

Zusammenfassung

In den Jahren 1988 und 1992 wurden im Eis des Porchabella-Gletschers (Gemeinde Bergün, Kanton Graubünden, Schweiz) Reste einer Frau und ihrer Ausrüstung gefunden. Im Gegensatz zu den gut erhaltenen Ausrüstungsgegenständen waren von den Weichteilen aber nur einzelne Partien (vor allem Schulter und linker Arm) erhalten, während die Skelettreste relativ vollständig vorlagen.

Die anthropologischen Untersuchungen ließen auf eine etwa 22 Jahre alte Frau schließen. Anhand ihrer Kleidung und Ausrüstung wird eine Herkunft aus Österreich postuliert. Zeitlich dürfte sie im 16. oder 17. Jahrhundert gelebt haben, wie aus der noch alten Durchbruchsfolge der Zähne hervorgeht und auch vom Schuhtyp her bestätigt wird. Anhand des Gletschertransportes kann diese Datierung stimmen; im ungünstigsten Falle wäre allerdings auch ein Alter von nur etwas über 200 Jahren möglich. Genauere Angaben erwarten wir von der endgültigen Bearbeitung, die noch 1994 erfolgen soll.

Résumé

En 1988 et en l992, on a découvert, ensevelis dans le glacier de Porchabella (Commune de Bergün, située dans les Grisons, Suisse) les restes mortels d' une femme ainsi que des effets lui appartenant.

Son équipement et les ossements étaient relativement en bon état de conservation, surtout l' épaule et le bras gauche, contrairement à certains tissus.

Les analyses anthropologiques ont demontré qu' il s'agissait d'une femme d'environ 22 ans. On peut déduire d' après ses vêtements et ses affaires personelles qu' elle venait d'Autriche.

Elle a vraisemblablement vécu au 16ème siècle ou au 17ème, comme semblent indiquer à la fois le perçage des dents et le genre des chaussures.

En disposant des données techniques concernant le trajet parcouru dans le cas que ce corps était étendu au sommet du glacier, dans l'hypothèse la plus avantageuse, on peut retenir une période de séjour d' environ 400 ans; dans le pire des cas, on peut l'éstimer à 200 ans.

L'étude définitive, s'achevant encore en l994, nous apportera des informations plus précises.

Riassunto

Nel 1988 e l992 si è scoperto seppellita nei ghiacci del ghiacciaio di Porchabella (Commune di Bergun, situato nei Grigioni, Svizzera) i resti mortali appartenenti ad una donna, come pure qualche effetti personali.

Questi effetti e le sue ossa erano relativamente ben conservate; soprattutto la spalla ed il braccio sinistro; contrariamente a certi tessuti.

Le analisi anthropologiche hanno dimostrato che si trattava di una donne di 22 anni. Si può dedurre dai suoi vestiti ma soprattutto dalle sue scarpe, ch'esse veniva dal Austria.

Sembra aver vissuto probabilmente nel 16 secolo, oppure nei diciasettesimo. Questo fatto sembra indicato, sia della sequenza di eruzione dei denti che dalle scarpe utilizzate.

Secondo i dati tecnici disponibili concerenti il tragetto percorso da quel corpo nel caso ch'esso steso sulla cima fosse; si può ritenere il tempo del suo soggiorno, in circa in 400 anni. Nel peggiore dei casi, si può stimare in 200 anini.

Lo studio definitivo, terminando nel 1994, ci portera delle informazioni più precise.

References

Bay, R. (1958), Das Gebiß des Neanderthalers, in: Hundert Jahre Neanderthaler, Böhlau-Verlag, Köln – Graz, p. 123–140 (=Beihefte der Bonner Jahrbücher, Bd. 7).

Ferembach, F. u. A. (1979), Empfehlungen für die Alters- und Geschlechtsdiagnose am Skelett. In: Homo 30, Anhang.

Correspondence: Dr. Bruno Kaufmann, Anthropologisches Forschungsinstitut, St. Jakobstrasse 30, CH-4147 Aesch, Switzerland.

Iceman research: current events

Iceman's last weeks

K. Spindler

Institut für Ur- und Frühgeschichte, Universität Innsbruck, Austria

We now have archaeological, medical and botanical information from which we can piece together the events of the last weeks in the life of the man from the glacier, that finally led to his lonely death in the rocky cleft on the Hauslabjoch.

From an archaeological point of view, the condition of the bow, the quiver and its contents are of utmost importance in this context.

The bow is made of yew-wood and approximately 1.82 m in length. The cross-section is in the form of a horse-shoe tapering towards both ends. The surface of the wood is completely covered with rough carving marks and examinations with the microscope have shown that there are no visible impressions at the two ends where the bowstrings could have been attached[1] (Fig. 1).

Thus, the bow differs greatly from other Neolithic and Bronze Age finds from Central Europe in general, but especially from bows from Alpine regions. Although these other finds are not very numerous, they are homogenous. I am referring here to the examples of the bows from Koldingen[2], from the Lötschenpaß[3] and from Thayngen-Weier[4], to which others can be added[5]. These other bows have, as the modern bow, a thickened grip in the middle of the bow which the archer holds. The shoulders of the bow are normally flattened at a right angle in the direction of shooting so that the bow bends well. Both ends are constructed to allow the attachment of the bow strings or so that the loops of the strings can be hung onto the ends. Moreover, the prehistoric bows known to us are generally well finished and the surfaces smoothed. None of these characteristics are to be found on the Hauslabjoch bow.

These observations lead us to the compelling conclusion that Iceman's bow was an unfinished implement. As it can be assumed that a man between 35 and 40 must have had other bows before this one, we can also assume that he must have either lost or broken such a bow.

The fact that the man's equipment included a piece of cord (Fig. 2a), which can almost certainly be regarded as a bowstring, also fits into the picture. The cord was kept with the pieces of unfinished arrows and other objects in the quiver[6]. This further indicates that the cord was an extremely important piece of equipment and that the man therefore kept it in a safe place.

X-ray pictures taken before the quiver was opened at the Roman Germanic Central Museum in Mainz showed the shape of what appeared to be a coil. When the quiver was opened an irregularly wound coil of 14.1 cm was revealed. The cord consisted of two threads. From the botanist's report we learn that the material used for the cord is bast, as opposed to all the other pieces of string amongst the find from the Hauslabjoch, which were all made out of intertwined pieces of grass. One end of the cord in question is knotted and at this point 7 mm

---

[1] Markus Egg, Zur Ausrüstung des Toten vom Hauslabjoch, Gem. Schnals (Südtirol). In: Frank Höpfel/Werner Platzer/Konrad Spindler (eds.), Der Mann im Eis, vol. 1. Veröffentlichungen der Universität Innsbruck 187 (1992) 254 plate 1.1–2.

[2] Klaus Beckhoff, Der Eibenbogen von Koldingen, Stadt Pattensen, Ldkr. Hannover. Nachrichten aus Niedersachsens Urgeschichte 46, 1977, 177–188.

[3] Werner Meyer, Der Sölder vom Theodulpaß und andere Gletscherfunde aus der Schweiz. In: Höpfel et al. (n. 1) 322 et seq. Fig. 1–2. – Werner Bellwald, Drei spätneolithisch/frühbronzezeitliche Pfeilbögen aus dem Gletschereis am Lötschenpaß. Archäologie der Schweiz 15, 1992, 166–171.

[4] Walter Ulrich Guyan, Bogen und Pfeil als Jagdwaffe im „Weier". In: Die ersten Bauern, vol. I. Pfahlbaufunde Europas. Forschungsberichte zur Ausstellung im Schweizerischen Landesmuseum Zürich (1990) 135–138.

[5] Klaus Beckhoff, Der Eibenbogen von Vrees. Die Kunde N. F. 15, 1964, 113–125. – John G. D. Clark, Neolithic Bows from Somerset, England, and the Prehistory of Archery in Northwestern Europe. Proceedings of the Prehistoric Society N. S. 29, 1963, 50–98. – Jürg Rageth, Der Lago di Ledro im Trentino und seine Beziehungen zu den alpinen und mitteleuropäischen Kulturen. Bericht der Römisch-Germanischen Kommission 55, 1974, 196 Fig. 13, plate 107.7–8. – Renato Perini, Scavi archeologici nella Zona palafitticola di Fiavè-Carera, vol. 2 (1987) 356 Fig. 174. – Renè Wyss, Wirtschaft und Technik. In: Ur- und frühgeschichtliche Archäologie der Schweiz, vol. 2. Die jüngere Steinzeit (1969) 127 Fig. 6.1–2.

[6] Egg (n. 1) 258 Fig. 4.4.

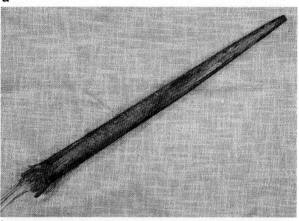

Fig. 1. (a) Bow from the Hauslabjoch, fragment, 1991 (photo by RGZM Mainz). (b) Bow from the Hauslabjoch, fragment, 1992 (photo by Hans Nothdurfter). (c) Detail of bow from the Hauslabjoch with carving marks (photo by Gerhard Sommer)

in diametre. At the other end the cord tapers to 3.5 mm in thickness. The full unwound length cannot be measured exactly, but is estimated to be between 1.9 and 2.1 m.

Nowadays we assume that bowstrings were made from animal sinews and the German term "Bogensehne" (bow sinew) implies this intrinsically. This assumption is further supported by Iceman in that two hamstrings from an animal of a size between that of a deer and that of a cow were also found among the equipment which he carried with him[7] (Fig. 2b). On the other hand, we have the ethnological evidence which proves that recent primitive peoples used and use vegetable fibrous materials as well as animal sinews for their bowstrings[8]. Moreover, the contemporary bow-maker Harm Paulsen used a twisted string of flax for a replica of a Late Neolithic-Early Bronze Age bow. Paulsen twined the string himself and offered no further explanation for his choice of material[9]. His choice was free as we do not possess proof from finds about the material of prehistoric bows.

There are few remaining doubts about the fact that the bundle of string found in Iceman's quiver was a bowstring. We may even go as far as to assume that it was the string from his old bow, which could still be used and which the man was intending to use again for his new bow.

The rough bow from the Hauslabjoch had obviously been first carved out from the trunk of a yew tree[10]. The finishing, that is to say, carving the basic piece of yew into a bow was then begun, but interrupted forever by the death of the Hauslabjoch man. In this context it is of importance to know how many hours' work had already been put into the making of the bow.

Rough data for the length of time needed to make a bow of yew wood have been provided by experimental archaeologists. The bow-maker Harm Paulsen made a replica of the bow from Koldingen as mentioned above[11]. However, as opposed to the case of the Hauslabjoch bow, Paulsen did not use split logs but used instead a thin trunk of a yew tree, whose diametre was just sufficient for carving out the bow. Thus, Paulsen saved himself a considerable amount of work as compared to the work involved in making the Hauslabjoch bow, for which split logs were cut from a larger tree. According to botanist's report, the Hauslabjoch bow must have been acquired from a trunk which was at least 20 cm in diameter. Therefore, the time necessary to fell a yew of this thickness, to cut the trunk to the desired length – in this case approximately 2 m – and the work involved in splitting the wood must be added to the working time calculated to make the bow. As yet no experiments with yew have been carried out, but there are data available for oak. Robert Pleyer took 40 minutes to fell a 25 cm thick

---

[7] Egg (n. 1) 258 Fig. 4.2.
[8] Walter Hirschberg/Alfred Janata (eds.), Technologie und Ergologie in der Völkerkunde. Ethnologische Paperbacks (³1986) 213.
[9] Harm Paulsen, Schußversuche mit einem Nachbau des Bogens von Koldingen, Ldkr. Hannover. In: Experimentelle Archäologie in Deutschland. Archäologische Mitteilungen aus Nordwestdeutschland, suppl. 4 (1990) 301 Fig. 1b, 4.
[10] Klaus Beckhoff, Eignung und Verwendung einheimischer Holzarten für prähistorische Pfeilbogen. Die Kunde N. F. 19, 1968, 85–101.
[11] Paulsen (n. 9).

oak tree using a stone axe with parallel shafts[12], Hermann Holsten and Kai Martens required 42 minutes to fell a 28 cm thick oak with a flintstone axe[13].

Approximately the same length of time is necessary when working with a bronze axe. Experiments with a copper axe, such as the one which was found with the Hauslabjoch man[14], have not as yet been carried out. However, we can assume that the time would be similar to that needed when working with a stone or a bronze axe.

It only took 15 minutes to cut the oak trunk to the desired length as it had been continually turned during the preparation so that it could be worked symmetrically and radially.

Split planks were also produced during these experiments and here Pleyer used an axe with a knee handle type of shaft with a blade made of stone, a so called "Schuhleistenkeil". He split the wood hacking at it from the front with short, hard blows. Then he split off the planks using wooden wedges. In this way he was able to produce planks which were 2 m in length, 25 cm in width and 5 cm in breadth[15]. Unfortunately he does not cite the time needed for this task. It can be assumed that not more than an hour was needed. On the other hand, he was using a straight oak trunk with no branches and freshly felled.

Paulsen worked the bow out of the log of the thin trunked yew tree, which he had chosen, using a stone axe. This took him three hours and 42 minutes. The contemporary bow-maker chose fragments of flintstone with straight edges for the finishing touches. Final polishing was not necessary as the paring and planing with flintstone produced surprisingly smooth surfaces. Another hour and 39 minutes were needed for finishing and completing the bow. This means that the complete working time is calculated at five hours and 21 minutes[16].

If we now transfer the results from the experimental archaeologists to the rough bow from the Hauslabjoch and add the additional work of making a bow from split logs, it becomes clear that a complete bow could be produced in a day at the most. The fact that a Neolithic man certainly had more practice working with these tools has not even been considered in the calculations.

The yew tree (*Taxus baccata*) grows in what is taken to be the environment of the Iceman, up to a height of approximately 1,600 m above sea-level. The place where he

Fig. 2. (a) Wound bast cord from the quiver from the Hauslabjoch (photo by RGZM Mainz). (b) Bundle of animal sinews from the quiver from the Hauslabjoch (photo by RGZM Mainz)

met his death lies at 3,210 m above sea-level, which means that there is at least 1,600 m difference between the level at which the yew tree grows and the Hauslabjoch. This distance can also be covered by a grown man with experience in the mountains in a day under normal circumstances. All things considered, we must estimate a time lapse of at least two days between acquisition of the wood for the new bow and the interruption of the work on the bow due to death.

It would appear justified to see the loss of the old bow in connection with events which caused damage to the quiver and its contents. The quiver consists of a narrow, rectangular leather bag which tapers slightly to the bottom and which is stiffened by a hazel switch on the side seam[17].

The quiver (Fig. 3) and contents were not recovered from the site until two days after the official recovery of the mummy. The recovery of the quiver was conducted with great care by Dr. Gernot Patzelt, Professor at the University of Innsbruck, and his team of glaciologists[18] (Fig. 4).

---

[12] Robert Pleyer, Holzarbeitung mit altneolithischem geschliffenem Steingerät. In: Experimentelle Archäologie (n. 9) 227–230.

[13] Hermann Holsten/Kai Martens, Die Axt im Walde – Versuche zur Holzbearbeitung mit Flint-, Bronze und Stahlwerkzeugen. In: Experimentelle Archäologie (n. 9) 231–243.

[14] Egg (n. 1) 260–262 Fig. 1, plate 6.

[15] Pleyer (n. 12) 230 Fig. 6.

[16] Paulsen (n. 9).

[17] Egg (n. 1) 254–258 plate 1.3, 2.

[18] Gernot Patzelt, Neues vom Ötztaler Eismann. Mitteilungen des Oesterreichischen Alpenvereins 47 (117) 1992, no. 2, 23 et seq. photo 3.

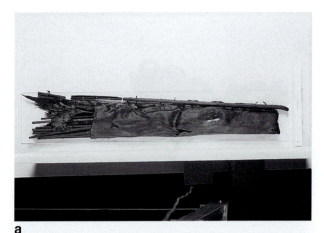

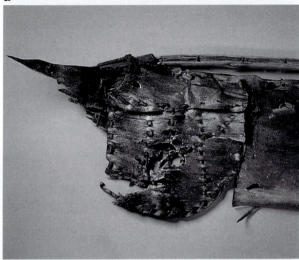

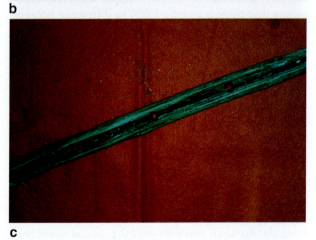

Fig. 3. (a) Quiver from the Hauslabjoch (photo by RGZM Mainz). (b) Quiver from the Hauslabjoch, detail of wing flap (photo by RGZM Mainz). (c) The piece of stiffening support of the quiver from the Hauslabjoch (photo by Roger Teissl).

Thus, it can be said with certainty that the state in which the quiver was found corresponds exactly to that in which it was placed in the cleft approximately 5,300 to 5,200 years ago. After the recovery of the quiver, it was discovered that it was considerably damaged and that this must have occurred in antiquity.

– The fastening lid, whose existence is proved by the remains of straps on the front of the quiver, is missing.

– The strap used for carrying the quiver, a broad band of leather on the back of the quiver, is torn at an angle above mouth of the quiver and is also missing.

– The front of the quiver ends with a semi-circular wing flap, which is hidden under the proper lid when the quiver is closed and which can be opened from the side. This is also torn off at an angle, but was found amongst the other pieces of equipment belonging to the glacial mummy. Due to the tumultuous circumstances of the recovery of the mummy, it is not possible to ascertain the exact location of this piece. However, it is almost certain that the part of the wing flap torn off in antiquity lay with the mummy itself at a distance of approximately 5 m from the quiver. It can, therefore, be assumed that Iceman kept this piece of the quiver on his body in order to repair the quiver when an opportunity arose.

– Similar observations have been made concerning the wooden supporting strut which was attached to side seams of the quiver with a fine leather strap. After the recovery of the find, it soon became apparent that the supporting strut was broken into three pieces. The full length of the strut was 92.2 cm and it was broken into a long bottom piece, a short middle piece (Fig. 3c) and a still shorter top piece. All three pieces fit together at the points of fracture and there are no pieces missing.

– As in the case of the wing flap, the middle piece of the strut was also found with the mummy itself and not with the quiver. It was thus discovered four days before the glaciologists recovered the quiver (Fig. 4)[19]. It is also true here that the damage must have taken place before the death of the man and that it was his intention to repair the strut. It is possible that the opposite end of the strap used for carrying the quiver was attached to the middle piece of the strut for which reason it had been detached from the seam of the quiver.

The archaeological analysis of the quiver shows that it had been badly damaged before the death of the man, that parts of the quiver – i. e. the fastening lid and the strap – had been lost and that other pieces – i. e. the torn off wing flap and the broken middle piece of the strut – were being kept to be repaired.

The fact that pieces of the quiver are missing indicates that the man could not find all the fragments of the quiver, whatever the circumstances were under which it was damaged. This leads to the conclusion that unusually violent events caused damage to the quiver and the loss of the bow.

The contents of the quiver were found in a similar state. The quiver only contained two arrows primed for

---

[19] Elisabeth Zissernig, Der Mann vom Hauslabjoch – Von der Entdeckung bis zur Bergung. In: Höpfel et al. (n. 1) 239.

shooting and twelve unfinished arrow shafts[20] (Fig. 5 a). It thus becomes apparent that the man had had to use a great number of arrows some time before his death and this in turn indicates a violent struggle with opponents.

Therefore, the man from the Hauslabjoch had to replace the arrows missing from his stock. For this purpose he chose the tough, long, straight shoots of the viburnum (*Viburnum lantana*)[21]. All in all he started to make twelve new arrows. He cut the shoots into lengths between 84.5 and 87.8 cm. Only one of these shafts is notably shorter, i. e. 69.2 cm in length, than the others. One end of this shaft is cut neatly and the other end appears to have been broken, as it is frayed. This is presumably a secondary point of fracture. The missing part was not found in the quiver, which means that the piece must have been broken off before the shaft was placed in the quiver. On the other hand, it is possible that the shortened shaft was intended for repairing a broken arrow, whose head (here a composite arrow) appeared usable (cf. Fig. 5 b). In this case the fibrous end of the shortened shaft would have had to be sharpened to a point or hollowed out so that the two shaft pieces could be fitted together. This would have explained the fact the one end of the shaft had not been cut straight.

The rough shafts had been stripped of bark and the small twigs also removed in the process. However, the careful smoothing of the completed arrows was not evident here. On the other hand, grooves, on average 2 cm deep, had been made at the thicker, bottom ends of the viburnum shoots to enable the arrowheads to be fitted. For this fine carving work the Iceman would have used that small laminated piece of flint which was found in the tool-set in his belt pocket[22] (Fig. 5 d).

When the contents were removed from the quiver in the workshops of the Roman Germanic Museum in Mainz, it was noted that the rough shafts were in good condition but that the two finished arrows were broken. In both cases the point of fracture was around the arrowhead. In the case of one of the arrows, the composite arrow, the breaking had been of such a violent nature as to break off the flintstone head and to cause the stem to protrude into the the front of the arrow shaft. There are other breaks in the shafts, in the one case rather in the middle and in the other case rather at the end of the shaft.

From these observations we can obviously make some conclusions about the chronology of events, for both the completed arrows must have been broken before the rough shafts were put into the quiver. Otherwise the

---

[20] Egg (n. 1) 254–258 plate 3.
[21] Sigmar Bortenschlager/Werner Kofler/Klaus Oeggl/Werner Schoch, Erste Ergebnisse der Auswertung der vegetabilischen Reste vom Hauslabjochfund. In: Höpfel et al. (n. 1) Fig. 3.
[22] Egg (n. 1) 258 plate 5.4.

Fig. 4. (a) Top of quiver during discovery on 25. September 1991 (photo by Gerhard Markl). (b) Quiver from the Hauslabjoch after exposure (photo by Gernot Patzelt)

rough shafts would have protected the arrows or have been broken themselves. Thus we can reconstruct the following chronology.

– The man's normal supply of arrows had been used down to the last two which were found in the quiver.

– The quiver with the rest of its contents was damaged in a violent way, as mentioned above, and this led to the last two arrows being broken. This event will be described as the "Disaster" in the following text. The loss or breakage of the old bow must have been connected with these events.

– The Iceman began an obviously chaotic retreat leaving behind him a number of pieces of equipment – the strap to which the quiver was attached and the lid of the quiver.

– As a consequence he made attempts to substitute the missing pieces of equipment.

– He found these within the area in which the yew and the viburnum grow. They both grow in what is accepted to be the man's environment, that is to say up to a height of 1,600 m.

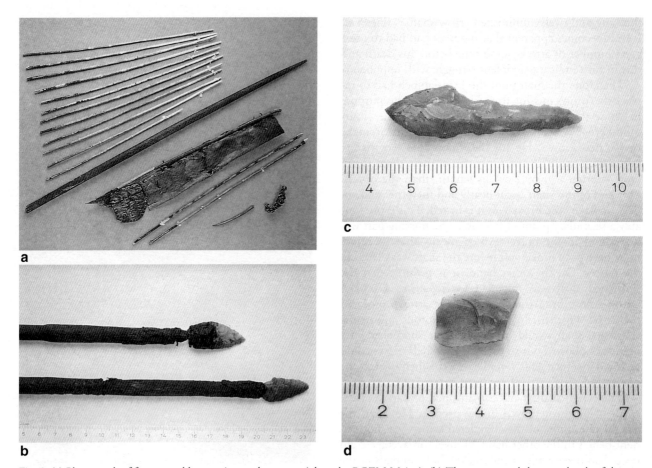

Fig. 5. (a) Photograph of fragmented bow, quiver and contents (photo by RGZM Mainz). (b) The area around the arrowheads of the two complete arrows from the Hauslabjoch; the lower arrow is a composite arrow (photo by RGZM Mainz). (c, d) Flintstone tools from the belt pocket of the man from the Hauslabjoch (photo by RGZM Mainz)

We also have figures provided by the experimental archaeologists for the length of time needed to make an arrow using prehistoric methods. According to their experiments it takes about two and a quarter hours to produce an arrow using Neolithic tools including the flintstone arrowhead, which had been dihedrally retouched, and the feathering[23]. Therefore, a man would need around 27 hours to make twelve arrows. These calculations are, however, of little use to us as the rough arrow shafts were only in the early stages of production. Admittedly, it would not have taken longer than a day to find suitable shoots from the viburnum in the woods, cut them to the required length, remove the bark and make the grooves to insert the arrowheads. Thus, the time span between the "disaster" and the death of the man in the chasm must have been at least three days, if we take the time into account which would be needed to collect the raw materials, to begin work on the arrows and to climb from the growth area of the yew and the viburnum up to the main Alpine ridge under normal circumstances.

We can, however, assume that the events surrounding the man's last days were not "normal", since he did not have any material for arrowheads on him, or at least not the usual supply.

From statistics calculated from finds, we can assume that the flintsone was the normal material for strengthening arrows in the Neolithic Age. On the other hand, there are some examples of arrowheads made of bone or antler on which jointing marks show without a doubt that they were used for the strengthening of arrows[24]. Here, however, it would appear that arrowheads of bone or antler were used when the normal raw material, i. e. flintsone, was not available in the required quantities. Josef Wininger has coined the German term "Ausweichmaterial" (alternative material) to explain this[25]. This being the case, the bundle of four pieces of antler, which the Iceman kept in his quiver together with

---

[23] Paulsen (n. 9) 301–303 Figs. 5, 8, 9.

[24] Josef Winiger, Beinerne Doppelspitzen aus dem Bieler See – Ihre Funktion und Geschichte. Jahrbuch der Schweizerischen Gesellschaft für Ur- und Frühgeschichte 75, 1992, 65–99.

[25] Winiger (n. 24) 92.

other tools, gains in importance[26]. These four pieces of antler are only roughly carved and are around 15 cm long, so that it would have been possible to make several arrowheads of the type from Lake Biel[27] (Fig. 6 b). The maximum number of arrowheads from each piece of antler would have been five, making a total of 20, since the Lake Biel arrowheads have a minumum length of approximately 3 cm.

We do not know for which purpose the pieces of antler were kept, but they certainly could have been kept as an alternative material for the strengthening of the man's arrows. Since flintstone does not naturally occur in the area anywhere around the Hauslabjoch – the Hauslabjoch is an area of Central Alpine primitive rock – this detail also points to irregularities in the man's ascent to the main Alpine Ridge.

One final archaeological aspect should be added. This concerns the lack of provisions. Only a portion of neck of what was probably dried ibex or rock goat meat[28] and one sloeberry *(Prunus spinosa)*[29] can be regarded as provisions.

Considering that the sloe (Fig. 7 c) ripens in September or October and only becomes edible after the first frost, the season of Iceman's death can be relatively reliably defined. Moreover, there are other indications which verify this[30]. The event occurred at the beginning of autumn, shortly before the first onslaught of winter snow, although it must be emphasized that during no month in the year is the region of the find completely certain to be free from snowfall. However, in the summer a possible covering of snow generally thaws when the weather improves.

The fact that the man ventured into the peak regions of the Alps without proper provisions at this time of year also points to unusual pressures.

The medical indications point even more convincingly than the archaeolgical indications to the fact that Iceman was in an extremely critical situation during his last journey.

An unusually high reduction in the body fat, which is responsible for the proper functioning of all the organs of the human body, is strinking. If it is not a case of post-mortal degeneration, then we might consider a period of starvation prior to decease in combination with a lack of reserves of fatty tissues as a cause of death[31].

A series of rib fractures on the right thorax, which had not healed, was decisive and certainly a cause of physical weakness[32]. The X-ray picture shows that the third to the sixth ribs are broken and that the points of fracture are slightly out of place without there being any signs of osseous healing. Since callus tissue begins to form on the ribs and can be diagnosed by X-ray around eight weeks after the occurrence of injury, the man from the Hauslabjoch cannot have sustained his injuries more than two months before his death.

Furthermore, the results from the X-rays show varying stages of calcification of the right and left humerus[33]. The right humerus, that is to say on the side of the injured thorax, is slightly decalcified in comparison to the left humerus, which displays normality. The roentgenologist sees this as a demonstration of inactivity, a phenomenon which manifests itself about two to three weeks after the injury. Therefore, the Iceman must have taken care of his right arm during the period after injury in order to keep the pain down to an endurable level and to accelerate the healing process. A series of fractures on the left thorax, which were well healed, demonstrate that he had had experience with rib fractures[34].

There is no good reason not to regard the physical injury and the loss and/or damage of parts the Iceman's equipment as intrinsically connected. The medical report implies that the "disaster" must have occurred about two months, at the most, before the death of the individual. The varied state of calcification of the humerus bones shows that the event must have taken place two to three weeks before decease. The man's will to survive is underlined by the fact that he strove to complete his equipment despite the fact that he was handicapped due to his injuries. From the archaeological point of view, the "disaster" must have occurred at least three days before the death of the man. If we now take the man's handicaps into account, it becomes clear that the length of time needed to start re-equipping himself, must have been very much longer than first surmised. His mobility and skill must have been greatly impeded. Thus, the information

---

[26] Egg (n. 1) 258 plate 4.1.
[27] Winiger (n. 24) 92.
[28] Angela von den Driesch/Joris Peters, Zur Ausrüstung des Mannes im Eis – Gegenstände und Knochenreste tierischer Herkunft. In: Konrad Spindler/Elisabeth Rastbichler-Zissernig/Harald Wilfing/Dieter zur Nedden/Hans Nothdurfter (eds.), Der Mann im Eis. Neue Funde und Ergebnisse. Veröffentlichungen des Forschungsinstitutes für Alpine Vorzeit der Universität Innsbruck (1995), pp. 59–66.
[29] Bortenschlager et al. (n. 21) 311.
[30] The report on the container, in which embers from ripe, freshly harvested Norway maple leaves were found, dates the event between the month of July and September: Sigmar Bortenschlager, orally. The lack of marks from bites from carrion feeders points to a continual covering of snow, which cannot occur before the end of September.

[31] Werner Platzer/Horst Seidler, orally.
[32] Dieter zur Nedden/Klaus Wicke/Rudolf Knapp/Horst Seidler/Harald Wilfing/Gerhard Weber/Konrad Spindler/William A. Murphy/Gertrud Hauser/Werner Platzer, New findings of the Tyrolean Iceman – archeological and CT-body analysis suggest personal disaster before death. Journal of Archaeological Science 21, 1994, 809–818.
[33] Dieter zur Nedden, personal communication.
[34] Spindler et al. (n. 32).

Fig. 6. (a) Bundle of four pieces of deer antler (photo by RGZM Mainz). (b) "Bone" arrowheads from various Swiss damp earth/soil settlements (from Winiger 1992; comp. note 24)

gained from varies sources fits together well. The Iceman must have survived the "disaster" for at least two weeks.

But what can have persuaded him to have ventured into such inhospitable regions shortly before the start of winter, with insufficient equipment and impeded by fractured bones? The answer is, of course, open to speculation. At the same time, there is apparently no question about the fact that the Iceman felt himself to be threatened following the "disaster", or, maybe, he was even "on the run". This situation would explain the incomplete state of his equipment in connection with his physical handicap.

In this connection, further indications about his activities during the last weeks of his life would be useful. It is, therefore, now necessary to turn our attention to the first results of botanical observations. Two grains of one-grained wheat (*Triticum monococcum*), still in the husks, were found in the hairs of his fur clothes[35] (Fig. 9). They were obviously not part of his provisions and must have got into his clothes unintentionally and transported up to the Hauslabjoch without the man's knowledge. Simi-

larly, fragments of corn from threshing were separated out from the insulating material of the ember container, one of the two birch bark vessels[36]. Here we have two fragments of husk and a rachis segment of one-grained wheat and a husk of *Triticum sp*.

This find implies that the man had been present during the threshing of corn within the last weeks of his life. On the one hand, this tells us that he was in contact with an agricultural community, or more exactly with a grain-growing community, and that he was present in this Late Neolithic village after the grain harvest, while the corn was being threshed. The insulating material in the ember container consisted of grass and fresh leaves, from which chlorophyll could still be extracted, and this must have been renewed not too long before his death. This shows that he had obviously not had the corn fragments from threshing on him for long[37]. They can only have become mixed with the insulating material by chance, and this process can have only occurred at the place where the threshing was carried out.

Thus, we can reconstruct one further event of considerable import which took place within the last weeks Iceman's life. Although the following order of events is not conclusive, it appears useful to list them in this way.

– The man is present in a community during the threshing of the grain harvest.
– The "disaster" occurs causing bodily injury and damage to equipment.
– The man flees in the direction of the main Alpine chain.

When possible a fugitive chooses a path, which should bring him to a particular goal, and where he has the advantage of knowing the terrain better than his pursuer.

The fact that the Iceman fled in the direction of the Hauslabjoch may appear incomprehensible at first glance, but it is in line with the facts which we have gained from the botanists.

In another paper, I have made a detailed attempt to show that the man's "native village" is most likely to have been south of the main Alpine chain, in an area favorable to settlements and known as the Val Venosta (Ger.: Vinschgau) region, in the Upper Adige Valley (Ger.: Etschtal). I use the term "native village" (Heimatdorf) to describe what is accepted to be the man's environment[38] (Fig. 10). Significant archaeological correlations between the Hauslabjoch find and Neolithic cultural phenomena of the Val Venosta and the Copper Age cultural area of Upper Italy, in this case especially the Remedello culture

---

[35] Bortenschlager et al. (n. 21) 310.

[36] Egg (n. 1) 266. – Konrad Spindler, Der Mann im Eis. Sandoz Bulletin 99, 1992, 27 with illustration. – Ibid., L'Homme du Glacier. Bulletin Sandoz 99, 1992, 27 with illustration.
[37] Bortenschlager et al. (n. 21) 311.
[38] Konrad Spindler, Der Mann im Eis (1993) 233 et seq.

Iceman's last weeks 257

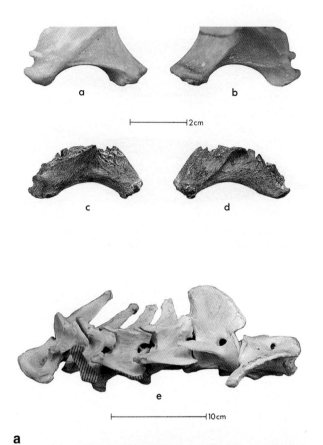

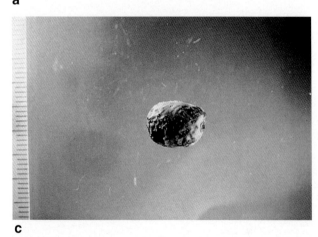

Fig. 7. (a) Cervical vertebra of ibex from the Hauslabjoch (photo by Inst. f. Paleoanatomie München). (b) Male ibex showing the position of the fragments of cervical vertebra found on the Hauslabjoch (from von den Driesch/Peters 1995; comp. note 28). (c) Sloe from the Hauslabjoch (photo by Inst. f. Botanik Innsbruck)

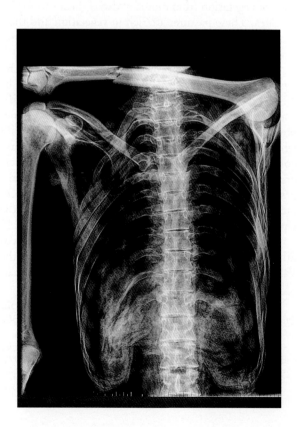

Fig. 8. X-ray of the thorax of the man from the Hauslabjoch showing the healed series of fractures of the left rib-cage and the unhealed series on the right (photo by Dieter zur Nedden)

of the northern Po Valley[39] (Fig. 11), are all-important for this theory.

Taking this human geographic situation into account, we then pose the question as to what could have caused a Neolithic inhabitant of the Val Venosta area to travel through the Val Senales (Ger.: Schnalstal) and over the passes of the main Alpine chain to the regions of the Upper Ötz Valley. There the tree-line runs at about 2,300 m above sea-level. Above the tree-line there is a broad belt of mountain pastures, which in turn gradually becomes a rocky area devoid of vegetation.

Since 1968 numerous pollen profiles from the Upper Ötz Valley have been analysed by the Department of Botany (under Dr. Sigmar Bortenschlager, Professor of the University of Innsbruck) in collaboration with the Department of Meteorology and Geophysics and the Department for Alpine Research of the University of Innsbruck. The analyses were carried out in connection with observations on the historical development of vegetation, glaciers and climate. During this project the pollen profile from the highest Alpine moorland on the Rofenberg at a height of 2,760 m above sea-level, which is exactly opposite the Hauslabjoch, was analysed[40].

The analyses show traces of anthropogenous influences in the regions above the tree-line since the end of the fifth millenium B. C., that is to say around one thousand years before the Iceman (Fig. 12). These influences can be observed from a clear change in structure of the herbal vegetation, to be exact from a change in the layer of vegetation from moors of dwarf shrubs to Alpine pastures. These pastures are rich in vegetation and there is a predominance of plants that indicate pasture, in this case especially the *Ligusticum mutellina*.

These observations prove that the high pasture regions of the Upper Ötz Valley had been cultivated by Neotithic man centuries before our Iceman. Due to the considerable climb from the settlements, transhumance obviously only occurred during the summer months[41], which for reasons of human geography could only have gone out from the Val Venosta region. The animals were driven from these regions over a number of days via the passes of the main Alpine chain (the Niederjoch, Hochjoch, Hauslabjoch, Tisenjoch) into the valleys of the Upper Ötz Valley (Rofental, Niedertal and Gurgler Tal).

The transhumance, which has its origins in the Neolithic Age, is still practised today[42]. The farmers from the Val Venosta region drive their sheep and few goats

Fig. 9. Grains of corn from the hair of the fur clothes of the man from the Hauslabjoch (photo by RGZM Mainz)

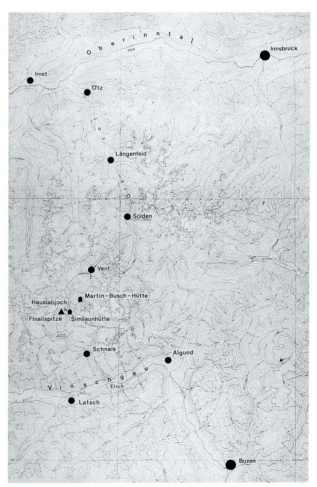

Fig. 10. Map of the mountainous terrain between the Upper Inn Valley and the Adige Valley showing important places mentioned in the text

---

[39] Raffaele C. de Marinis, La più antica metallurgia nell' Italia settentrionale. In: Höpfel et al. (n. 1) 389–409. – Spindler (n. 38).
[40] Bortenschlager et al. (n. 21).
[41] Wolfgang Dehn, „Transhumance" in der westlichen Späthallstattkultur? Archäologisches Korrespondenzblatt 2, 1972, 125–127.
[42] Hans Haid, Aufbruch in die Einsamkeit – 5000 Jahre Überleben in den Alpen (1992).

through the Val Senales and over the same passes into the Upper Ötz Valley, where they still possess rights of pasture.

In this connection one must point to the results of economic research from Neolithic Switzerland[43]. They show that the keeping of domestic animals varies according to the proximity to mountainous regions (in this case French and Swiss Jura). The proportion of small animals (sheep and goats) increases noticeably in comparison to that of larger animals, especially cattle, otherwise the most important animal in Neolithic agriculture[44]. Although there are no data determinating animal bones from Neolithic settlements, comparable data can be applied to the region of the Upper Adige Valley, which lies between the Ötz Valley Alps in the north and the massif of the Ortler in the south. Here, too, the keeping of sheep and goats would have been a characteristic feature in the economy of the prehistoric village. At the same time, this behaviour shows the significant social position of the herdsman.

If we consider our Iceman's entire equipment[45] (clothing, utensils, weapons, containers, therapeutic agents, lighters, back-carrier, substitute materials) under this aspect, we cannot avoid the impression that it had been put together excellently for intended longish periods of absence.

For these reasons, the opinion was often expressed during the early stages of interpretation of the find, that the man was an outlaw, who had lived alone, without contacts with his fellow humans. That this cannot have been the case is proved by the fragments of grain, which have already been mentioned and which prove his presence in a settlement during the preparations for winter. The state of his clothes indicates the same thing[46]. The clothes had obviously been made with great care, even skill. Moreover, the fact that we can observe skilled repairs alongside very amateur mending (Fig. 13) shows that more than one hand had worked on the clothes. At least one other person had made the clothes, whereas the amateur mending must have been done by Iceman himself at a time when he had been out of the village for a certain time. Finally, the tattoos[47] show that he was under some kind of medical care, for at least those on his back (Fig. 14a) cannot have been done by his own hand. There is certainly a connection between the marks, found especially on the joints (spine, knee, ankle) (Fig. 14b), and the arthritic condition of these joints shown by X-ray pictures[48].

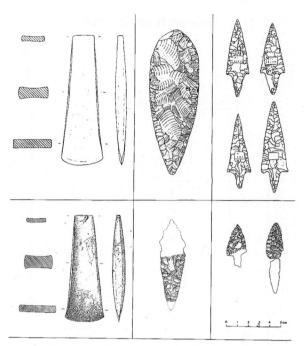

Fig. 11. Comparison of relevant pieces of equipment of the man from the Hauslabjoch (bottom) with inventory items from Remedello Sotto, Grave 102 (top)

All the aspects mentioned above indicate that the man from the Hauslabjoch was indeed intergrated in a community, but that he was equipped to spend periods outside the community. If we now take into account the economic model for the Neolithic inhabitants of the Val Venosta area, which has been described here in a very brief form, the idea that the man was a wandering herdsman becomes convincing. This means, and no part of the evidence is inconsistent with this theory, that our man had spent the summer before his death with herds of sheep and goats in the high pastures of the Upper Ötz Valley. He may have been with other herdsmen. In September he then gathered his animals together, looked for lost lambs and kids and began to drive them down to the valley, which would take several days.

Everything seems to indicate that he reached his native village in the Val Venosta area safely. The inhabitants were in the middle of their harvest, and at this time of year everyone in a farming community is extremely busy. It is certain that the fragments from threshing got into his fur clothes and into the insulating material of the ember-container at this point. One of the tasks of the Hauslabjoch man was doubtlessly to select the animals, well-fed at this time of year, for slaughter, to conserve the

---

[43] Jörg Schibler/Peter J. Suter, Jagd und Viehzucht im schweizerischen Neolithikum. In: Die ersten Bauern (n. 4) 91–103.

[44] Angela von den Driesch/Joris Peters/Marlies Stork, 7000 Jahre Nutztierhaltung in Bayern. In: Bauern in Bayern – Von den Anfängen bis zur Römerzeit. Katalog des Gäubodenmuseums Straubing, no. 19 (1992) 157–164.

[45] Egg (n. 1).

[46] Egg (n. 1) 266–268.

[47] Luigi Capasso/Arnaldo Capelli/Luigi Frati/Renato Mariani-Costantine, Notes on the Paleopathology of the Mummy from Hauslabjoch/Val Senales (Southern Tyrol, Italy). In: Höpfel et al. (n. 1) 209–213.

[48] Dieter zur Nedden/Klaus Wicke, Der Eismann aus der

Sicht der radiologischen und computertomographischen Daten. In: Höpfel et al. (n. 1) 133 et seq., 142. – Ibid., The Similaun Mummy as Observed from the Viewpoint of Radiological and CT Data. In: Frank Höpfel/Werner Platzer/Konrad Spindler (eds.), The Similaun Mummy, vol. 1. University of Innsbruck Publication 187 (1992) 3–19 (separate print).

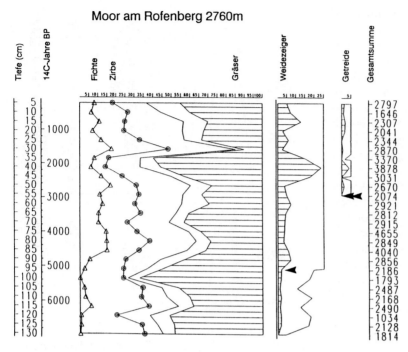

Fig. 12. Simplified pollen profile of the moor on the Rofen Mountain (from Bortenschlager et al., 1992; comp. note 24)

meat and to prepare the skins for tanning. It is possible that he was responsible for sheep-shearing, if sheep's wool was just used at this period. This question has not yet been definitively clarified in terms of domestic history[49]. In any case, there is no doubt about the fact that provisions for winter were made at this time of year and that Iceman was actively involved. Thus, further notable details about his last weeks emerge.

At some point during these days the incident, which we have termed as the "disaster", must have occurred. In any case there was a violent confrontation, which lead to the man being physically injured and to pieces of his equipment being abandoned or damaged. It is justified to theorize on the causes of the "disaster", but it is not possible to prove anything. During his absence of several months from his village, there may have been changes in the situation of his family. The "disaster" need not necessarily have been triggered off dramatically by jealousy; there may have been a confrontation in connection with the hierarchal order (a struggle for power). Another possible cause for his personal "disaster" might have been connected with his own misbehaviour, the loss of sheep or goats or an infringement of the communal rules.

On the other hand, we must bear one factor in mind which affected all village communities. It is certainly not by chance that numerous New Stone Age settlements are either fortified or in a position which can be easily defended. This shows that the inhabitants reckoned with threats. A particularly good booty would be expected shortly after the bringing-in of the harvest. The mass grave of Talheim in Württemberg from the band ceramics period, in which 34 heads were found buried, demonstrates that massacres of the entire population of villages occurred in the Neolithic Age. In this case children, fertile women and old men were violently murdered and heedlessly buried together[50]. As it was obviously the case that not even small children were spared, it would appear that the opponents would have been enraged that a grown man, even if injured, had escaped the pogrom. If we continue in this vein, it appears to me that certain suppositions are justified, which could contribute to the clarification of the fate of our Iceman.

In my opinion, the above considerations render a flight and pursuit lasting days if not weeks understandable. The fact that the hunted man chose the route through the Val Senales to the Hauslabjoch would be explained by the fact that the route and the terrain were familiar to him because of annual transhumance over the main Alpine chain. It is only too understandable that he attemped first and foremost to make up that part of the equipment which was necessary for survival, i. e. the hunting weapons. The quiver was left to be repaired later. Under the assumption that he was an itinerant herdsman, he would have been one of the few men of his day who were familiar with high Alpine regions. Due to the approach of winter, he must have been hoping to

---

[49] von den Driesch et al. (n. 44) 160 et seq.

[50] Joachim Wahl/Günter König, Anthropologische Untersuchung der menschlichen Skelettreste aus dem bandkeramischen Massengrab bei Talheim, Kreis Heibronn. Fundberichte aus Baden-Württemberg 12, 1987, 67–193.

Fig. 13. (a) Detail of the skin clothing (the hairs have fallen out secondarily) of the man from the Hauslabjoch showing the seams sewn carefully with animal sinew garn (photo by RGZM Mainz). (b) Ditto with amateur mending of grass garn (photo by RGZM Mainz)

 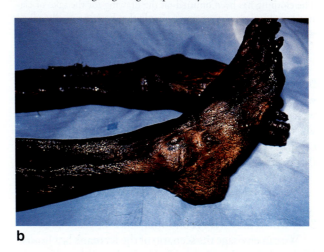

Fig. 14. (a) Tattoo on back of the man from the Hauslabjoch (photo by Hans Unterdorfer). (b) Ditto. Tattoo of bundle of strokes on the right foot (photo by Gerhard Sommer)

find a hiding-place in the summer pasturing area after shaking off his pursuers.

Archaeological field-work under Dr. Walter Leitner, Professor at the University of Innsbruck, in June 1993 led to the discovery of a storage place in the Upper Ötz Valley, only 10 km linear distance from the Hauslabjoch (Fig. 15a). On the bottom of a small semi-cave ("Abri") flintsone artefacts were found, which prove the presence of prehistoric man in the area above the tree-line, in the middle of the belt of high pastures. It is true that the shelter cannot be dated exactly at the moment (Epi-palaeolithic to Neolithic), but its existence does clearly prove human presence in the region at least back to the time of the Iceman, if not much earlier. The shelter was either a hunting post or shelter for herdsmen. Iceman may indeed have had such an aim, when he had crossed the main Alpine chain and made his way towards the Upper Ötz Valley.

It is not known whether he knew that his last hour had come when he had reached the chasm below the Hauslabjoch and only 80 m from the summit ridge. In any case, the trough offered some protection from the weather conditions, which were obviously an immediate threat to the man. In these days of the thirty-third century before Christ, the chasm must have been completely or nearly free from snow and ice. The man was in a condition of exhaustion, when he was surprised by an early snow storm. He layed his axe, bow and back-carrier on the rock ledge, squatted down and ate the last of his food supplies – a piece of dried ibex meat.

During the 1991 dig, fragments of the transverse apophyses of the forth and fifth cervical vertebra of a male ibex were found at the site[51]. Since the points of fracture of the

---

[51] von den Driesch/Peters (n. 28).

262   K. Spindler

Fig. 15. (a) Epipalaeolithic-Neolithic (Abri) Shelter in the upper Ötz Valley (photo by Walter Leitner). (b) Reconstruction of the position in which the man from the Hauslabjoch died (top) and of the position slightly changed by the movement of ice (bottom). The clothing has been omitted on the drawing for the sake of clarity, but this does not, of course, imply that he was actually naked in the ice (drawing by Michael Schick)

bones appear to have be cut or hacked with a stone utensil, we can assume that they are from pieces of meat which were taken out of the neck of the animal when the venison was being prepared. The protruding apophyses from the cervical vertebra were probably torn off accidently during this process. The meat is dried in portions and after this process the dried meat is a nutritious source of food, which can be stored and easily transported. The man knawed carefully at the bones and then spat them onto the rockface, where they were found thousands of years later.

We can envisage the scenario of the Iceman's last hours. He knew that sleep meant death. To keep himself warm and awake he trudged a few paces up and down. His quiver fell to the ground, 5 m away from the rock ledge. He must have staggered forward, completely exhausted and frozen, he then tripped on a rock after taking five steps and could not get back onto his feet. His cap fell off, so that he was now bare-headed. With his last ounce of strength he turned himself onto his left side, the least painful position for his injured rib-cage[52] (Fig. 15 b). Then he fell into the sleep from which he was to awaken no more. Snow covered his body. The fact that the body showed no traces of bites from animals proves that the man had been surprised by snow and that his body was completely covered during the night of his death. Even in high Alpine regions a dead body is discovered and attacked by carrion feeders, especially eagles, vultures and other birds of prey within a few hours. The good state of the body shows that it must have remained covered by a layer of light snow, which being in the area of permanent frost, made the dry mummification of the body possible. The covering of snow was more and thicker in the winters that followed without the thawing rates of the year of his death ever being reached again[53]. This did not happen again until several years of little snow in the winter of 1991. The thawing rates of that same year were increased additionally by a meteorological event between the fifth and the eighth of April. A strong southern pressure area brought a large amount of Sahara dust into the area, which settled as a dirty-yellowish-brownish layer on the snowfields of the Alps[54]. Thus the glacier melted vertically at a rate 10 cm a day in September 1991. On about the 16th of September the mummy began to emerge. It was discovered three days later. The first snow fell during the

---

[52] zur Nedden et al. (n. 32).

[53] Patzelt (n. 18).

[54] Gernot Patzelt, Gletscherbericht 1990/91 – Sammelbericht über die Gletschervermessungen des Oesterreichischen Alpenvereins im Jahre 1991. Mitteilungen des Oesterreichischen Alpenvereins 47 (117) 1992, no. 2, 17–22.

night between the 22nd and 23rd of September. Altogether the mummy would only have been visible for about six days. In the winter of 1991/91 approximately 700 cm of snow fell on the Hauslabjoch and the following summer melted this down to about 100 cm. The large amounts of snow during the winter of 1993 further increased the covering of snow so that the Alpine glaciers will have a period of recovery. The fact that after between 5,300 and 5,200 years similar climatic conditions freed the man from the Hauslabjoch, giving us insight into an otherwise hidden Neolithic fate, is almost a miracle.

Summary

The author presents the archaeological, botanical and anatomical of medical evidence relating to the events of the last few days of the Iceman's life. The unfinished arrows and the half-completed bow indicate that he had lost his weapons and was in the process of re-arming himself. The quiver and the two primed arrows show clear signs of damage that has been proved to originate from before entombment in the ice of Hauslabjoch. An intravital series of fractured ribs and atrophic changes to the humerus on the same side of the body are also indicative of a violent conflict. The presence of threshing and winnowing fragments proves that, shortly before his death, the Iceman spent some time in a human settlement in which the grain crop was threshed. The theory is therefore proposed that shortly before his death the Iceman suffered some personal catastrophe involving damage to his possessions and physical injury. He fled in the direction of the inner Ötz Valley, a region of high alpine pastures he may have been familiar with from summer transhumance. Just beyond the ridge of the main Alpine chain he was caught by a sudden fall in temperature and snowfall, which he did not survive.

Zusammenfassung

Im vorliegenden Beitrag werden Hinweise aus archäologischer, anatomisch-medizinischer und botanischer Sicht zusammengestellt, die Auskünfte über Ereignisse in den letzten Tagen und Wochen des Mannes vom Hauslabjoch geben können. Die unfertigen Pfeile ebenso wie der in Arbeit befindliche Bogen belegen Verlust und Wiederbeschaffungsversuch dieser Waffen. Der Köcher und die beiden sogenannten schußbereiten Pfeile zeigen erhebliche Beschädigungen, die nachweislich bereits vor ihrer Deponierung am Hauslabjoch eingetreten sein müssen. Ein intravital erlittener Serienrippenbruch, verbunden mit Atrophieerscheinungen des Humerus an der gleichen Körperseite, deutet ebenfalls auf eine gewalttätige Auseinandersetzung hin. Druschreste und Worfelabfälle belegen, daß der Mann vom Hauslabjoch kurz vor seinem Tode in einer menschlichen Siedlung anwesend war, in der die Getreideernte gedroschen wurde. Es wird daher die Vermutung ausgesprochen, daß dem Mann kurz vor seinem Tode ein „Desaster" mit Körperverletzung und Sachbeschädigung widerfuhr. Er flüchtete in Richtung hinteres Ötztal, eine Hochweideregion, die ihm möglicherweise von einer sommerlichen Wanderweidewirtschaft („Transhumance") her vertraut war. Kurz hinter dem Alpenhauptkamm überraschte ihn ein Wettersturz mit Schneefall, den er nicht überlebte.

Résumé

L'article présent résumé des références du point de vue archéologique, anatomique-médical et botanique, qui peuvent donner des renseignements sur les événements des derniers jours et semaines dans la vie de l'homme du Hauslabjoch. Les flèches non-terminées ainsi que l'arc en fabrication témoignent de leur perte et de la tentative de s'approvisionner de nouveau de ses armes. Le carquois et les deux flèches prêtes à l'usage montrent des endommagements considérables qui se sont passés avant qu'ils soient déposés au Hauslabjoch comme on peut en apporter la preuve. Une infraction intravitale de plusieurs côtes liée à l'apparence atrophiée de l'humérus du même côté du corps portent à croire qu'un dispute violant a eu lieu. Des restes de battage et des déchets venant de l'action de vanner donnent la preuve que l'homme du Hauslabjoch a été présent, peu avant sa mort, dans une agglomération humaine dans laquelle on battait le blé après sa récolte. Ainsi peut-on faire la supposition que l'homme a subi un "désastre" avec blessure du corps et endommagement des biens. Il s'est enfui en direction de l'arrière partie de la vallée de l'Ötztal, une région de haute alpage qu'il a probablement connue d'une transhumance en été. Peu après la chaîne montagneuse de l'Alpenhauptkamm il a été surpris par une brusque chute de température accompagnée des chutes de neige auxquelles il n'a pas survécu.

Riassunto

Il presente articolo riassume le indicazioni archeologiche, anatomiche e botaniche che possono delucidare sugli eventi delle ultime settimane nella vita dell'uomo dell'Hauslabjoch. Le frecce non ultimate, nonchè l'arco ancora in stato di fabbricazione testimoniano della perdita e dei tentativi di recupero di tali armi. La faretra e le due frecce pronte all'uso presentano notevoli danni evidentemente verificatisi giá prima della loro deposizione sull'Hauslabjoch. L'ipotesi di un episodio di violenza viene confermata anche dalla presenza di una serie intravitale di fratture costali, connesse ad atrofie dell'umero sullo stesso lato del corpo. Dai residui della trebbiatura e della spulatura trovati sul corpo dell'uomo dell' Hauslabjoch si può evincere che quest'ultimo, prima della sua morte, si sia trovato in un insediamento umano in cui veniva trebbiato il grano. Viene tuttavia formulata l'ipotesi che l'uomo prima della sua morte abbia subito uno scontro provocandogli lesioni fisiche e danni al suo materiale. Egli fuggì in direzione della valle posteriore dell'Ötz, una regione di transumanza che gli era probabilmente nota. Dietro la catena principale alpina fu sorpreso da una improvvisa caduta di neve alla quale non sopravvisse.

Correspondence: Dr. K. Spindler, Institut für Ur- und Frühgeschichte, Universität Innsbruck, Innrain 52, A-6020 Innsbruck, Austria.

Post-mortem alterations of human lipids – part I: evaluation of adipocere formation and mummification by desiccation

T. L. Bereuter[1], E. Lorbeer[1], C. Reiter[2], H. Seidler[3], and H. Unterdorfer[4]

[1] Institute for Organic Chemistry, University of Vienna, Austria
[2] Institute for Forensic Medicine, University of Vienna, Austria
[3] Institute for Human Biology, University of Vienna, Austria
[4] Institute for Forensic Medicine, University of Innsbruck, Austria

Introduction

The discoveries of well-preserved corpses have fascinated scientists, and have been subjects of many investigations. Preservation at the macroscopic level reflects a certain degree of preservation of original structural components of the tissues such as scleroproteins. Nevertheless, transformation processes have to render the body components from the steady state during lifetime to a stable state after death. A stable state is reached if the reaction products are thermodynamically stable or kinetically inert. In addition, these products have to be unattractive for scavengers and microorganisms. This has, indeed, been observed for the three fundamental preservation types, namely (i) mummification by desiccation, (ii) transformation to adipocere, and (iii) "tanning" in the acidic bog environment.

This article focuses on adipocere formation. The active mechanisms responsible for the formation of adipocere are yet not fully understood. Systematic investigations are difficult since suitable samples are scarce. Nevertheless, research has been given an important stimulus by the discovery of the Late Neolithic corpse known as Iceman. As he was released by the glacier, experienced forensic scientists would have expected to find adipocere. Instead, macroscopic investigations indicate that rapid desiccation caused the observed state of preservation. Therefore, we investigate both adipocere formation as well as natural mummification by desiccation.

Adipocere

The name adipocere derives from the Latin words *adeps*, meaning fat, and *cera*, meaning wax. Translated into English it is called *fat wax* or *fatty wax*[1]. In analogy, the German word for adipocere is *Fettwachs*. In both cases, the name is misleading because it is neither a fat nor a wax. Adipocere is formed by the post-mortal conversion of body fat into a lipid mixture with wax like consistency and of greyish-white colour. Chemical analysis showed that it consists mainly of fatty acids.

Adipocere was first mentioned in 1661 and again, in more detail, in 1722 and 1744. As those reports were not noticed, Foucroy and Thouret are regarded as the discoverers of adipocere. In 1789, they found corpses in common graves with nearly unchanged intravital anatomy.

The formation of adipocere takes usually about three to six months, although the shortest observed time for partial conversion was six weeks. A complete transformation of the whole body lasts years. The conditions under which this process can occur are manifold. None of them has to be fulfilled strictly, except for the humid and anaerobic conditions which are a prerequisite for adipocere formation. Most common is the transformation of water-logged bodies which is usually restricted to the torso with the other parts macerated. Adipocere formation can also be found in humid graves without air access (e.g. in loam soil or bog). Furthermore, corpses released by glaciers also show adipocere formation. The extent to which these rare findings are preserved is variant. In some cases, the corpses are crushed by the shearing forces and by the pressure of the glacier ice (Meyer 1992), whereas in other cases bodies are fully conserved (Ambach 1991). Between these two extremes, all grades of destruction and preservation may be observed (Ambach 1992).

---

[1] A further expression for adipocere is grave wax.

The major constituents of adipocere are free fatty acids with even numbers of carbon atoms.[2] Saturated fatty acids such as myristic acid, palmitic acid (the predominant acid), and stearic acid are present in characteristic relations (Table 1). In some cases, hydroxy-fatty acids such as 10-hydroxy-stearic acid are found to be main components. These saturated fatty acids as well as the hydroxy-fatty acids are solid at room temperature. Unsaturated fatty acids are present in smaller total amounts; owing to their low melting points, they are liquid. Further components are sterols.

Different theories have been developed to explain adipocere formation (Döring 1973). The theories of (i) *fat migration*, (ii) *saponification*, and (iii) *hydrogenation* are described here:

(i) *Fat migration theory:*

The fat, which is set free by decomposition of structural elements, migrates into the surrounding tissues. The neutral fats (human subcutaneous fat mainly consists of triacylglycerols; oleic acid (48 %) and palmitic acid (26 %) are the main acids bound in the triacylglycerols) are cleaved and the liquid fission products such as oleic acid and glycerol are lost by diffusion/gravitational separation.

(ii) *Saponification theory:*

The fragmentation of the body fat into fatty acids and glycerol by hydrolysis is assumed to be the main reaction.

Table 1. Fatty acid pattern of human subcutaneous fat before and after the transformation to adipocere (Berg 1975)

Fatty acids (FA)		m. p. [°C]	"Fresh" subcutaneous fat [%]	"Artificial" adipocere [%]
common name	short notation			
saturated FA				
lauric acid	C12:0	45	2.0	2.0
myristic acid	C14:0	54	6.0	7.0
pentadecanoic acid	C15:0	52	0,5	0.5
palmitic acid	C16:0	62	26.0	53.0
heptadecanoic acid	C17:0	61	1.0	1.0
stearic acid	C18:0	70	7.5	9.0
arachidic acid	C20:0	76	1.5	1.5
unsaturated FA[a]				
palmitoleic acid	C16:1	1	5,0	2,0
oleic acid	C18:1	16	48,5	22,0
gadoleic acid	C20:1	23	2,0	2,0

[a] The double bond is not specified in the literature but is probably the cis isomer, and located at the 9th carbon atom.

In a basic reaction medium containing alkaline-earth metals, water insoluble soaps of the saturated fatty acids would be formed. However, the associated reaction products such as glycerol and oleic acid are at all only present in small amounts. This is not explained by the theory. The theory was developed at an early stage of research when soaps were believed to be the main components. Nevertheless, only low concentrations of fatty acid salts were found in adipocere and the saponification process, therefore, may not have the importance as believed.

(iii) *Hydrogenation theory:*

This theory (Figure 1) is supported by microbiological experiments. In subcutaneous fat deposits, large amounts of fatty acids are set free by fission of the triacylglycerols. The surrounding tissues are soaked with the liquid mixture containing mainly unsaturated fatty acids. These will undergo the postulated hydrogenation process to form saturated fatty acids. In the case of oleic acid, stearic acid would be produced. As palmitic acid is the major component of adipocere, the loss of a C_2-unit by a single step β-oxidation must be assumed. The solid nature of palmitic acid and stearic acid means that a tissue will harden by their accumulation.

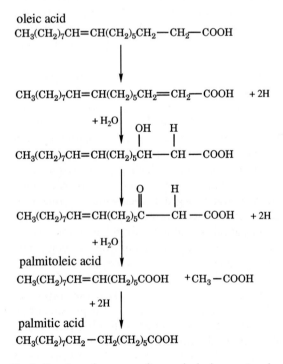

Fig. 1. Reaction scheme according to the hydrogenation theory

Mummification

This kind of preservation is mainly based on desiccation of the body tissues. In the case of artificial mummification, the desiccation process is accelerated by, e.g.,

---

[2] About 10 % of the total fatty acids are described to be calcium and magnesium soaps.

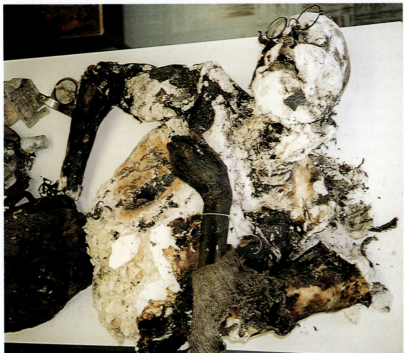

Fig. 2. Case 1

Fig. 3. Thigh of Case 2

Fig. 4. Case 3

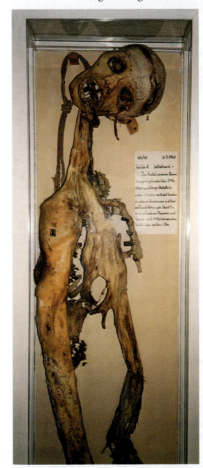

Fig. 5. Case 4

forced air circulation, smoking, or salts. Other treatments such as embalming often are more deleterious than advantageous. The different kinds of artificial mummification have been topics of chemical archaeological investigations, but largely concerned with the means of mummification and not with the results of the lipid transformation. To our knowledge, no detailed chemical analysis has been performed to determine the lipid alteration of natural mummified tissues about which we report here. Although there is no sharp boundary between natural and artificial mummification, the first is almost free of interferences in the gas chromatograms caused by artefacts.

If dry, cool, and, in particular, airy conditions prevail, desiccation may be fast enough to prevent putrefaction and above all the development of maggots. The presence of firmly rooted hair is characteristic for this so-called primary mummification. On the other hand, cutis and hairs are lost during secondary mummification. In this case, putrefaction and maggots are active at an early stage. Their activities reduce the water content in addition to that brought about by other desiccation processes. This loss of water can stop the destruction processes, and, therefore, preserves the last stage of decomposition. As a rule of thumb, one can state that the faster the process of desiccation is, the better the degree of macroscopic preservation will be.

In our own studies, we correlate the lipid composition with environmental conditions which evoke the specific kind of lipid transformation observed in adipocere and desiccated tissues. The lipid components were partly identified by the use of standards and by the coupled technique of gas chromatography – mass spectrometry.

Experimental

Instrumentation

High resolution gas chromatograph Mega 5300 (Fisons) equipped with:

- on-column injector with secondary cooling
- flame ionisation detector
- computer data station
- glass capillary column (25 m × 0.32 mm) coated with a 0.1 μm thick film of OV 1701-OH (Blum 1991)

Operating conditions

- on-column injection of 1 μl sample at an oven temperature of 70 °C after secondary cooling for 0.5 min
- temperature program: from 70 °C to 170 °C at 25 °C/min, and to 380 °C at 10 °C/min, with an isothermal hold at maximum temperature for 15 min
- detector temperature: 390 °C
- hydrogen as carrier gas: column head pressure of 1 bar, flow of 4 ml/min

Scheme of analysis

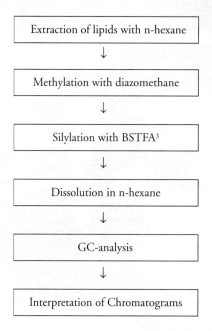

Case reports

Fatty wax type of mummification

Case 1: The body of a 62 year-old identified male alpinist (Figure 2) was found in August 1991 after 57 years of submersion in the Sulztalferner (glacier of the Stubai Alps) at an altitude of 2700 m. The circumstances of the accident are unknown (Ambach 1992). Damage was caused by shearing forces of the glacier and by maceration in the occipital skull area. Well preserved soft tissues are observed in the area of the trunk by the fatty wax type of mummification and in the distal area of the extremities by desiccation type of mummification (Spindler 1993). Samples of the desiccated tissue were taken from the chin and of adipocere from the upper arm, respectively.

Case 2: The corpse of a 28 year-old identified female alpinist (Figure 3) was found in August 1952 after 29 years of submersion in the Madatschferner (glacier of the Oeztal Alps). Severe bone fractures (Ambach 1991), maceration of the forearm, and extended transformation of the body by the fatty wax type of mummification were observed. The sample was taken from the adipocere of the thorax.

Case 3: The torso of a 30 year-old woman (Figure 4) was found in the Achensee (a mountain lake in the north-east of Innsbruck) in September 1989. The corpse was discovered in a motor car in which the skeleton of a man was also found. It was a case of multiple suicide. Both bodies were submerged for 50 years and 50 m below the water surface at about 4 °C. Nevertheless, only the soft tissues of the torso of the woman were transformed to adipocere. It was postulated that the physiologically higher fat content of women generally and the adiposity of this woman particularly were responsible for the different states of preservation (Rabl 1990). Samples were taken from the upper arm.

---

[3] N,O-bis(trimethylsilyl)-trifluoroacetamide containing 1 % v/v trimethylsilyl chloride.

Desiccation type of mummification

Case 4: A man who had been hung from the neck was found at a tree in the Ahrntal (a valley in the south of Innsbruck) after 3 months post-mortal in July 1940 (Figure 5). Preservation of the head, parts of the arms, shoulders, and torso by the desiccation type of mummification was observed. The other parts of the body were scattered, and found completely macerated on the ground below the corpse. The sample was taken from the upper arm.

Case 5: The corpse of a 41 year-old identified man (Figure 6) was found in April 1993 after 3.5 years lying in an apartment. The circumstances of his death are unknown. The onset of putrefaction and fly-maggots damaged the corpse. In this state, the body was preserved by the desiccation type of mummification. The sample was taken from the abdominal wall.

Case 6: The body of a 47 year-old identified man (Figure 7) was found in April 1993 after 4 years lying in an apartment. Once again, the circumstances of death are unknown. Beginning putrefaction lead to a state in which the body was preserved by the desiccation type of mummification. Samples were taken from the back and the thigh.

Reference tissue

Case 7: Fresh subcutaneous fat was taken three hours after death from the thigh of a 35 year-old man who had been killed by a shot to the head. The corpse showed no symptoms of any disorder in lipid metabolism. The sample was freeze dried prior to analysis.

The samples were taken from outer tissue layers of corpses, and stored at -20 °C until required for analysis. Those from the upper arm (case 1, 4, and 5) were fractionated by cuts horizontally to the surface. For the samples from case 6 and 7, fractionation was done on a sledge microtom after freezing of the samples. The resulting layers were about 0.5 to 1.0 mm (outer layer) and 2 to 3 mm (inner layers) thick. These portions underwent the procedures described in the scheme of analysis.

By using high performance high temperature capillary gas chromatography, we 'fingerprinted' the lipophilic extracts of adipose tissues. Our screening procedure shows a great potential for resolution of complex mixtures of lipid components including fatty acids, acylglycerols, and sterols in a single chromatographic step. Thus, quantities of sample in the lower mg range are quite sufficient.

Results and discussion

Chemical analysis versus macroscopic investigation

The corpse of an alpinist (case 1, Figure 2) has been released from a glacier. This finding showed several different kinds of post-mortem alterations. Total macer-

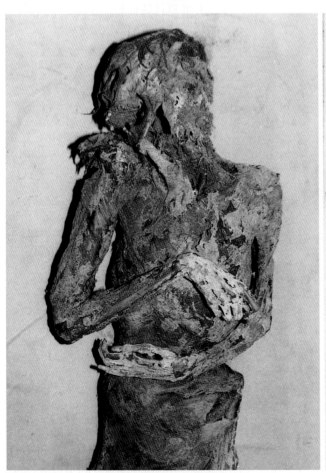

Fig. 6. Case 5

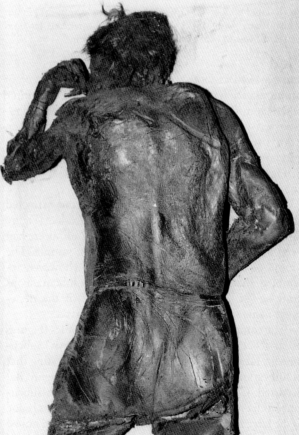

Fig. 7. Case 6

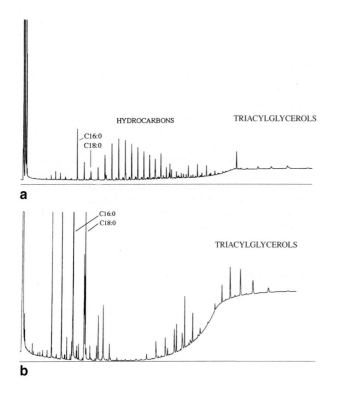
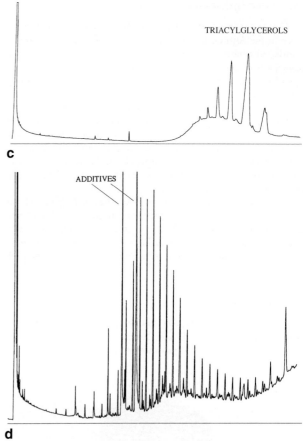

Fig. 8. Lipid profiles. (a) Desiccated tissue from case 1. (b) Homogeneous fatty wax from case 2. (c) Fresh subcutaneous fat from case 7. (d) Cream with an ointment base of paraffin

ation as well as fatty wax type and desiccation type of mummification were observed. We investigated a sample taken from the chin region, where evidently no adipocere was present. In addition, subcutaneous fat taken from a recently deceased man (reference tissue, case 7) and homogeneous adipocere from a human corpse (case 2, Figure 3), released from a glacier, were analysed for comparison.

The fatty acid pattern (Figure 8a) of the extracted chin tissue (case 1) was almost identical to that (Figure 8b) of the adipocere (case 2), although no adipocere had been observed macroscopically in the chin sample. The profile of the triacylglycerols (Figure 8a), however, resembled that (Figure 8c) of the "fresh" subcutaneous fat (case 7). The original triacylglycerols of the desiccated tissue seem to be at least partly preserved.

The comparison of the triacylglycerol profiles (Figures 8b and 8c) of the adipocere and of the "fresh" subcutaneous fat showed significant differences indicating that modification of the original triacylglycerols (Figure 8c) occurred. Chromatographic analysis of the fresh subcutaneous fat resulted in broad peaks (Figure 8c); this is due to the large number of triacylglycerols which could not be resolved by the method. It is clear that the analysis of acylglycerols needs further development of the current technique. The problem can be demonstrated by a simple calculation: in the column for fresh subcutaneous fat of Table 1 are five fatty acids listed which have relative concentrations larger than one percent. If we assume that these acids can be bound to either one of the hydroxyl groups of glycerol through an ester linkage, 125 triglycerides are possible. The total number of triglycerides is reduced to 75 provided that none of the optical isomers are distinguished and to 35 if no isomers are specified. The sharp peaks observed for the triacylglycerols (Figure 8b) indicate that only certain isomers resisted degradation. Chemical alterations in the chains of fatty acids bound in the triacylglycerols may be responsible for this selective resistance, especially if the degradation processes are catalysed by enzymes which are specific for certain fatty acids.

In addition to the natural lipids, we found a homologous series of hydrocarbons in the chin sample (Figure 8a). Probably, the chin was greased with a sun protection cream on paraffin basis (Figure 8d). The paraffins were coextracted by the apolar solvent, and caused in the chromatogram signals of hydrocarbons. This explanation for the artefact is supported by the finding of a sun protection cream in the equipment of the alpinist.

These results demonstrate the power of gas chromatography in the analysis of lipid components extracted from human body tissues. Furthermore, autopsy supplemented by chemical analysis gives a more accurate description of the kind and state of mummification than mere macroscopic inspection.

Lipid transformation under different conditions

The free fatty acid patterns of the outer tissue layers taken from three human corpses were also compared. One of the corpses examined was the alpinist (case 1, Figure 2), the second body selected had been submerged in a fresh water lake (case 3, Figure 4), and the third corpse was a man (case 4, Figure 7) who had been hung. Informations in detail about duration and conditions of the post-mortem storage, but also physical descriptions, are available for each case. The corpse released from the glacier showed the presence of adipocere, and the body found in the lake also underwent fatty wax type of mummification. In contrast, the hanged man was mummified due to desiccation by dry winds. Samples were taken from the upper arm of the corpses; the visible features of the samples of the desiccation type mummy and of the fatty wax type corpses were quite different. However, samples of the water corpse and the glacier corpse appeared similar in that they possessed adipocere. In addition, it was observed that the water corpse had lost all structural differentiation, whereas the glacier corpse showed layers of differing colour and consistency[4]. The samples were cut up parallel to the dermis. In the case of the glacier body, the five zones visible were separated into five fractions for separate analysis. As the sample of the water corpse was homogeneous, the division into three portions was arbitrary. Nevertheless, chemical analysis showed no differences in the constitution of the fatty acids between the fractions of the inhomogeneous sample of the glacier body (Figure 9) and, in contrary, differences in the fractions of the homogeneous sample of the water body (Figure 10). The sample of the hanged man could be separated into muscle and dermis plus subcutaneous fat (Figure 11), also revealing different fatty acid constitutions.

All samples exhibited decomposition of the triacylglycerols, domination of palmitic acid as the major fatty acid, and presence of sterols. The cyclic carbon skeleton of sterols appears not to be affected by chemical or microbial alteration. It is of interest, however, that sterols are observed only by certain analysts (Berg 1975). We suppose that this is due to the different analytical

---

[4] Light microscopical evaluation displayed that samples of the glacier body had lost the epidermis and that from the dermis mainly the reticular layer has been preserved. The dermis is followed by subcutaneous fat. The layer underneath the subcutaneous fat possibly contained remnants of skeletal muscles with fibrous sheaths (Plenk 1993).VVV

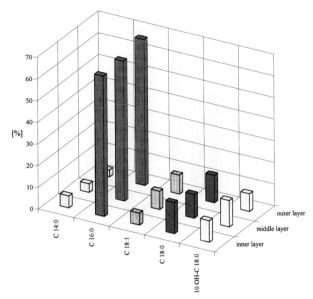

Fig. 9. Fatty acid distribution – Case 1

methods used, especially the method of derivatization. Characteristics in the fatty acid patterns of the three corpses have been observed. The adipocere of the glacier corpse (case 1) mainly consisted of saturated fatty acids, but oleic acid and 10-hydroxy-stearic acid were also present in larger amounts (Figure 9). In contrast, the adipocere of the water corpse (case 3) contained only smaller amounts of oleic acid (Figure 10). The ratio of oleic acid to stearic acid and the amount of modified di- and triacylglycerols can be used to differentiate between the two environments of water and ice during the conservation process. For the desiccation type mummy (case 4), oleic acid was found to be one of the major constituents of the lipid extract. In the deeper layer of the tissue oleic acid is about 50 % of the relative concentration of palmitic acid. A declining gradient can be observed between the deeper tissue (probably muscle) and the outer layer (Figure 11). This supports the theory that oleic acid is lost by diffusion. The gas pressure which develops during putrefaction may accelerate these losses. The hydroxy-fatty acid, which was present at a significant concentration in the water corpse and the glacier body, did not occur in the mummy of the desiccation type. The hydroxy-fatty acid may, therefore, be crucial for the properties of the adipocere.

Desiccation type of mummification

Desiccation as a natural type of mummification is a well-known process of soft tissue preservation in human corpses (Berg 1975). However, reliable chemical descriptions of the lipid composition are not available. The free fatty acid patterns of the outer tissue layers taken from

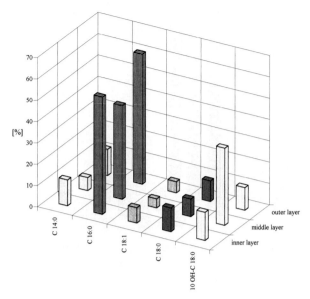

Fig. 10. Fatty acid distribution – Case 3

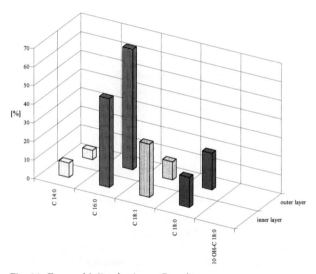

Fig. 11. Fatty acid distribution – Case 4

three human corpses were determined. Two of the corpses investigated (case 5 and 6) had been found lying in their apartments. The third corpse is the hanged man (vide supra, case 4), who had lost all liquefied parts of the body. Although samples were taken from different parts of the corpses, the optical aspects were quite similar. The macroscopic result of the medical and forensic investigation was the same, namely the desiccation type of mummification. Detailed case reports are available.

Significant differences between the fatty acid patterns of the samples taken from the hanged corpse and the other two corpses were observed. Most remarkable are the various ratios of oleic acid to palmitic acid. The main fatty acid of the hanged corpse was palmitic acid, whereas in the other samples of the apartment bodies oleic acid is also dominant. The loss of liquid oleic acid observed in the sample of the hanged man is possibly due to diffusional/gravitational separation, as it is assumed in the fat migration theory. Furthermore, it was found that sampling at different sites can lead to different results. This can be the result of various micro-environments which are generated, for example, by the ground on which the corpses were lying while part of the body was surrounded by circulating air.

Conclusion

The results obtained show that macroscopic investigations should be followed by chemical analysis in order to obtain reliable and detailed information about the kind and state of preservation. Gas chromatography is well suited for this purpose. None of the theories described, which should explain the process of adipocere formation, is able to account for all the observations conclusively. An accurate separation into subclasses of adipocere (e.g. according to the different conditions under which it was formed) seems to be necessary. This applies also to the desiccation type of mummification. The differences between each one of the analysed samples may be characteristic for the special conditions under which the corpses have been stored. This has to be confirmed by the investigation of samples with similar histories.

Summary

Adipocere formation and mummification by desiccation, both natural modes of body preservation, are discussed. Lipid extracts from outer tissues of naturally preserved bodies were analysed by high-temperature gas chromatography in combination with microanalytical transformations. The analytical investigation of the lipid composition of tissues from desiccated mummies is described here for the first time.

Zusammenfassung

Fettwachsbildung und Trockenmumifikation als natürliche Arten der Leichenkonservierung werden diskutiert. Die analytische Untersuchung von Lipidextrakten der äußeren Gewebsschichten natürlich konservierter Leichen erfolgte durch Hochtemperatur-Gaschromatographie in Kombination mit mikroanalytischen Transformationen. Vergleichbare Untersuchungen für trockenmumifizierte Leichen wurden bisher nicht beschrieben.

Résumé

La formation de l'adipocire et la momification par dessèchement, deux modes naturels de conservation du cadavre, sont discutées. Les extraits de lipides des tissus externes des cadavres naturellement conservés ont été analysés par chromatographie en phase gazeuse à haute température en combinaison avec des transformations micro-analytiques. Les recherches sur la composition en lipide des tissus des cadavres momifiés secs sont décrites ici pour la premiére fois.

Riassunto

Vengono discusse la formazione di adipocera e la mummificazione tramite essicamento, due tipi naturali di conservazione di cadaveri. Gli esami analitici degli estratti di lipidi dei tessuti esterni di cadaveri, conservati in modo naturale, furono effettuati tramite gas-cromatografia ad alta temperatura in combinazione trasformazioni micro-analitiche. Simili studi su mummie sono stati eseguiti per la prima volta.

Acknowledgments

W. Rabl and H. Schuler[5] are thanked for their support in the course of sampling from case 1–4.

References

Ambach 1991 – E. Ambach, W. Tributsch, R. Henn, Fatal accidents on glaciers: forensic, criminological, and glaciological conclusions. *J. Forensic Sci.* **36** (1991) 1469–1473.

---

[5] Institute for Forensic Medicine, University of Innsbruck.

Ambach 1992 – W. Ambach, E. Ambach, W. Tributsch, R. Henn, H. Unterdorfer, Corpses released from glacier ice: glaciological and forensic aspects. *J. Wilderness Med.* **3** (1992) 372–376.

Berg 1975 – S. Berg, Leichenzersetzung und Leichenzerstörung. In: B. Mueller (ed.), *Gerichtliche Medizin*, Springer-Verlag, Berlin–Heidelberg–New York, 1975, pp. 62–106.

Blum 1991 – W. Blum, R. Aichholz, *Hochtemperatur Gaschromatographie*, Hüthig-Verlag, Heidelberg, 1991.

Döring 1973 – G. Döring, Chemische Altersveränderungen des Leichenlipids, thesis, University of Göttingen, 1973.

Meyer 1992 – W. Meyer, Der Söldner vom Theodulpaß und andere Gletscherfunde aus der Schweiz. In: F. Höpfel, W. Platzer, K. Spindler (eds.), *Der Mann im Eis*, vol. 1, University of Innsbruck, 1992, pp. 321–333.

Rabl 1990 – W. Rabl, E. Ambach, W. Tributsch, Leichenveränderungen nach 50 Jahren Wasserzeit. *Beitr. Gerichtl. Med.* **49** (1990) 85–89.

Plenk 1993 – H. Plenk jr., K. Groszschmidt, H. Wilfing, H. Seidler, Light microscopical observations on a skin-tissue sample from the forearm of a human glacier corpse. Poster presented at the International Mummy-Symposium, September 1993, Innsbruck, Austria.

Nedden 1994 – D. zur Nedden, K. Wicke, R. Knapp, H. Seidler, H. Wilfing, G. Weber, K. Spindler, W. A. Murphy, G. Hauser, W. Platzer, New findings on the Tyrolean "Iceman": archaeological and CT-body analysis suggest personal disaster before death. *J. Archaeol. Sci.* **21** (1994) 809–818.

Correspondence: Mag. Thomas L. Bereuter, Institute of Organic Chemistry, University of Vienna, Währinger Strasse 38, A-1090 Vienna, Austria.

Post-mortem alterations of human lipids – part II: lipid composition of a skin sample from the Iceman

T. L. Bereuter[1], C. Reiter[2], H. Seidler[3], and W. Platzer[4]

[1] Institute for Organic Chemistry, University of Vienna, Austria
[2] Institute for Forensic Medicine, University of Vienna, Austria
[3] Institute for Human Biology, University of Vienna, Austria
[4] Institute for Alpine Prehistory, University of Innsbruck, Austria

Introduction

For the different types of mummification see Part I (Bereuter 1996, in this volume, pp. 265–273). We present here the preliminary results obtained by investigation of a small part of skin tissue from the Iceman. The sample was found during the second excavation at the place of discovery of the Iceman near the Tiesenjoch. It is part of the skin from the left pelvic region.

Experimental

A 1.5 mm thick strip was taken from the sample (Figure 1) and divided on a freeze microtom into four layers parallel to the skin surface (Figure 2), each in the low mg range. These four fractions were analyzed separately by capillary gas chromatography with flame ionization detection (GC-FID), but also with mass selective detection (GC-MS). The sample pretreatment and the GC-FID conditions are described in Part I (Bereuter 1996). For identification of the signals in the chromatograms, the sample solutions were derivatized, spiked with standard substances, and the retention times compared. The major constituents are confirmed by mass spectrometry.

Results and discussion

Sample description

The macroscopic aspect of the tissue (Figure 1) shows the corium without visible parts of epidermis, but with traces of subcutaneous fat (Figure 3).

The histological investigation (Figure 4) exhibits collagen bundles of connective tissue arranged in parallel layers with underlying honeycombed layers of soft connective tissue as well as with dense tubular structure residues, maybe ducts of glands. No cell nucleus could be stained. The sample shows mineral deposits on both surfaces. The epidermis was found to be completely lost, further to the Iceman's loss of hair and nails previously described (Lippert 1991).

Chemical analysis

Almost all ester bonds of the triacylglycerols in the skin sample of the Iceman are cleaved so that fatty acids (Figure 5) and glycerin are set free. These facts exclude the speculations of immediate freeze drying of the body after death followed by imbeddement in glacier ice without any significant temperature increase till the time of discovery in 1991.

Oleic acid (C18:1) is the major component of the triacylglycerols in human fat. The shifting to palmitic acid in Iceman's lipids as the main constituent of the fatty acids without functionalities in the carbon chain and the low content of oleic acid (Figure 6) correspond with previous analysis of adipocere. The shifting to 10-hydroxy stearic acid – being a marker for adipocere – as the main constituent of the total fatty acids (Figure 5) demonstrates a virtual complete transformation of the original fat to adipocere. These results correspond with the macroscopic finding of the so-called état mammeloné (Berg 1975) in the skin of Iceman's face (Figure 7). État mammeloné is the accumulation of adipocere producing light colored spots on the skin.

The comparison of the pattern of hydroxy fatty acids in the four layers (Figure 8) displayed no gradients from the inner to the outer side of the skin. Nevertheless, a slight increase in unsaturated fatty acids (palmitoleic acid, C16:1, m.p.=1 °C; oleic acid, C18:1, m.p.=16 °C) was observed in the outer layer (Figure 6). This can be explained by the fat migration theory which postulates a

Fig. 1. Iceman's skin tissue from which the sample was taken. ×15

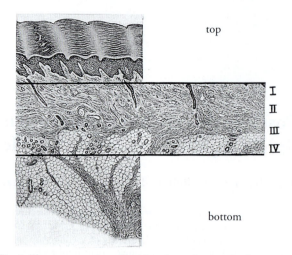

Fig. 2. Transverse section of epidermis, corium, and subcutaneous fat. The part pointed out corresponds to the sample of the Iceman with numbering of the four layers investigated

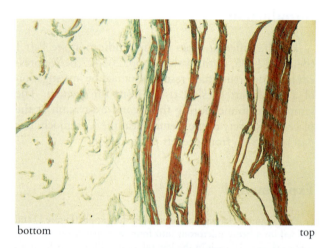

Fig. 3 (above). View on the bottom side of the skin tissue from the Iceman. ×25. Fig. 4 (right). Histological appearence of the skin tissue from the Iceman (Masson trichrome stain). ×100

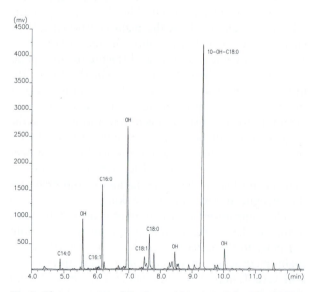

Fig. 5. Chromatogram of the fatty acids extracted from layer II (see also Fig. 2). The various hydroxy fatty acids are marked with OH

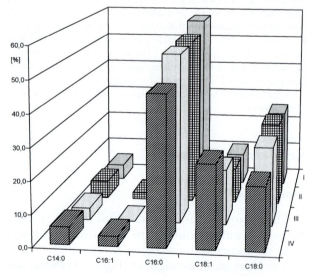

Fig. 6. Fatty acid profiles. IV is the external layer of the corium and I is the internal layer of the corium with traces of subcutaneous fat (see also Fig. 2)

migration process of more mobile fatty acids caused by diffusion and gravitation.

The lipid extract of our sample is nearly exclusively composed of fatty acids. About 86 % of them are fatty acids with hydroxy functionality in the carbon chain. Only 3 % are unsaturated fatty acids. 97 % are, therefore, stable compounds. This state can be seen as the final state of the transformation process of the fat components.

The high proportion of hydroxy fatty acids in the total lipids is the result of a high oxidational state. Possibly premortal highly oxidative stress could have led to the production of enzymes able to autocatalyze postmortem oxidation of unsaturated fatty acids. Experiments at humans have shown to increase significantly the oxidation of lipids by cold exposure (Vallerand 1990). Samples from humans who died through exposure after prolonged exercise would, therefore, be of great interest for further investigation.

The high state of oxidation is in contradiction to the hydrogenation theory that states reductive conditions under which double bonds of unsaturated fatty acids are hydrated.

As we have found only small amounts of fatty acids with uneven carbon numbers, traces of branched fatty acids, and no trans-fatty acids we have no indications for high bacterial activities. This is based on the assumption that the hydroxy fatty acids are not produced microbially.

During the first stages of putrefaction, the epidermis, hair, and nails (Berg 1975, Capasso 1995) can peel off as in case of secondary mummification. However, the good state of the body and the chemical findings described above do not correspond to secondary mummification. Therefore, in the case of the Iceman we assume the cause to be skin wrinkling, also known as washerwoman's hands (Püschl 1985), in which the epidermis soaked with water strips off. The slightest movement in the water or gas pressure produced by putrefaction can lead to a complete loss of the epidermis (Figure 9). Apart from the epidermis, hair and nails are also lost by this process. If rapid desiccation preserves embedded hair and nails – as in primary mummification – the connective tissue looses its typical swelling capacity in water. Therefore, hair and nails remain firmly rooted, but epidermis is desquamated by water (Reiter 1996). The finding of diatom frustules and algal (chrysophycean) stomatocysts on Iceman's cloak and foot wear normally indicate a water storage, although it is possible that these organisms could have been harvested together with the grass used for production of the clothes or caused by mere fossil contamination of the archeological site (Rollo 1995). Our chemical findings described above as well as investigations of muscle fibres from the Iceman (Galler 1995) point to a certain period of submersion in water.

Fig. 7. Oral area of the Iceman with état mammeloné spots

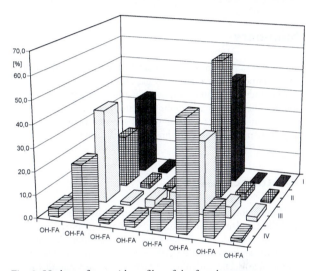

Fig. 8. Hydroxy fatty acid profiles of the four layers

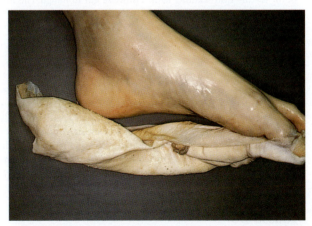

Fig. 9. Foot of a body submerged in water for 25 days at 15 °C and stripped off epidermis sock

These facts all together show that the body was lying in water for at least several months before the desiccation process started.

Conclusion

Histological and chemical investigations of Iceman's skin sample show characteristic postmortal alterations. The loss of epidermis as well as transformation of fat into adipocere indicate submersion in water for at least several months and desiccation afterwards. Results of other research projects confirm this hypothesis.

Final results of the chemical lipid analysis of the sample from the Iceman will be published soon. We are at present carrying out artificial mummification experiments to study the influence of time and temperature on the postmortal processes in water, in the light of which the presented preliminary results will have to be rediscussed.

Summary

A skin sample from the Iceman is described histologically. Its lipid composition is analyzed by gas chromatography. Loss of epidermis and formation of adipocere in a high oxidational state were found. The preliminary results indicate that the Iceman was submerged in water for at least several months before desiccation. There are no indications for high bacterial activities.

Zusammenfassung

Eine Hautprobe des Eismannes wurde histologisch untersucht und die Lipidzusammensetzung gaschromatographisch analysiert. Festgestellt wurde ein Zustand nach Abgehen der Oberhaut und eine Umwandlung des Körperfettes in Fettwachs von hohem Oxidationsstatus. Die vorläufigen Ergebnisse weisen darauf hin, daß der Eismann zumindest mehrere Monate im Wasser gelagert war, bevor es zur Austrocknung kam. Hinweise auf umfangreiche mikrobielle Aktivitäten wurden nicht gefunden.

Résumé

Un énchantillon de la peau de l'homme de la glace a été décrit histologiquement et la composition de ses lipides a été analysée par chromatographie an gaz. On a constaté la perte de l'épidermie et la transformation de graisse en adipocire qui est dans un état élevé d'oxidation. Ces premiers résultats indiquent que l'homme de la glace a été sous l'eau au moins pendant quelques mois avant son desséchement. Rien n'indique par contre une grande activité bactérielle.

Riassunto

Un campione di cute dell'uomo del ghiaccio è stato oggetto di esami istologici e la sua composizione lipidica è stata rilevata a mezzo di una cromatografia a gas. Si è osservato che l'epidermide si era staccata e che si era formata adipocera in uno stato altamente ossidato. I risultati provvisori degli studi finora effettuati indicano che il cadavere si era trovato per diversi mesi sommerso da aqua prima che si verificasse la mumificazione a secco. Non ci sono segni di rilevanti attività batteriche.

Acknowledgment

This work was supported from the jubilee funds (project number 4818) of the Austrian National Bank (OeNB).

References

Bereuter 1996 – T. L. Bereuter, E. Lorbeer, C. Reiter, H. Seidler, H. Unterdorfer, Post Mortem Alterations of Human Lipids – Part I: Evaluation of Adipocere Formation and Mummification by Desiccation. In: K. Spindler, H. Wilfing, E. Rastbichler-Zissernig, D. zur Nedden, H. Nothdurfter (eds.), *Human Mummies*, Springer-Verlag, Wien New York, pp. 265–273. 1996.

Berg 1975 – S. Berg, Leichenzerstörung und Leichenzersetzung. In: B. Mueller (ed.), *Gerichtliche Medizin*, Springer-Verlag, Berlin Heidelberg New York, 1975, S 62–106.

Capasso 1995 – L. Capasso, Ungueal Morphology and Pathology of the "Iceman". In: K. Spindler, E. Rastbichler-Zissernig, H. Wilfing, D. zur Nedden, H. Nothdurfter (eds.) Der Mann im Eis. Neue Funde und Ergebnisse. Springer-Verlag, Wien New York, 1995, pp. 231–239.

Galler 1995 – S. Galler, Noch Kraft in den Muskeln des Tiroler Eismanns? Eine physiologische Zustandsanalyse der Muskulatur. In: K. Spindler, E. Rastbichler-Zissernig, H. Wilfing, D. zur Nedden, H. Nothdurfter (eds.) Der Mann im Eis. Neue Funde und Ergebnisse. Springer-Verlag, Wien New York, 1995, pp. 253–268.

Lippert 1991 – A. Lippert, K. Spindler, Die Auffindung einer frühbronzezeitlichen Gletschermumie am Hauslabjoch in den Ötztaler Alpen, *Österreichische Archäologie* **2** (1991) 11–17.

Püschel 1985 – K. Püschel, A. Schneider, Die Waschhautbildung im Süß- und Salzwasser bei unterschiedlichen Temperaturen, *Z. Rechtsmed.* **95** (1985) 1–18.

Reiter 1996 – publication in preparation.

Rollo 1995 – F. Rollo, S. Antonini, M. Ubaldi, W. Asci, The Neolithic Microbial Flora of the Iceman's grass: Morphological Description and DNA Analysis. In: K. Spindler, E. Rastbichler-Zissernig, H. Wilfing, D. zur Nedden, H. Nothdurfter (eds.) Der Mann im Eis. Neue Funde und Ergebnisse. Springer-Verlag, Wien New York, 1995, pp. 107–114.

Vallerand 1990 – A. L. Vallerand, I. Jacobs, Influence of Cold Exposure on Plasma Triglyceride Clearance in Humans, *Metab. Clin. Exp.* **39** (1990) 1211–1218.

Correspondence: Mag. Thomas L. Bereuter, Institute for Organic Chemistry, University of Vienna, Währinger Strasse 38, A-1090 Vienna, Austria.

Comparison of the lipid profile of the Tyrolean Iceman with bodies recovered from glaciers

A. Makristathis[1], R. Mader[2], K. Varmuza[3], I. Simonitsch[4], J. Schwarzmeier[2], H. Seidler[5], W. Platzer[6], H. Unterndorfer[7], and R. Scheithauer[7]

[1] Department of Clinical Microbiology, Hygiene Institute, University of Vienna
[2] Department of Internal Medicine I and L. Boltzmann Institute for Cytokine Research, University of Vienna
[3] Institute of General Chemistry, Technical University of Vienna
[4] Department of Clinical Pathology, University of Vienna
[5] Institute of Human Biology, University of Vienna
[6] Institute of Anatomy, University of Innsbruck, Austria
[7] Institute of Forensic Medicine, University of Innsbruck, Austria

In September 1991, an approximately 5000 year-old frozen mummy of a man was found in the Tyrolean Alps. The Tyrolean Iceman is a unique find, which has been the subject of several studies since then. Even from frozen anthropologic samples, decomposition of macromolecules is one of the major obstacles in the analysis of proteins or nucleic acids (1, 2). Small molecules present at high concentrations in tissue, such as fatty acids, should have a better chance to evade the decomposition process. We therefore analyzed the lipid profile of samples from the skin, trabecular bone, nose cavity and paranasal sinus of the tyrolean ice man and other human corpses, buried in glaciers (skin, muscle from calf and thigh, cardial muscle, lung, liver, bone marrow), by gas-liquid chromatography / mass spectrometry. The lipid profile of these samples was compared with the corresponding tissue samples from fresh human corpses. Additionally, three tissue samples from a well preserved female body, recovered from a lake after 50 years, could be analyzed.

Briefly, samples were evaluated with respect to fatty acids including hydroxy acids according to the following method (3): homogenized tissue was treated for 30 min at 100 °C with NaOH and methanol in order to saponify the lipid material. The sodium salts of the free fatty acids were converted to their methyl esters by heating with methanol and hydrochloric acid at a temperature of 80 °C, and were then extracted with n-hexane and t-butylethylether. The extracts were quantitatively analyzed by gas-liquid chromatography (Hewlett Packard 5890), using phenyl methyl silicone as stationary phase. Compounds were identified by gas chromatography/mass spectrometry (GC/MS, Finnigan 8200) using the database Masslib.

Independent of their origin, the lipid profile of the skin biopsies from fresh human corpses (Fig. 1) was dominated by fatty acids of even numbered carbon chains. High concentrations of palmitate and stearate were present together with unsaturated fatty acids of the same length (e.g. palmitoleate, oleate and linoleate). Minor concentrations of laurate, myristate, and myristoleate were observed. Fatty acids with odd numbered carbon chains were rarely and only in small amounts detected. Muscle, lung and liver were distinguished from skin (with attached fat) and bone marrow by a prominent signal for arachidonate.

In contrast, samples from a female and a male body, buried in glaciers for 29 and 57 years, respectively, had similar lipid profiles and were characterized by a highly decreased concentration of unsaturated fatty acids. The characteristic pattern of arachidonate and related compounds was absent. A large single peak of hydroxylated fatty acids, mainly stearate but in some specimens also palmitate, was observed (Fig. 2). This pattern could also be confirmed in tissue from the body recovered from sweat water with the exception of a prominent signal for myristate and a lower one for pentadecanoate.

The four biopsies from the Tyrolean Iceman showed a less homogeneous lipid profile. In three of four biopsies, abiotic decomposition was observed by the presence of ramified fatty acids with carbon chains of 19 C-atoms. In the same three biopsies hydroxylated palmitate could be detected. When compared with the other corpses buried in glaciers, hydroxylated stearate was present in similar concentrations in all samples of the Tyrolean Iceman (range: 15–48 % of the total fatty acids in the Tyrolean Iceman vs. 9.8–49 % in the other corpses).

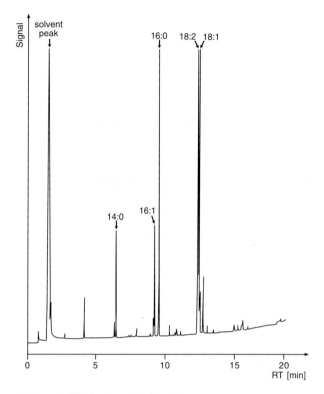

Fig. 1. Lipid profile of the skin (with attached fat) from a fresh human corpse

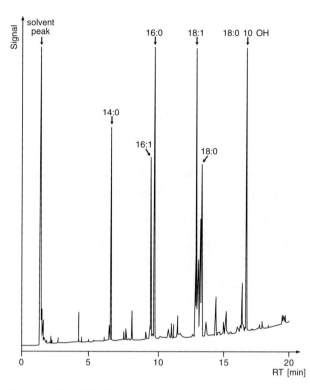

Fig. 3. Lipid profile of the trabecular bone from the Tyrolean Iceman

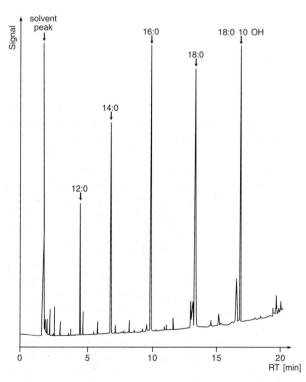

Fig. 2. Lipid profile of the liver from a female body recovered from a glacier; burial time: 29 years

The differences between the lipid profile of the Tyrolean Iceman and that of other corpses recovered from glaciers included:
i) the absence of laurate;
ii) the high concentration of oleate (range: 8.7–22 % of the total fatty acids; the lowest concentration was observed in skin and the highest level in the trabecular bone; Fig. 3);
iii) the sporadic presence of palmitoleate;
iv) the presence of isononadecanoate in specimens of the tyrolean ice man.

Preliminary results of chemometric evaluations based on the prinicipal component analysis (n=26 samples including 16 features) indicate a higher similarity between fresh samples and the Tyrolean Iceman when compared with other bodies recovered from glaciers.

We have shown that GC/MS-analyses of fatty acids directly from corpses enclosed in glaciers is feasible and offers insights in the degradation processes of human tissue. This preliminary data suggest mainly two events to occur during the millennial conservation process: unsaturated fatty acids were converted to saturated ones, and oxidation to hydroxy fatty acids took place. The oxidation of unsaturated fatty acids to hydroxylated fatty acids can be attributed to hydroxyl radicals and other reactive oxygen species (4), easily generated in the environment of glaciers due to the strong ultraviolet radiation in high altitude.

The higher heterogeneity of the samples from the Tyrolean Iceman suggests that their condition of residence under the microaerophilic environment was different. Owing to the fact that the precise position of the Tyrolean Iceman could not be reconstructed (5) and that we have no knowledge concerning the storage conditions of the corpse, it is difficult to explain the partly inhomogenous lipid profile. It was surprising, however, that a substantial part of the fatty acids of the Tyrolean Iceman remained still unaltered during 5000 years enclosed in the glacier.

On a quantitative base, this data will be analyzed in detail by chemometric methods in order to correlate the lipid profile of the respective samples with parameters such as origin, age and storage conditions.

Summary

The lipid profile of biopsies from the Neolithic corpse discovered in a Tyrolean alpine glacier in 1991 has been evaluated by gas-liquid chromatography/mass spectrometry. Furthermore, we analyzed tissue specimens from human bodies, buried for some decades in glaciers, as well as corresponding samples from fresh human corpses. The lipid profiles of the prehistoric ice man and other bodies recovered from glaciers were characterized by the presence of hydroxylated fatty acids, especially hydroxylated stearate. In comparison with other glacier corpses, the higher concentration of unsaturated fatty acids in specimens from the Tyrolean Iceman, indicates the good preservation of this body over a period of at least 5000 years. This could also be confirmed by the preliminary results of chemometric analyses.

Zusammenfassung

Das Lipidprofil von Biopsien des 1991 im Eis der Tiroler Alpen entdeckten neolithischen Körpers wurde mittels Gaschromatographie/Massenspektrometrie untersucht. Als Vergleich haben wir Gewebsproben humaner Gletscherleichen mit einer Verweildauer von einigen Jahrzehnten im Eis sowie korrespondierendes Gewebe frischer Leichen analysiert. Das Lipidprofil des prähistorischen „Mannes im Eis" und der anderen Gletscherleichen war durch das Vorhandensein von hydroxylierten Fettsäuren, vor allem hydroxyliertem Stearat, charakterisiert. Die höhere Konzentration an ungesättigten Fettsäuren im Vergleich zu wesentlich kürzer konservierten Gletscherleichen weist auf den guten Erhaltungszustand dieser Leiche über die Zeitdauer von mindestens 5000 Jahren hin. Die Besonderheit dieses Fundes konnte auch durch die vorläufigen Ergebnisse der chemometrischen Analysen bestätigt werden.

Résumé

Le profil lipidique des biopsies des tissus de l'homme néolithique découvert dans un glacier des alpes tyroliennes en 1991 a éte évalué au moyen de la chromatographie au gaz liquide resp. de la spectométrie des masses.

En outre nous avons analysé des échantillons de tissus de cadavres ayant séjourné pendant plusieurs décades dans un glacier ainsi que des échantillons de cadavres de glaciers receamment découverts. Les profils des lipides de l'homme préhistoirique du glacier ainsi que ceux des autres cadavres récents étaient caractérisés par la présence de lipo-acides hydroxylatés et en particulier de stéarates hydroxylatés. Par rapport aux cadavres plus récents la majeure concentration de lipo-acides non saturés dans les échantillons de tissus de l'homme néolithique démontre un bon état de préservation depuis au moins 5000 ans. Ce fait peut aussi être confirmé par les résultats préliminaires des analyses chémométriques.

Riassunto

Il profilo lipidico delle biopsies del cadavere neolitico ritrovato nel ghiaccio delle alpi tirolesi nel 1991 è stato rilevato per mezzo della cromatografia con gas liquido rispi spettrometria delle masse inoltre abbiamo analizzato campioni di tessuti umani di cadaveri umani, coperti da ghiacciai per alcuni decenni, nunché campioni di tessuti. Di persone appena de cedute e ritrovate su ghiacciai il profilo lipidico delli uomo neolitico del ghiacciaio nonchè delle altre salme ritrovate su ghiacciai era caratterizzato dalla presenza di acidi lipidici idrossilati ed in particolare di stearati idrussilati. La maggiore concentrazione di acidi lipidici insature nei campioni di tessuto nell'uomo del similaun rispetto a quelli di altre salme conservatesi meno a lungo in ghiacciai indica. Il buona stato di conservazione di questo cadavere durante un periodi di almeno 5000 anni. Questo fatto si è potuto confermare anche per mezzo dei risultati provvisori delle analisi chimometriche.

References

1. Lubec, G., M. Weninger, S. R. Anderson (1994): Racemization and oxidation studies of hair protein in the Homo tirolensis. FASEB J. **8:** 1166–1169.
2. Handt, O., M. Richards, M. Trommsdorff, C. Kilger, J. Simanainen, O. Georgiev, K. Bauer, A. Stone, R. Hedges, W. Schaffner, G. Utermann, B. Sykes, S. Pääbo (1994): Molecular genetic analyses of the Tyrolean ice man. Science **264:** 1775–1778.
3. Miller, L. T. (1982): Single derivatization method for routine analysis of bacterial whole-cell fatty acid methyl esters, including hydroxy acids. J. Clin. Microbiol. **16:** 584–586.
4. Gourmelon, M., J. Cillard, M. Pommepuy (1994): Visible light damage to Escherichia coli in seawater: oxidative stress hypothesis. J. Appl. Bacteriol. **77 (1):** 105–112.
5. Seidler, H., W. Bernhard, M. Teschler-Nicola, W. Platzer, D. zur Nedden, R. Henn, A. Oberhauser, T. Sjovold (1992): Some anthropological aspects of the prehistoric tyrolean ice man. Science **258:** 455–457.

Correspondence: Dr. A. Makristathis, Department of Clinical Microbiology, Hygiene Institute, University of Vienna, Währinger Gürtel 18–20, A-1090 Vienna, Austria.

Trace element contents of the Iceman's bones
Preliminary results

C. Kralik[1], W. Kiesl[1], H. Seidler[2], W. Platzer[3], and W. Rabl[4]

[1] Institut für Geochemie, Universität Wien, Austria.
[2] Institut für Humanbiologie, Universität Wien, Austria.
[3] Institut für Anatomie, Universität Innsbruck, Austria.
[4] Institut für gerichtliche Medizin, Universität Innsbruck, Austria.

Trace elements in bones

Trace element concentrations of excavated human bones can be used for the reconstruction of ancient diet, the palaeoenvironment or as an indicator for long term exposure to toxic elements (Grupe, 1991). The approach is limited by some factors as the influence of the burial environment, the inhomogeneous distribution of trace elements in bone, and the susceptibility of bone to diagenetic contamination.

Interaction of bone with soil can drastically raise the level of various elements such as Fe, Al, Mn, and V (Hancock et al., 1987; Lambert et al., 1991), thus erasing the original trace element signature, while other elements may be leached by percolating water, thus reducing their respective abundances. Problems are also encountered when comparing bones from different skeletons. Even in modern bones from a single skeleton trace elements are inhomogeneously distributed within one bone as well as between canellous and cortical bone, due to different metabolic turnover rates during lifetime. Trace elements were found to be more homogeneously distributed in the shafts of long limb bones than in any other type of bone (Brätter et al., 1977). Moreover, cortical bone is usually less susceptible to diagenetic contamination than cancellous bone. For the study of trace element contents of archaeological human remains, samples should therefore be extracted from the shafts of long limb bones, excluding the inner and outer surfaces. When sampling different skeletons, samples should be taken from the same type of bone and from the same area of the bone (Grupe, 1988).

Following these precautions, trace element data of bones can yield valuable information about an individual's diet and health status.

Samples

Bone fragments of the Iceman were found at the Hauslabjoch-site during the 1992 excavation by the Institute of Prehistory, University of Innsbruck/Austria. We studied a small piece of cancellous bone and a fragment of cortical bone, both from the femur of the Iceman.

As a reference we analyzed a piece of cortical bone of a non-soil-buried male femur, dating from the beginning of this century, and a clavicle fragment of a female body found in a lake after 50 years of 'burial' in this wet environment.

Analytical techniques

The bone samples of the Iceman, weighing 6 and 12 mg respectively, were repeatedly sonicated in deionized water to remove superficial contamination by rock particles. A piece of the clavicle of the body from the lake was clipped off and cleaned in the same way. About 5 cm of the shaft of the non-soil-buried femur were removed by a surgical steel saw. The sample was crushed and particles free of any interior or exterior bone surface were handpicked under the microscope. The sample was cleaned as described above. About 200 mg of this material were used for analysis.

The samples were analyzed by instrumental neutron activation analysis (INAA). Synthetic multielement standards were used as standards (Koeberl, 1993). Four biological reference materials (GSH-1: Human Hair; NIST 1566a: Oyster Tissue; NIST 1577a: Bovine Liver; IAEA I-155: Whey Powder) were irradiated together with the samples and standards to check accuracy and precision of the method at the low abundance levels of most elements present in the bone matrix. Samples and standards were irradiated for 24 hours at a flux of $1.7 \cdot 10^{12}$ n cm^{-2} s^{-1}.

Samples were counted twice on a high sensitivity HPGe detector (48 % rel. efficiency; resolution <2 keV at 1332 keV) after different cooling periods. Following this procedure we were able to determine As, Au, Ba, Br, Ce, Co, Cr, Cs, Eu, Fe, Hg, K, La, Lu, Na, Rb, Sb, Sc, Se, Sm, Sr, Tb, Th, U, Yb, and Zn in the samples.

Results

Elemental abundances determined in the bone samples are given in Table 1. Two analyses of the non-soil-buried femur were done to check for the variation of elemental concentrations within a single bone. For Fe, Zn and Sr differences between the two replicates are small, whereas Se concentrations are different by a factor of 3 in the two samples. Elements present at the ppb level show variations in abundance ranging from a factor of 2 in the case of Co to almost 6 for Eu. Nevertheless, the range of concentrations defined by this non-soil-buried femur can serve as a reference for diagenetic alteration of bone samples, because data do agree well with concentration ranges given for cortical bone in previous studies (Grupe, 1988; Hancock et al., 1987; Lambert et al., 1991).

Table 1. Trace elemental composition of bone samples analyzed by INAA

		Iceman femur cancellous bone	Iceman femur cortical bone surface	Body from lake clavicle	Non-soil-buried femur cortical bone	Non-soil-buried femur cortical bone
Na	ppm	3673	3898	3211	n.d.[a]	5058
K	ppm	<181	<178	<6000	n.d.	<487
Sc	ppb	4.2	18.7	63.4	7.1	3.8
Cr	ppm	<0.5	0.42	1.71	<0.2	<0.6
Fe	ppm	355.0	2928.0	2234.2	16.1	25.2
Co	ppb	1580	3370	486	12	24
Zn	ppm	122.2	240.6	104.3	102.4	112.7
As	ppb	120	308	947	n.d.	<182
Se	ppm	<0.7	1.27	0.28	0.13	0.41
Br	ppm	0.45	0.56	2.03	n.d.	0.62
Rb	ppm	<0.63	0.52	0.32	<0.21	<0.56
Sr	ppm	82	88	156	220	229
Sb	ppb	<29	29	637	n.d.	<24
Cs	ppb	9	<28	<6	<3	<18
Ba	ppm	14	24	10	<14	16
La	ppb	234	2470	543	n.d.	<32
Ce	ppb	<663	5153	971	<174	<670
Sm	ppb	95.7	442.2	71.5	n.d.	7.9
Eu	ppb	24.9	87.5	3.1	1.2	7.9
Tb	ppb	23	46	6	<3	<12
Yb	ppb	<52	106	43	66	<48
Lu	ppb	<14	<19	7.0	<90	<26
Au	ppb	<0.1	0.11	0.82	n.d.	<0.4
Hg	ppb	62	<170	99	<100	137
Th	ppb	<51	34.6	38.1	<16	<53
U	ppb	<10	<10	2000	n.d.	<10

[a] n.d. Not analyzed

Diagenetic alteration of bone samples is easily detected using their rare earth element (REE) content. The REE content of geological material is more than three orders of magnitude higher than that of biological material. Moreover, the REE have typical distribution patterns in rocks. Contamination of the bone samples of the Iceman and of the body from the lake is indicated by their 'sedimentary type' chondrite-normalized REE patterns paralleling the REE pattern of average sedimentary rock (Fig. 1). The sample of cortical bone is much more affected by contamination than the cancellous bone sample of the Iceman and the clavicle sample of the body from the lake. REE abundances of the non-soil-buried femur are well below the detection limit of the method, except for Sm, and Eu.

Abundance ratios of the samples normalized to the respective abundances in the non-soil-buried femur also reflect alteration by the burial environment (Fig. 2). Elements as Fe, Co, and the REE, which are sensitive to diagenetic additions in soil-buried bones, also show distinctive higher concentration levels in the Iceman samples and in the clavicle of the body from the lake. Ratios for Sc, Fe, Co, Sm, and Eu are considerably higher in the Iceman's cortical bone sample than in his cancellous bone. No net influx from the burial environment was observed for Zn, Sr, Ba, and Hg in the Iceman samples and in the clavicle of the body from the lake.

Sodium, showing only small variations between cortical and cancellous bone according to literature data (cortical bone: 5081 ppm Lambert et al., 1991; cancellous bone: 5100 ppm Brätter et al., 1977), is slightly depleted in the Iceman's bone samples and in the sample of the body from the lake. This is probably due to leaching of Na during burial (Price et al., 1992).

Arsenic is slightliy enriched in the Iceman's cortical bone compared to his cancellous bone. However, the As concentration in the clavicle of the body from the lake is more than three times higher.

Discussion

Elemental data of the Iceman's bone samples as well as of the clavicle bone of the body from the lake clearly indicate that contamination by the burial environment is by far exceeding any leaching effects by glacier or lake waters.

Previous studies found that cancellous bone is more easily affected by diagenetic alteration than cortical bone, because of the larger surface (Brätter et al., 1977; Grupe, 1988; Hancock et al., 1987; Lambert et al., 1991). In this study, however, REE patterns as well as el-

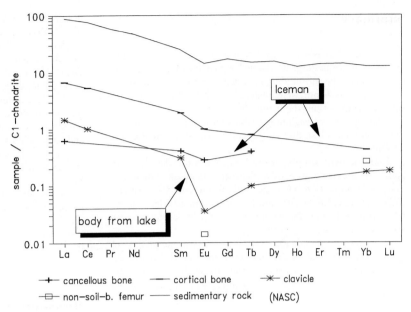

Fig. 1. Rare earth element patterns of bone samples (samples normalized to C1-chondrites). The REE pattern of average sedimentary rock is given as a reference

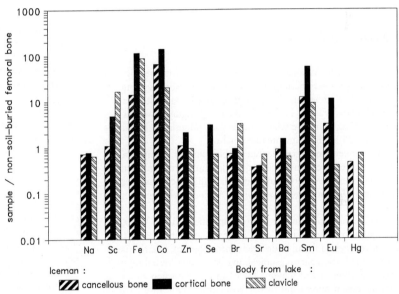

Fig. 2. Trace element abundances in bone samples of the Iceman and the body from the lake normalized to abundances in cortical bone of the non-soil-buried femur

ement ratios indicate that the Iceman's cortical bone was more contaminated by the burial environment than his cancellous bone. In previous studies contamination was found to be concentrated at bone surfaces (Badone & Farquhar, 1982; Lambert et al., 1991). This probably points to the cortical bone fragment originating from a surficial part of the femur.

Concentration of As in the clavicle of the body from the lake is 3 to 8 times higher than in the Iceman's bones. The high As abundance in the Iceman's samples is therefore more likely due to contamination by the burial environment than due to (long term) exposure to this toxic element during his lifetime (e.g. by ore smelting and refining).

Drill core samples of cortical bone from the femur or humerus of the Iceman will probably be less affected by the burial environment and will therefore be more promising for further studies on his health and nutritional status.

Acknowledgements

The authors wish to thank the Institute of Prehistory, University of Innsbruck, the Austrian Academy of Sciences, and the Institute of Forensic Medicine, University of Innsbruck, for providing the samples.

Summary

We wanted to assess the alteration of the trace element signature of the bones of the glacier-buried Iceman by leaching and contamination by the surrounding rock debris. Small bone fragments of the Iceman's femur were analyzed by INAA for 24 trace elements. Samples from the femur shaft of a non-soil-buried skeleton dating from the beginning of this century, and one clavicle of a body found in a lake after 50 years of 'burial' time were used as a reference.

Both the bone samples of the Iceman and the clavicle of the body from the lake show massive diagenetic alteration of their trace elemental composition. They have 'sedimentarytype' REE patterns and elevated Fe, Co, Sc, Sm, and Eu concentrations relative to the non-soil-buried femur. No net influx from the burial environment was observed for Zn, Sr, Ba, and Hg.

Trace element abundances of the Iceman's bone samples as well as of the clavicle bone of the body from the lake clearly indicate that, except for a few elements, contamination by the burial environment is by far exceeding any leaching effects by glacier or lake waters.

Zusammenfassung

Die mögliche Veränderung des Spurenelementmusters in den Knochen der Gletschermumie sowohl durch Auswaschung als auch durch Eintrag aus der Lagerungsumgebung sollten erfaßt werden. Dazu wurden Proben der Femurspongiosa und der Femurkompakta der Gletschermumie vom Hauslabjoch mittels INAA auf 24 Spurenelemente untersucht.

Als Vergleichsmaterial dienten Proben aus der Kompakta des Femurschaftes eines vom Beginn dieses Jahrhunderts stammenden, nie erdgelagerten Skelettes und eine Clavicula einer Wasserleiche, die nach 50 Jahren Liegezeit aus einem Alpensee geborgen worden war.

Die Knochenproben der Gletschermumie und der Wasserleiche weisen im Vergleich zu dem nie erdgelagerten Femur starke Veränderungen in ihrem Spurenelementmuster auf. Die Kontamination durch die Lagerungsumgebung ist an den Verteilungsmustern der Seltenen Erdelemente (SEE) in den Knochen zu erkennen, die denen durchschnittlicher Sedimentgesteine ähnlich sind. Die massive Veränderung des Elementmusters während der Liegezeit ist an den Knochenproben von Gletschermumie und Wasserleiche auch an den erhöhten Konzentrationen von Fe, Co, Sc, Sm und Eu erkennbar. Für die Elemente Zn, Sr, Ba und Hg ist dagegen in den Knochen der Gletschermumie und der Wasserleiche keine Erhöhung der Elementgehalte im Vergleich zu dem nie erdgelagerten Femur festzustellen.

Die erhöhten Konzentrationen an fast allen untersuchten Elementen in den Knochenproben der Gletschermumie und der Wasserleiche zeigen, daß die Kontamination durch das Lagerungsmilieu die Auslaugung der Knochen durch Gletscherwasser bzw. Wasser des Sees bei weitem überwiegt.

Résumé

L'alteration de la signature des élements traces dans les os du cadavre de glaciers par extraction et contamination par les roches environnentes est le thème de l'étude présente. 24 élements traces d'échantillons de fémurspongiosa et fémurcompacta du cadavre de glaciers ont été analysés par INAA. Comme référence, des échantillons du manche du fémur d'un squelette non enterré originant du début du ciècle, et d'une clavicula d'un cadavre des eaux, qui a passé 50 ans dans un lac alpin, ont été utilisés.

Les échantillons des os du cadavre du glacier et de la clavicule du cadavre des eaux montrent une alteration significative dans la composition des éléments traces: La signature REE ressemblent à celle des roches sédimentaires. Les concentrations de Fe, Co, Sc, Sm et Eu sont élevées relativement au fémur non enterré. Aucune influence de l'environnement n'a été enregistré pour Zn, Sr, Ba et Hg.

A l'exeption de certains éléments, la contamination par l'environnement est de beaucoup plus significative que les effects d'extraction par les glaciers ou lac. Ceci est indiqué par le contenu d'élements traces dans les os du cadavre de glacier et de la clavicule du cadavre des eaux.

Riassunto

Per provare l'alterazione della matrice oligoelementare nelle ossa della mummia del Similaun, dovuta a demineralizzazione da dilavamento ed a mineralizzazione da contaminazione dall'ambiente roccioso circostante, si sono analizzati, tramite INAA 24 elementi-traccia di campioni della spongiosa e della compacta femoris della mummia.

Come materiale di riferimento ci se è avvalsi di campioni della compacta corporis femoris di uno scheletro mai inumato, risalente all'inizio del nostro secolo, e dalla clavicola di un cadavere ritrovato dopo 50 anni di permanenza in un lago alpino.

I campioni ossei della mummia ritrovata sul ghiacciaio e del cadavere ritrovato nel lago presentarono grandi alterazioni diagenetiche della matrice oligominerale con matrici REE simili a rocce sedimentari nonchè elevate concentrazioni di Fe, Co, Sc, Sm ed Eu che il femore mai inumato invece non presenta. Nessun apporto dall'ambiente circostante è state invece osservato quanto ad una maggiore presenza degli clementi Zn, Sr, Ba e Hg rispetto al femore mai inumato.

Concentrazioni più abbondanti di pressochè tutti gli elementi presi in esame nei campioni ossei della mummia del ghiacciaio e del cadavere ritrovato nel lago indicano chiaramente che, ad eccezione di alcuni pochi elementi, la contaminazione dall'ambiente circostante supera di gran lunga il dilavamento dovuto all'effetto del ghiacciaio o dell'acqua.

References

Badone, E., Farquhar, R. M. (1982): Application of neutron activation analysis to the study of element concentration and exchange in fossil bones. J. Radioanal. Chem., **69**, 291–311.

Brätter, P., Gawlik, D., Lausch, J., Rösick, U. (1977): On the distribution of trace elements in human skeletons. J. Radioanal. Chem., **37**, 393–403.

Grupe, G. (1988): Impact of the choice of bone samples on trace element data in excavated human skeletons. J. Arch. Sci., **15**, 123–129.

Grupe, G. (1991): Anthropogene Schwermetallkonzentrationen in menschlichen Skelettfunden als Monitor früher Umweltbelastungen. Z. Umweltchemie und Ökotoxikologie, **3**, 226–229.

Hancock, R. G. V., Grynpas, M. D., Alpert, B. (1987): Are archaeological bones similar to modern bones? An INAA assessment. J. Radioanal. Nucl. Chem., **110**, 283–291.

Koeberl, C. (1993): Instrumental neutron activation analysis of geochemical and cosmochemical samples. J. Radioanal. Nucl. Chem., **168**, 47–60.

Lambert, J. B., Xue, L., Buikstra, J. E. (1991): Inorganic analysis of excavated human bone after surface removal. J. Arch. Sci., **18**, 363–383.

Price, T. D., Blitz, J., Burton, J., Ezzo, J. A. (1992): Diagenesis in Prehistoric Bone: Problems and Solutions. J. Arch. Sci., **19**, 513–529.

Correspondence: Dr. C. Kralik, Seidlgasse 1/13, A-1030 Vienna, Austria.

Remarks on the anatomy of a mummified cat regarding the extent of preservation

J. Weisgram[1], H. Splechtna[1], H. Hilgers[1], M. Walzl[1], W. Leitner[2], and H. Seidler[3]

[1] Institut für Zoologie der Universität Wien, Austria
[2] Institut für Ur- und Frühgeschichte der Universität Innsbruck, Austria
[3] Institut für Humanbiologie der Universität Wien, Austria

Introduction

On July 24, 1994, the remains of a cat were found on the Stubai-Glacier, 2,900 m above sea-level at the bottom of the "Schaufelferner" facing northeast. Haim & Nosko (1993) have given inaccurate information concerning the locality.

The corpse was completely air dried, bearing numerous marks of insect feeding. Investigations indicate the "Dresdnerhütte", situated at 2,302 m above sea-level, as a possible provenance of the animal.

Radiocarbon dating conducted at the Swiss Federal Institute of Technology, Zürich, have shown that the corpse dates either from the period 1959–1962 or that of 1980–1983. The first date is more likely, since locals remember keeping pets in that area about 30 years ago.

Results

External aspect of the right side of the body

When found, the right side of the mummy lay upon the ground. The outer ear is dried up and curled on its periphery. The orbita shows a sharp border (max. diameter 16 mm) and is empty, except of some dry remains of the eye-bulb, probably parts of the sclera. The lips are dried and shrinked, the incisors and the canine tooth are exposed. A small number of vibrissae can be found above the orbita. A larger number of vibrissae are situated on the upper lip, some of them are preserved in full lenght, others are obviously broken (Fig. 2).

The skin has a parchment-like appearance with numerous holes, probably caused by insect-feeding. High damage can be seen in the middle trunk region (Fig. 2) around the border thorax-abdomen, the last three ribs are exposed (Fig. 3). The skin is missing on a ventrolateral aspect of the lower jaw, the mandible lies free. A further lesion of the skin on the right upper thigh exposes the distal part of the femur and the knee.

The anal region is completely destroyed and opens to the body cavity, the fully mazerated right Tuber ischiadicum of the pelvis lies free.

The tail is preserved at full length, it lies on the right side of the body and runs caudad and dorsad to the aerea of caudal vertebra 7, where it changes its direction immediately to rostral and in the region of caudal vertebra 15 it turns caudad again (Fig. 2). The forelimbs are crossed, the left one is positioned above the right.

The genital region shows remains of the scrotum, which is open at the right side, and of the preputium (Fig. 4). The testes and the Corpus penis are missing. The sex of the corpse can be clearly determined as male.

External aspect of the left side of the body

The outer ear has its natural shape, is raised and shows only slight shrinkage. The orbita (max. diameter 11 mm) contains remains of the eye, probably parts of the sclera (Fig. 1). Three vibrissae are preserved above the orbita. The lips are elevated and expose the incisors and the canine tooth, several vibrissae are situated on the upper lip (Fig. 1).

The skin is dry, parchment-like and shows some holes (probably caused by insect feeding) but no major destructions (compared to the right side of the body). A small lesion exposes the knee joint. In microscopic view single hair can be found all over the skin, tufts of hair remains exist on the midventral region (Fig. 12) as well as on the upper thigh.

Skeleton

The skull is in good condition, no fractures are visible in the X-ray shots (Fig. 8). All permanent teeth are preserved, the second dentition is completely developed.

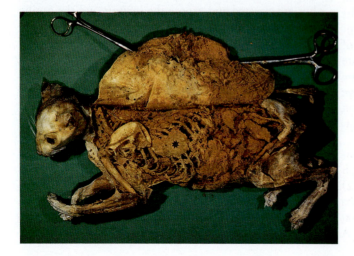

Fig. 1. Head in frontal view, note the vibrissae above the orbitae and on the upper lips. The different diameter of the orbitae results from the better preservation of the skin of the left side. Photo: Walzl

Fig. 2. Right side; the flattened skin of the neck indicates that it lay upon a flat stone; note the damage of the middle trunk region and the position of the tail. Photo: Walzl

Fig. 3. Detail from Fig. 2 to show the destroyed body wall of the right side. The body cavity is open, the ends of the three caudalmost ribs (white arrows) can be seen from outside. Photo: Walzl

Fig. 4. Genital region seen from left, remains of the scrotum (white arrow) and the preputium (white triangle) can be seen. Photo: Walzl

Fig. 5. Left side, skin open, note that the inner organs are almost rotten. A remain of the external thoracic musculature is marked by an asterisk. Photo: Walzl

Fig. 6. Detail from Fig. 5 to show the grave wax (white arrow) in the hindlimb-region, the stomach and parts of the intestine (white dot) are partly transformed into adipocere. A remain from the liver (small white arrow) can be identified by its form. Photo: Walzl

The vertebral column shows a dorsad dislocation of the second cervical vertebra (Fig. 8). As there is no indication, that this dislocation happend post mortem, it could be the cause of death as well.

Two more fractures exist between caudal vertebrae 7 and 8, as well as caudal vertebrae 15 and 16 (Fig. 11). In the area of the tail-fractures the intervertebral disks (Fig. 11) are detached from the vertebral bodies.

The pelvis is intact, the *Tuberculum ischiadicum* is still separated from the Os ischium by a cartilagineous plate. The bones of the hindlimbs are undamaged, the Patella and the Ligamentum patellae as well as the tendon of the M. quadriceps femoris and the tendon of Achilles are preserved and can be seen clearly in the X-ray shots (Fig. 10). In the distal end of the tibia the epiphyseal plate is visible.

The bones of the forelimb and the shoulder are fully intact. In the left Scapula the caudal cartilagineous rim (*Cartilago scapulae*) is detached, it is well preserved but shifted approx. 6 mm to cranial (Fig. 7). In the right Scapula the Cartilago scapulae is in situ.

The epiphyseal plate can be seen on the proximal end of both humeri.

All ribs, costal cartilages and sternebrae are preserved and in correct position (Fig. 9).

The degree of ossification and the state of the teeth indicate, that the animal had an age between 8 and 13 month at death.

Abdominal cavity

The organs of the abdominal cavity are half rotten and hardly identifiable. The stomach and probably parts of the intestine are transformed into a clot of grave wax (adipocere), a strong layer of adipocere can be found also

Fig. 7. Left side, shoulder region, the *Cartilago scapulae* (white arrow) is dislocated to cranial, probably by shrinking. Photo: Walzl

in the position of the ventral abdominal muscles, just under the skin (Fig. 6). A part of the liver could be identified by its form.

Thoracic cavity

The organs of the thoracic cavity are completely rotten, neither the heart nor the lungs could be identified. One remain of external thoracic musculature lies outside the ribs of the left side of the body (Fig. 5).

Cervical region

The musculature of the neck is completely rotten, along the caudolateral region of the cervical vertebrate column a net of nerves and probably blood vessels is preserved. Parts of the esophagus exist but the larynx and the trachea could not be found.

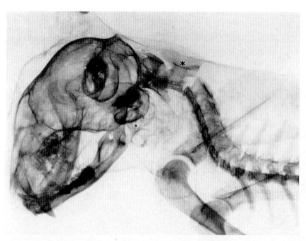

Fig. 8. Radiogram of the head and neck to show the dislocated second cervical vertebra (asterisk). X-ray: Geres. Photo: Weisgram

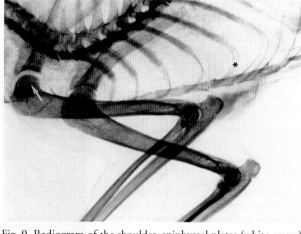

Fig. 9. Radiogram of the shoulder, epiphyseal plates (white arrow) can be seen as well as the preserved ribs and costal cartilages (black asterisk). X-ray: Geres. Photo: Weisgram

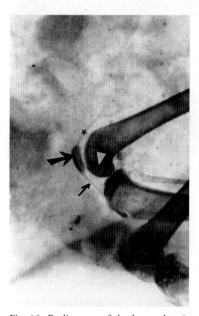

Fig. 10. Radiogram of the knee, showing the patella (thick black arrow), the *Ligamentum patellae* (thin black arrow) the tendon of the M. quadriceps femoris (black asterisk) and the distal epiphyseal plate of the femur (white triangle). X-ray: Geres. Photo: Weisgram

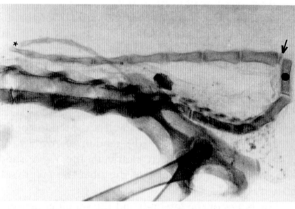

Fig. 11. Radiogram of the tail to show the fractures between caudal vertebrae 7 (black dot) and 8, and 15–16 (asterisk) a dislocated intervertebral disk (black arrow) can be seen. X-ray: Geres. Photo: Weisgram

Histology

A part of the skin of the left shoulder was examined. Epidermal layers are completely lacking, only dermis and subcutis are present (Fig. 13). Groups of hair follicles are located within the dermis. The cavities in the deeper layers of the dermis and subcutis may be remains of fat cells.

Acknowledgements

We kindly thank Dr. Geres, Institut für Röntgenologie der Veterinärmedizinischen Universität Wien (Head: Prof. Dr. Mayrhofer) for making the X-ray shots. L. Rudoll is acknowledged for preparing the histological sections.

Summary

The anatomy of the mummified cat, found on the Stubai-glacier in 1992, was studied by gross dissection and by X-ray shots. The surface of the skin was photographed by a stereomicroscope Wild M 420. For histological sectioning the skin was treated with a softening fluid and a special dehydrating sequence after Sandison (1955) and embedded in Araldite. The semithin sections (2 µm) were stained with Methylenblue-azur II after Richardson and photographed by a Reichert Polyvar.

The mummy shows a parchment-like skin with numerous vibrissae and remains of hair and is highly damaged by insect-feeding, especially on the right side of the body. The skeleton is undamaged except of a dislocation of the second cervical vertebra and fractures between caudal vertebrae 7 and 8, as well as 15 and 16.

The inner organs are mostly rotten and hardly identifiable, a few clots of adipocere were found in the abdominal cavity and below the skin. The sex of the cat could be clearly determined as male.

The degree of ossification of the skeleton as well as the state of the teeth indicate an age at death between 8 and 13 month.

Zusammenfassung

Die Anatomie einer 1992 auf dem Stubaier Gletscher gefundenen mumifizierten Katze wurde anhand von Sektionen und Röntgenaufnahmen untersucht. Die Hautoberfläche wurde mit einem Stereomikroskop Wild M 420 photographiert. Für histologische Schnitte wurde die Haut mit einem Aufweichungsmittel und einem speziellen Dehydratisierungsverfahren nach Sandison (1955) behandelt und in Araldit eingebettet. Die Semidünnschnitte (2 μm) wurden nach Richardson mit Methylenblau-Azur II gefärbt und mit einem Reichert Polyvar aufgenommen.

Die Mumie hat eine pergamentartige Haut, weist noch zahlreiche Schnurrhaare und Fellreste auf, und ist, besonders auf der rechten Seite des Körpers, stark durch Insektenfraß beschädigt. Das Skelett ist unversehrt, abgesehen von einer Luxation des zweiten Halswirbels und Frakturen zwischen dem 7. und 8. sowie dem 15. und 16. Schwanzwirbel.

Die inneren Organe sind größtenteils verwest und kaum zu identifizieren; in der Bauchhöhle und unter der Haut fanden sich einige Klumpen Leichenwachs. Das Geschlecht des Tieres konnte eindeutig als männlich festgestellt werden.

Der Verknöcherungsgrad des Skeletts und der Zustand der Zähne deuten auf ein Lebensalter des Tieres von 8 bis 13 Monaten.

Fig. 12. Micrograph of the skin of the left midventral body region with remains of hair tufts, scale bar = 1 mm. Photo: Walzl

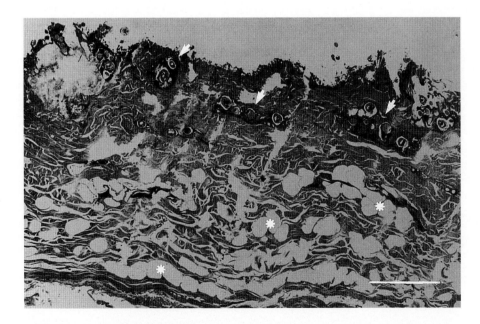

Fig. 13. Section of skin, showing groups of hair follicles (white arrows) and clusters of presumed fat cells (white asterisks), scale bar = 200 μm. Photo: Hilgers

Résumé

L'anatomie d'un chat momifié, trouvé en 1992 sur le glacier de la vallée de Stubai en Autriche, a été examinée à l'aide d'une autopsie complète et de radioscopies. La peau a été photographiée avec un stéréomicroscope du type Wild M 420. En préparation aux coupes histologiques, la peau avait été traitée avec un agent amollissant, ensuite déshydratée avec un procédé spécial introduit par Sandison (1955) et finalement couverte d'araldite. Les coupes fines (2 µm) ont été colorées avec du bleu methylène-azur II suivant la méthode de Richardson et reprises avec un appareil photographique du modèle Reichert Polyvar.

La peau de la momie est parcheminée, et on y trouve encore de nombreux poils de moustache et des restes du pelage. Le cadavre, et sourtout sa partie gauche, est fortement detruit par les insectes. Le squelette est intact, à part une luxation de la deuxième vertèbre cervicale ainsi que des fractions entre la 7ème et la 8ème et entre la 15ème et la 16ème vertèbre de la queue.

Les viscères sont en grande partie putréfiés et donc difficiles à identifier; dans la cavité abdominale et sous la peau ont été trouvées des masses compactes d'adipocire. Le sexe de l'animal a pu être clairement déterminé comme masculin. Le degré d'ossification et l'état des dents permettent de situer l'âge de l'animal entre 8 et 13 mois.

Riassunto

L'anatomia di un gatto mummificato rinvenuto nel 1992 sul ghiacciaio dello Stubai è stata analizzata in base ad una sezione grossolana ed ad una radiografia. La superficie della pelle è stata fotografata con uno stereomicroscopio Wild M 420. Per la sezione istologica la pelle è stata trattata con un fluido ammorbidente, sottoposta ad un procedimento desidratante speciale secondo il metodo Sandison (1955) ed immersa in araldite. Le sezioni semisottili (2 µm) sono state colorate secondo il metodo Richardson con blue di metilene-Azur II e fotografate con una Reichert Polyvar.

La pelle della mummia ha un aspetto simile alla pergamena, presenta numerosi baffi e peli ed è fortemente corrosa da insetti sulla parte destra del corpo. Lo scheletro è intatto salvo una lussazione della seconda vertebra cervicale e delle fratture fra la settima e la ottava nonchè la quindicesima e la sedicesima vertebra caudiale.

Gli organi interni sono per lo più distrutti e difficilmente identificabili. Nella cavità addominale e sotto la pelle sono stati trovati alcuni agglomerati di adipocera. Si tratta evidentemente di un gatto maschile. Il grado dell'ossificazione dello scheletro nonchè lo stato della dentatura suggeriscono un'età dai 8 ai 13 mesi.

References

Bloom, W. and Fawcett, D. W.: A Textbook of Histology. W. B. Saunders Company, Philadelphia - London - Toronto 1975.

Heim, M. and Nosko, W.: Die Ötztalfälschung. Rowohlt Verlag GmbH, Hamburg 1993.

Popesko, P.: Atlas der topographischen Anatomie der Haustiere. F. Enke Verlag, Stuttgart 1979.

Romer, A. S.: The Vertebrate Body. W. B. Saunders Company, Philadelphia - London - Toronto 1970.

Sandison, A. T.: The histological examination of mummified material. Stain Technology 30, 6: 277–283 (1955).

Wischnitzer, S.: Atlas and Dissection Guide for Comparative Anatomy. W. H. Freeman and Company, Oxford - New York 1979.

Correspondence: Dr. J. Weisgram, Institut für Zoologie der Universität Wien, Althanstrasse 14, A-1090 Vienna, Austria.

SpringerNewsHumanbiology

Veröffentlichungen des Forschungsinstituts für Alpine Vorzeit der Universität Innsbruck, The Man in the Ice
edited by H. Moser, W. Platzer, H. Seidler, K. Spindler

Each volume will comprise 200-350 pages and contain approximately 100 illustrations in colour and 250 in black-and-white. Each volume will be cloth-bound and have a dustwrapper. The volumes are, with the exception of the first two, to appear in English, each article having in addition a summary in German, English, French and Italian. The volumes will appear at irregular intervals (once to twice a year).

A discount of 20 % will be granted on a subscription to the complete work, at present planned to consist of 9 volumes. The whole work will begin with Vol. 2 of the series "The Man in the Ice". The first volume having been published privately by the University of Innsbruck.

Plan of the Edition

K. Spindler et al. (Hrsg.), Der Mann im Eis. Neue Funde und Ergebnisse.
(The Man in the Ice, Vol. 2)
S. Bortenschlager (ed.), Biogenous Finds discovered along with the Iceman. Palaeobotanical results. (The Man in the Ice, Vol. 4)
T. Sjovold et al. (eds.), Iceman's Tattoos – Significance, Biological and Cultural Aspects. (The Man in the Ice, Vol. 5)
W. Platzer et al. (eds.), Radiological and Anatomic Investigations on the Iceman.
(The Man in the Ice, Vol. 6)
K. Spindler et al. (eds.), Mummies of the World.
(The Man in the Ice, Vol. 7)
K. Spindler et al. (eds.), Fossil Record at the Iceman's Site –
Archaeological Findings. Archaeological and pre-historical interpretations.
(The Man in the Ice, Vol. 8)
K. J. Irgolic (ed.), Ancient DNA and Trace Elements –
the Palaeoecological Interpretations. (The Man in the Ice, Vol. 9)

SpringerWienNewYork

P.O.Box 89, A-1201 Wien • New York, NY 10010, 175 Fifth Avenue
Heidelberger Platz 3, D-14197 Berlin • Tokyo 113, 3-13, Hongo 3-chome, Bunkyo-ku

*Springer-Verlag
and the Environment*

WE AT SPRINGER-VERLAG FIRMLY BELIEVE THAT AN international science publisher has a special obligation to the environment, and our corporate policies consistently reflect this conviction.

WE ALSO EXPECT OUR BUSINESS PARTNERS – PRINTERS, paper mills, packaging manufacturers, etc. – to commit themselves to using environmentally friendly materials and production processes.

THE PAPER IN THIS BOOK IS MADE FROM NO-CHLORINE pulp and is acid free, in conformance with international standards for paper permanency.